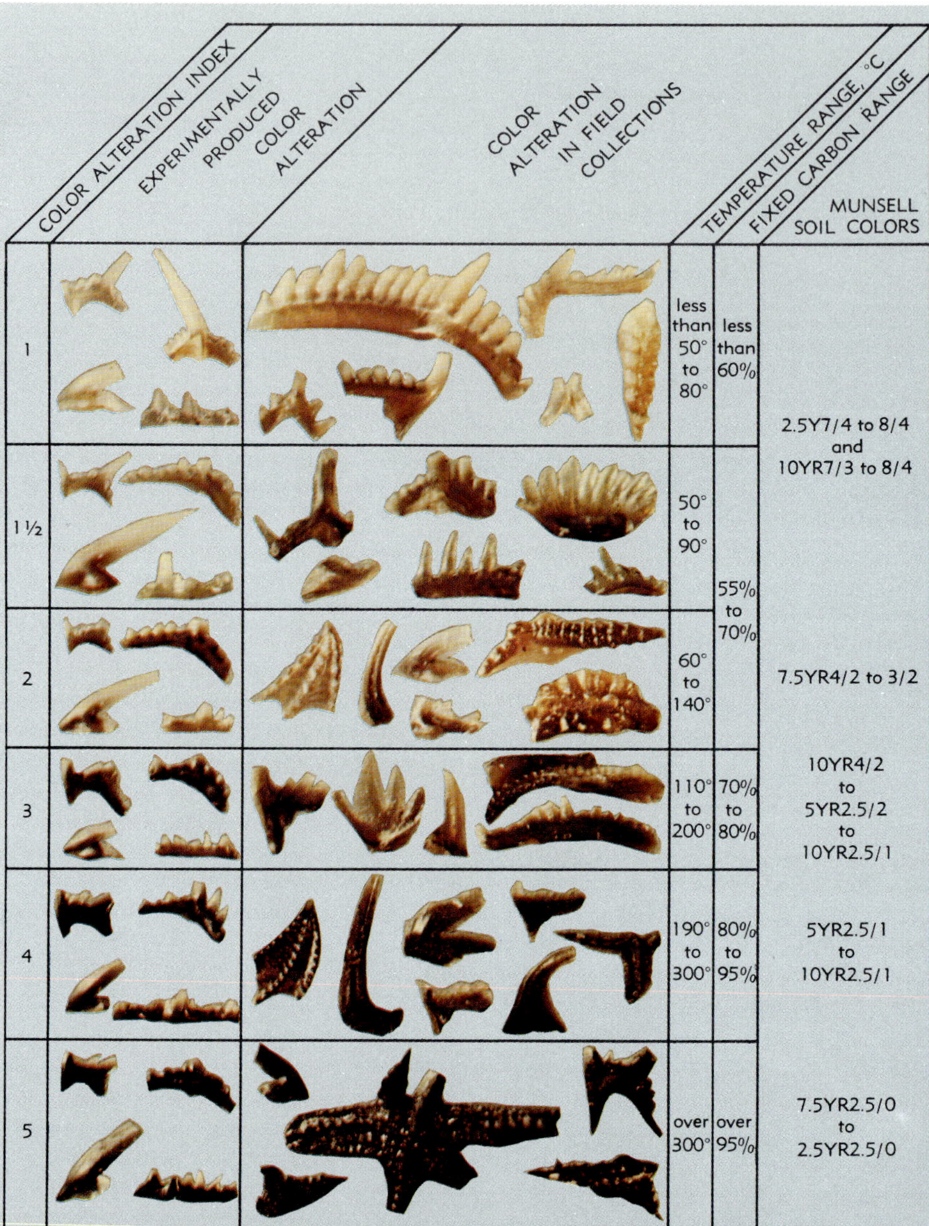

Frontispiece. Chart showing experimentally produced and naturally occurring colors in conodont elements together with the geological temperature and fixed carbon ranges for each Color Alteration Index (after Epstein, Epstein, & Harris, 1977). For further explanation, see section on Color and Alteration.

[Funds to reproduce this color illustration were provided by General Crude Oil Company.]

TREATISE ON INVERTEBRATE PALEONTOLOGY

Prepared under Sponsorship of
The Geological Society of America, Inc.

The Paleontological Society *The Society of Economic Paleontologists and Mineralogists*
The Palaeontographical Society *The Palaeontological Association*

RAYMOND C. MOORE
Founder

R. A. ROBISON
Editor

VIRGINIA ASHLOCK, JACK KEIM, ROGER B. WILLIAMS
Assistant Editors

Part W
MISCELLANEA

SUPPLEMENT 2
CONODONTA

By DAVID L. CLARK, WALTER C. SWEET, STIG M. BERGSTRÖM, GILBERT KLAPPER, RONALD L. AUSTIN, FRANK H. T. RHODES, KLAUS J. MÜLLER, WILLI ZIEGLER, MAURITS LINDSTRÖM, JAMES F. MILLER, and ANITA G. HARRIS

THE GEOLOGICAL SOCIETY OF AMERICA, INC.
and
THE UNIVERSITY OF KANSAS
BOULDER, COLORADO, and LAWRENCE, KANSAS

1981

Library of Congress Catalogue Card Number: 53-12913
ISBN 0-8137-3028-7

Published 1981

Distributed by the Geological Society of America, Inc., 3300 Penrose Place, Boulder, Colo. 80301, to which all communications should be addressed.

The *Treatise on Invertebrate Paleontology* has been made possible by (1) funding principally from the National Science Foundation of the United States, from the Geological Society of America through the bequest of Richard Alexander Fullerton Penrose, Jr., and from the bequest of Raymond C. and Lilian Moore; (2) contribution of the knowledge and labor of specialists throughout the world, working in cooperation under sponsorship of the Geological Society of America, the Paleontological Society, the Society of Economic Paleontologists and Mineralogists, the Palaeontographical Society, and the Palaeontological Association; and (3) acceptance by the University of Kansas of publication without any financial gain to the University.

TREATISE ON INVERTEBRATE PALEONTOLOGY

PARTS

Parts of the *Treatise* are distinguished by assigned letters with a view to indicating their systematic sequence while allowing publication of units in whatever order each is made ready for the press. Copies are available on orders sent to the Publication Sales Department, The Geological Society of America, 3300 Penrose Place, Boulder, Colorado 80301.

VOLUMES ALREADY PUBLISHED

Part A. INTRODUCTION, xxiii + 569 p., 371 fig., 1979.
Part C. PROTISTA 2 (Sarcodina, chiefly "Thecamoebians" and Foraminiferida), xxxi + 900 p., 5,311 fig., 1964.
Part D. PROTISTA 3 (chiefly Radiolaria, Tintinnina), xii + 195 p., 1,050 fig., 1954.
Part E. ARCHAEOCYATHA, PORIFERA, xviii + 122 p., 728 fig., 1955.
Part E, Volume 1. ARCHAEOCYATHA, Second Edition (Revised and Enlarged), xxx + 158 p., 871 fig., 1972.
Part F. COELENTERATA, xvii + 498 p., 2,700 fig., 1956.
Part F. COELENTERATA (Supplement 1). Rugosa and Tabulata, xl + 762 p., 3,317 fig., 1981.
Part G. BRYOZOA, xii + 253 p., 2,000 fig., 1953.
Part H. BRACHIOPODA, xxxii + 927 p., 5,198 fig., 1965.
Part I. MOLLUSCA 1 (Mollusca General Features, Scaphopoda, Amphineura, Monoplacophora, Gastropoda General Features, Archaeogastropoda, mainly Paleozoic Caenogastropoda and Opisthobranchia), xxiii + 351 p., 1,732 fig., 1960.
Part K. MOLLUSCA 3 (Cephalopoda General Features, Endoceratoidea, Actinoceratoidea, Nautiloidea, Bactritoidea), xxviii + 519 p., 2,382 fig., 1964.
Part L. MOLLUSCA 4 (Ammonoidea), xxii + 490 p., 3,800 fig., 1957.
Part N. MOLLUSCA 6 (Bivalvia), Volumes 1 and 2 (of 3), xxxviii + 952 p., 6,198 fig., 1969; Volume 3, iv + 272 p., 742 fig., 1971.
Part O. ARTHROPODA 1 (Arthropoda General Features, Protarthropoda, Euarthropoda General Features, Trilobitomorpha), xix + 560 p., 2,880 fig., 1959.
Part P. ARTHROPODA 2 (Chelicerata, Pycnogonida, Palaeoisopus), xvii + 181 p., 565 fig., 1955.
Part Q. ARTHROPODA 3 (Crustacea, Ostracoda), xxiii + 442 p., 3,476 fig., 1961.
Part R. ARTHROPODA 4 (Crustacea exclusive of Ostracoda, Myriapoda, Hexapoda), Volumes 1 and 2 (of 3), xxxvi + 651 p., 1,762 fig., 1969.
Part S. ECHINODERMATA 1 (Echinodermata General Features, Homalozoa, Crinozoa, exclusive of Crinoidea), xxx + 650 p., 2,868 fig., 1967 [1968].
Part T. ECHINODERMATA 2 (Crinoidea), Volumes 1-3, xxxviii + 1027 p., 4,833 fig., 1978.
Part U. ECHINODERMATA 3 (Asterozoans, Echinozoans), xxx + 695 p., 3,485 fig., 1966.
Part V. GRAPTOLITHINA, xvii + 101 p., 358 fig., 1955.

Part V. GRAPTOLITHINA, Second Edition (Revised and Enlarged), xxxii + 163 p., 507 fig., 1970.

Part W. MISCELLANEA (Conodonts, Conoidal Shells of Uncertain Affinities, Worms, Trace Fossils, Problematica), xxv + 259 p., 1,058 fig., 1962.

Part W. MISCELLANEA (Supplement 1). Trace Fossils and Problematica (Revised and Enlarged), xxi + 269 p., 912 fig., 1975.

THIS VOLUME

Part W. MISCELLANEA (Supplement 2). Conodonta, xxviii + 202 p., frontis., 858 fig., 1981.

VOLUMES IN PREPARATION

Part B. PROTISTA 1 (Chrysomonadida, Coccolithophorida, Charophyta, Diatomacea, etc.).

Part J. MOLLUSCA 2 (Gastropoda, Streptoneura exclusive of Archaeogastropoda, Euthyneura).

Part M. MOLLUSCA 5 (Coleoidea).

Part R. ARTHROPODA 4, Volume 3 (Hexapoda).

Part X. ADDENDA, INDEX.

Part E, Volume 2. PORIFERA (revised edition).

Part G. BRYOZOA (revised edition).

Part L. MOLLUSCA 4 (Ammonoidea, revised edition).

Part O. ARTHROPODA 1 (supplement 1, Trilobita).

Part Q. ARTHROPODA 3 (supplement 1, Ostracoda).

Part W. MISCELLANEA (supplement 3, worms).

CONTRIBUTING AUTHORS

[Arranged by countries and institutions; an alphabetical list follows.]

AUSTRALIA

South Australia Geological Survey (Adelaide)
N. H. Ludbrook

University of Adelaide
M. F. Glaessner

University of Queensland (Brisbane)
Dorothy Hill

AUSTRIA

Universität Wien (Paläontologisches Institut)
Adolf Papp

BELGIUM

Unattached
Charles Grégoire (Bruxelles)

Université de Liège
Georges Ubaghs

CANADA

Geological Survey of Canada (Ottawa)
J. A. Jeletzky, D. J. McLaren

Institute of Sedimentary & Petroleum Geology (Geological Survey of Canada, Calgary)
A. W. Norris

University of British Columbia (Vancouver)
V. J. Okulitch

University of Saskatchewan (Saskatoon)
W. A. S. Sarjeant

FRANCE

Université de Paris
Colette Dechaseaux (Laboratoire de Paléontologie des Vertébrés), Geneviève Lutaud (Laboratoire de Cytologie)

Université Paris-Sud (Orsay)
Michel Roux

GERMAN DEMOCRATIC REPUBLIC

Bergakademie Freiberg (Fachbereich Geowissenschaften)
A. H. Müller

GERMANY, FEDERAL REPUBLIC OF

Friedrich Wilhelms Universität (Bonn)
H. K. Erben, K. J. Müller

Natur-Museum und Forschungs-Institut Senckenberg (Frankfurt)
Herta Schmidt, Wolfgang Struve

Philipps Universität (Marburg)
Gerhard Hahn, Maurits Lindström, Willi Ziegler

Unattached
Hertha Sieverts-Doreck (Stuttgart-Möhringen)
Universität Münster
Helmut Hölder
Universität Tübingen
Jürgen Kullmann, Adolf Seilacher
Universität Würzburg
Klaus Sdzuy

ITALY

Universitá Modena
Eugenia Montanaro Gallitelli
Universitá di Roma
Franco Rasetti

JAPAN

Saito Ho-on Kai Museum of Natural History (Sendai)
Kotora Hatai
University of Tokyo
Tetsuro Hanai

NETHERLANDS

Rijksmuseum van Natuurlijke Historie (Leiden)
L. B. Holthuis
Vrije Universiteit Amsterdam
A. J. Breimer, M. J. S. Rudwick

NEW ZEALAND

Auckland Institute and Museum
A. W. B. Powell
Dominion Museum (Wellington)
R. K. Dell
New Zealand Geological Survey (Lower Hutt)
C. A. Fleming
Unattached
John Marwick (Havelock North)

NORWAY

Unattached
Tron Soot-Ryen (Hosle)
Universitet Oslo
Gunnar Henningsmoen (Institutt for Geologi)

POLAND

Pánstwowe Wydawnictwo Naukowe (Warszawa)
Gertruda Biernat, Adolf Riedel

SWEDEN

Naturhistoriska Riksmuseet Stockholm
Valdar Jaanusson
Universitet Lund
Gerhard Regnéll
Universitet Stockholm
Ivar Hessland
Universitet Uppsala
R. A. Reyment

SWITZERLAND

Universität Basel
Manfred Reichel

UNITED KINGDOM

British Museum (Natural History) (London)
P. L. Cook, M. K. Howarth, C. P. Nuttall
British Petroleum Company (Middlesex)
F. E. Eames
Institute of Geological Sciences (London)
Raymond Casey, R. V. Melville
Iraq Petroleum Company (London)
G. F. Elliott
Open University (Milton Keynes)
S. Conway Morris
Queen's University of Belfast
Margaret Jope, R. E. H. Reid, A. D. Wright
Unattached
Dennis Curry (Middlesex), Sir James Stubblefield (London), R. P. Tripp (Sevenoaks, Kent), C. W. Wright (Dorset), Sir Maurice Yonge (Edinburgh)
University of Birmingham
Anthony Hallam
University of Cambridge
H. B. Whittington
University College London
J. H. Callomon, D. T. Donovan
University College of Swansea (Swansea, Wales)
D. V. Ager
University of Glasgow
W. D. I. Rolfe, John Weir, Alwyn Williams
University of Hull
M. R. House
University of Manchester
E. R. Trueman
University of Southampton
R. L. Austin

UNITED STATES OF AMERICA

Academy of Natural Sciences of Philadelphia (Pennsylvania)
A. A. Olsson, Robert Robertson

American Museum of Natural History (New York)
R. L. Batten, N. D. Newell

Brown University (Providence, Rhode Island)
R. D. Staton

California Academy of Sciences (San Francisco)
Eugene Coan, A. G. Smith

Cornell University (Ithaca, New York)
W. S. Cole, F. H. T. Rhodes, J. W. Wells

Exxon Production Research Company (Houston, Texas)
H. H. Beaver, R. M. Jeffords, S. A. Levinson, D. B. Macurda, L. A. Smith

Field Museum of Natural History (Chicago)
Fritz Haas

Geological Society of America (Boulder, Colorado)
A. R. Palmer

Getty Oil Company (Houston, Texas)
Lavon McCormick

Harvard University (Cambridge, Massachusetts)
F. M. Carpenter, W. J. Clench, H. B. Fell, Ruth Turner

Illinois State Geological Survey (Urbana)
M. L. Thompson

Indiana Geological Survey (Bloomington)
R. H. Shaver

Joint Oceanographic Institutions, Inc. (Washington, D.C.)
W. W. Hay

Kansas Geological Survey (Lawrence)
D. E. Nodine Zeller

Kent State University (Kent, Ohio)
A. H. Coogan

Louisiana State University (Baton Rouge)
W. A. van den Bold, H. B. Stenzel

New Mexico Institute Mining & Technology (Socorro)
Christina Lochman-Balk

New York State Museum (Albany)
D. W. Fisher

Ohio State University (Columbus)
S. M. Bergström, Aurèle La Rocque, W. C. Sweet

Oklahoma Geological Survey (Norman)
T. W. Amsden, R. O. Fay

Oregon State University (Corvallis)
A. J. Boucot, J. G. Johnson

Paleontological Research Institution (Ithaca, New York)
K. V. W. Palmer

Princeton University (Princeton, New Jersey)
A. G. Fischer

Professional Geophysics, Inc. (Oklahoma City, Oklahoma)
J. A. Eyer

Queens College of the City of New York (Flushing)
R. M. Finks

Radford College (Radford, Virginia)
R. L. Hoffman

St. Mary's College of California (St. Mary's)
A. S. Campbell

San Diego Natural History Museum (San Diego, California)
George Radwin

San Francisco State University (San Francisco, California)
Y. T. Mandra

Smithsonian Institution (Washington, D.C.)
F. M. Bayer, R. H. Benson, R. S. Boardman, A. H. Cheetham, A. H. Clarke, Jr., T. G. Gibson, R. E. Grant, P. M. Kier, R. B. Manning, David Pawson

Southern Illinois University (Carbondale)
John Utgaard

Southern Methodist University (Dallas, Texas)
A. L. McAlester

Southwest Missouri State University (Springfield)
J. F. Miller

Stanford University (Stanford, California)
A. Myra Keen

Syracuse University (Syracuse, New York)
J. C. Brower

Unattached
R. Wright Barker (Bellaire, Texas), J. W. Hedgpeth (Santa Rosa, Calif.), H. S. Puri (Tallahassee, Florida)

United States Geological Survey (Washington, D.C.)
J. M. Berdan, R. C. Douglass, A. G. Harris, J. E. Hazel, O. L. Karklins, K. E. Lohman, N. F. Sohl, I. G. Sohn, E. L. Yochelson
University of Alaska (Fairbanks)
C. D. Wagner
University of California (Berkeley)
J. W. Durham
University of California (Los Angeles)
A. R. Loeblich, Jr., Helen Tappan
University of California (San Diego, La Jolla)
R. R. Hessler, W. A. Newman
University of Cincinnati (Ohio)
K. E. Caster, D. L. Meyer
University of Colorado (Boulder)
E. G. Kauffman, J. A. Van Couvering
University of Florida (Gainesville)
H. K. Brooks
University of Illinois (Urbana)
D. B. Blake, Philip Sandberg, H. W. Scott
University of Indiana (Bloomington)
N. Gary Lane
University of Iowa (Iowa City)
W. M. Furnish, B. F. Glenister, Gilbert Klapper, H. L. Strimple
University of Kansas (Lawrence)
R. A. Robison, A. J. Rowell
University of Massachusetts (Amherst)
C. W. Pitrat
University of Michigan (Ann Arbor)
R. V. Kesling, C. P. Morgan
University of Minnesota (Minneapolis)
F. M. Swain
University of Missouri (Columbia)
R. E. Peck
University of North Carolina (Wilmington)
V. A. Zullo
University of Rochester (Rochester, New York)
Curt Teichert
University of Texas (Arlington)
B. F. Perkins
University of Texas (Austin)
J. T. Sprinkle
University of Wisconsin (Madison)
D. L. Clark
University of Wyoming (Laramie)
D. W. Boyd
Western Reserve University (Cleveland, Ohio)
F. G. Stehli
Western Washington University (Bellingham)
C. A. Ross, J. R. P. Ross
Wichita State University (Kansas)
Paul Tasch
Woods Hole Oceanographic Institute (Massachusetts)
W. A. Berggren
Wright State University (Dayton, Ohio)
T. S. Wood

DECEASED

W. J. Arkell, R. S. Bassler, H. Boschma, C. C. Branson, O. M. Bulman, André Chavan, L. R. Cox, Harriet Exline, D. L. Frizzell, Julia Gardner, G. D. Hanna, Walter Häntzschel, H. J. Harrington, W. H. Hass, L. G. Hertlein, H. V. Howe, B. F. Howell, L. H. Hyman, J. B. Knight, Bernhard Kummel, M. W. deLaubenfels, Marius Lecompte, S. M. Manton, A. K. Miller, R. C. Moore, H. M. Muir-Wood, Alexander Petrunkevitch, Chr. Poulsen, H. Wienberg Rasmussen, Emma Richter, Rudolf Richter, O. H. Schindewolf, W. K. Spencer, L. W. Stephenson, Leif Størmer, E. C. Stumm, P. C. Sylvester-Bradley, Johannes Wanner, J. M. Weller, T. H. Withers, Arthur Wrigley.

Alphabetical List

Ager, D. V., London (Univ. College of Swansea)
Amsden, T. W., Norman, Okla. (Oklahoma Geol. Survey)
Arkell, W. J. (deceased)
Austin, R. L., Southampton, Eng. (Univ. Southampton)
Barker, R. W., Bellaire, Texas (unattached)
Bassler, R. S. (deceased)
Batten, R. L., New York (American Museum Nat. History)
Bayer, F. M., Washington, D.C. (Smithsonian Inst.)
Beaver, H. H., Houston, Texas (Exxon Production Research Company)

Benson, R. H., Washington, D.C. (Smithsonian Inst.)
Berdan, J. M., Washington, D.C. (U.S. Geol. Survey)
Berggren, W. A., Woods Hole, Mass. (Woods Hole Oceanographic Inst.)
Bergström, S. M., Ohio (Ohio State Univ.)
Biernat, Gertruda, Warszawa, Poland (Pánstwowe Wydawnictwo Naukowe)
Blake, D. B., Urbana, Ill. (Univ. Illinois)
Boardman, R. S., Washington, D.C. (Smithsonian Inst.)
Bold, W. A. van den, Baton Rouge, La. (Louisiana State Univ.)
Boschma, H. (deceased)
Boucot, A. J., Corvallis, Ore. (Oregon State Univ.)
Boyd, D. W., Laramie, Wyo. (Univ. Wyoming)
Branson, C. C. (deceased)
Breimer, A. J., Amsterdam, Netherlands (Inst. Aardwetensch. Vrije Univ.)
Brooks, H. K., Gainesville, Fla. (Univ. Florida)
Brower, J. C., Syracuse, N.Y. (Syracuse Univ.)
Bulman, O. M. B. (deceased)
Callomon, J. H., London (Univ. College)
Campbell, A. S., St. Mary's, Calif. (St. Mary's College)
Carpenter, F. M., Cambridge, Mass. (Harvard Univ.)
Casey, Raymond, London (Inst. Geol. Sciences)
Caster, K. E., Cincinnati, Ohio (Univ. Cincinnati)
Chavan, André (deceased)
Cheetham, A. H., Washington, D.C. (Smithsonian Inst.)
Clark, D. L., Madison, Wis. (Univ. Wisconsin)
Clarke, A. H., Jr., Washington, D. C. (Smithsonian Inst.)
Clench, W. J., Cambridge, Mass. (Harvard Univ.)
Coan, Eugene, San Francisco, Calif. (California Acad. Sci.)
Cole, W. S., Ithaca, N.Y. (Cornell Univ.)
Conway Morris, S., Milton Keynes, Eng. (Open Univ.)
Coogan, A. H., Kent, Ohio (Kent State Univ.)
Cook, P. L., London (British Museum Nat. History)
Cox, L. R. (deceased)
Curry, Dennis, Middlesex, Eng. (unattached)
Dechaseaux, Colette, Paris (Laboratoire de Paléontologie des Vertébrés)
Dell, R. K., Wellington, N.Z. (Dominion Museum)
Donovan, D. T., London, Eng. (Univ. College)
Douglass, R. C., Washington, D.C. (U.S. Geol. Survey)
Durham, J. W., Berkeley, Calif. (Univ. California)
Eames, F. E., Middlesex, Eng. (British Petroleum Co.)
Elliott, G. F., London (Iraq Petroleum Co.)
Erben, H. K., Bonn, West Germany (Friedrich Wilhelms Univ.)
Exline, Harriet (deceased)
Eyer, J. A., Oklahoma City, Okla. (Professional Geophysics, Inc.)
Fay, R. O., Norman, Okla. (Oklahoma Geol. Survey)
Fell, H. B., Cambridge, Mass. (Harvard Univ.)
Finks, R. M., Flushing, N.Y. (Queens College)
Fischer, A. G., Princeton, N.J. (Princeton Univ.)
Fisher, D. W., Albany, N.Y. (New York State Museum)
Fleming, C. A., Lower Hutt, N.Z. (New Zealand Geol. Survey)
Frizzell, D. L. (deceased)
Furnish, W. M., Iowa City, Iowa (Univ. Iowa)
Gardner, Julia (deceased)
Gibson, T. G., Washington, D.C. (Smithsonian Inst.)
Glaessner, M. F., Adelaide, S. Australia (Univ. Adelaide)
Glenister, B. F., Iowa City, Iowa (Univ. Iowa)
Grant, R. E., Washington, D.C. (Smithsonian Inst.)
Grégoire, Charles, Bruxelles, Belgium (unattached)
Haas, Fritz, Chicago, Ill. (Field Museum Nat. History)
Hahn, Gerhard, Marburg (Philipps Univ.)
Hallam, Anthony, Oxford, Eng. (Univ. Birmingham)
Hanai, Tetsuro, Tokyo (Univ. Tokyo)
Hanna, G. D. (deceased)
Häntzschel, Walter (deceased)
Harrington, H. J. (deceased)
Harris, A. G., Washington, D.C. (U.S. Geol. Survey)
Hass, W. H. (deceased)
Hatai, Kotora, Sendai, Japan (Saito Ho-on Kai Museum Nat. History)
Hay, W. W., Washington, D.C. (Joint Oceanographic Inst., Inc.)
Hazel, J. E., Washington, D.C. (U.S. Geol. Survey)
Hedgpeth, J. W., Santa Rosa, Calif. (unattached)
Henningsmoen, Gunnar, Oslo (Univ. Oslo)
Hertlein, L. G. (deceased)
Hessland, Ivar, Stockholm, Sweden (Univ. Stockholm)
Hessler, R. R., La Jolla, Calif. (Scripps Inst. Oceanography)
Hill, Dorothy, Brisbane, Australia (Univ. Queensland)
Hoffman, R. L., Radford, Va. (Radford College)
Hölder, Helmut, Münster, Germany (Univ. Münster)
Holthuis, L. B., Leiden, Netherlands (Rijksmuseum van Natuurlijke Historie)
House, M. R., Kingston upon Hull, Eng. (Univ. Hull)
Howarth, M. K., London (British Museum Nat. History)
Howe, H. V. (deceased)
Howell, B. F. (deceased)
Hyman, L. H. (deceased)
Jaanusson, Valdar, Stockholm (Naturhistoriska Riksmuseet)
Jeffords, R. M., Houston, Texas (Exxon Production Research Company)

Jeletzky, J. A., Ottawa, Ontario, Canada (Geol. Survey Canada)
Johnson, J. G., Corvallis, Ore. (Oregon State Univ.)
Jope, Margaret, Belfast, N. Ireland (Queen's Univ. of Belfast)
Karklins, O. L., Washington, D.C. (U.S. Geol. Survey)
Kauffman, E. G., Boulder, Colo. (Univ. Colorado)
Keen, A. Myra, Stanford, Calif. (Stanford Univ.)
Kesling, R. V., Ann Arbor, Mich. (Univ. Michigan)
Kier, P. M., Washington, D.C. (Smithsonian Inst.)
Klapper, Gilbert, Iowa City, Iowa (Univ. Iowa)
Knight, J. B. (deceased)
Kullmann, Jürgen, Tübingen, W. Germany (Univ. Tübingen)
Kummel, Bernhard (deceased)
Lane, N. Gary, Bloomington, Ind. (Univ. Indiana)
La Rocque Aurèle, Columbus, Ohio (Ohio State Univ.)
Laubenfels, M. W. de (deceased)
Lecompte, Marius (deceased)
Levinson, S. A., Houston, Texas (Exxon Production Research Company)
Lindström, Maurits, Marburg, Germany (Philipps Univ.)
Lochman-Balk, Christina, Socorro, N. Mex. (New Mexico Inst. Mining & Technology)
Loeblich, A. R., Jr., Los Angeles, Calif. (Univ. California)
Lohman, K. E., Washington, D.C. (U.S. Geol. Survey)
Ludbrook, N. H., Adelaide, S. Australia (South Australia Geol. Survey)
Lutaud, Geneviève, Paris (Laboratoire Cytologie, Univ. Paris)
McAlester, A. L., Dallas, Texas (Southern Methodist Univ.)
McCormick, Lavon, Houston, Texas (Getty Oil Company)
McLaren, D. J., Ottawa, Ontario, Canada (Geol. Survey Canada)
Macurda, D. B., Houston, Texas (Exxon Production Research Company)
Mandra, Y. T., San Francisco, Calif. (San Francisco State Univ.)
Manning, R. B., Washington, D.C. (Smithsonian Inst.)
Manton, S. M. (deceased)
Marwick, John, Havelock North, N.Z. (unattached)
Melville, R. V., London (Inst. Geol. Sciences)
Meyer, D. L., Cincinnati, O. (Univ. Cincinnati)
Miller, A. K. (deceased)
Miller, J. F., Springfield, Mo. (Southwest Missouri State Univ.)
Montanaro Gallitelli, Eugenia, Modena, Italy (Univ. Modena)
Moore, R. C. (deceased)
Morgan, C. P., Ann Arbor, Mich. (Univ. Michigan)
Muir-Wood, H. M. (deceased)
Müller, A. H., Freiberg, German Democratic Republic (Fachbereich Geowiss.)
Müller, K. J., Bonn, West Germany (Friedrich Wilhelms Univ.)
Newell, N. D., New York (American Museum Nat. History)
Newman, W. A., La Jolla, Calif. (Scripps Inst. Oceanography)
Norris, A. W., Calgary, Alberta, Canada (Geol. Survey Canada)
Nuttall, C. P., London (British Museum Nat. History)
Okulitch, V. J., Vancouver, Canada (Univ. British Columbia)
Olsson, A. A., Coral Gables, Fla. (Acad. Nat. Sci. Philadelphia)
Palmer, A. R., Boulder, Colo. (Geological Society of America)
Palmer, K. V. W., Ithaca, N.Y. (Paleont. Research Inst.)
Papp, Adolf, Wien, Austria (Univ. Wien)
Pawson, David, Washington, D.C. (Smithsonian Inst.)
Peck, R. E., Columbia, Mo. (Univ. Missouri)
Perkins, B. F., Arlington, Texas (Univ. Texas at Arlington)
Petrunkevitch, Alexander (deceased)
Pitrat, C. W., Amherst, Mass. (Univ. Massachusetts)
Poulsen, Chr. (deceased)
Powell, A. W. B., Auckland, N.Z. (Auckland Inst. & Museum)
Puri, H. S., Tallahassee, Fla. (unattached)
Radwin, George, San Diego, Calif. (San Diego Nat. History Museum)
Rasetti, Franco, Rome, Italy (Univ. Roma)
Rasmussen, H. Wienberg (deceased)
Regnéll, Gerhard, Lund, Sweden (Univ. Lund)
Reichel, Manfred, Basel, Switz. (Univ. Basel)
Reid, R. E. H., Belfast, N. Ireland (Queen's Univ. Belfast)
Reyment, R. A., Uppsala, Sweden (Univ. Uppsala)
Rhodes, F. H. T., Ithaca, N.Y. (Cornell Univ.)
Richter, Emma (deceased)
Richter, Rudolf (deceased)
Riedel, Adolf, Warszawa, Poland (Pánstwowe Wydawnictwo Naukowe)
Robertson, Robert, Philadelphia, Pa. (Acad. Nat. Sci.)
Robison, R. A., Lawrence, Kans. (Univ. Kansas)
Rolfe, W. D. I., Glasgow, Scot. (Univ. Glasgow)
Ross, C. A., Bellingham, Wash. (Western Washington Univ.)
Ross, J. R. P., Bellingham, Wash. (Western Washington Univ.)
Roux, Michel, Orsay, France (Univ. Paris-Sud)
Rowell, A. J., Lawrence, Kans. (Univ. Kansas)
Rudwick, M. J. S., Amsterdam (Vrije Univ.)
Sandberg, Philip, Urbana, Ill. (Univ. Illinois)
Sarjeant, W. A. S., Saskatoon, Canada (Univ. Saskatchewan)
Schindewolf, O. H. (deceased)
Schmidt, Herta, Frankfurt, Germany (Natur Museum u. Forsch.-Inst. Senckenberg)

Scott, H. W., Urbana, Ill. (Univ. Illinois)
Sdzuy, Klaus, Würzburg, Germany (Univ. Würzburg)
Seilacher, Adolf, Tübingen, West Germany (Univ. Tübingen)
Shaver, R. H., Bloomington, Ind. (Indiana Geol. Survey & Univ. Indiana)
Sieverts-Doreck, Hertha, Stuttgart-Möhringen, Germany (unattached)
Smith, A. G., San Francisco, Calif. (California Acad. Sci.)
Smith, L. A., Houston, Texas (Exxon Production Research Company)
Sohl, N. F., Washington, D.C. (U.S. Geol. Survey)
Sohn, I. G., Washington, D.C. (U.S. Geol. Survey)
Soot-Ryen, Tron, Hosle, Nor. (unattached)
Spencer, W. K. (deceased)
Sprinkle, J. T., Austin, Texas (Univ. Texas)
Staton, R. D., Providence, R.I. (Brown Univ.)
Stehli, F. G., Cleveland, Ohio (Western Reserve Univ.)
Stenzel, H. B., Baton Rouge, La. (Louisiana State Univ.)
Stephenson, L. W. (deceased)
Størmer, Leif (deceased)
Strimple, H. L., Iowa City, Iowa (Univ. Iowa)
Struve, Wolfgang, Frankfurt, Germany (Natur-Museum u. Forsch.-Inst. Senckenberg)
Stubblefield, Sir James, London (unattached)
Stumm, E. C. (deceased)
Swain, F. M., Minneapolis, Minn. (Univ. Minnesota)
Sweet, W. C., Columbus, Ohio (Ohio State Univ.)
Sylvester-Bradley, P. C. (deceased)
Tappan, Helen, Los Angeles, Calif. (Univ. California)
Tasch, Paul, Wichita, Kans. (Wichita State Univ.)
Teichert, Curt, Rochester, N.Y. (Univ. Rochester)
Thompson, M. L., Urbana, Ill. (Illinois State Geol. Survey)
Tripp, R. P., Sevenoaks, Kent, Eng. (unattached)
Trueman, E. R., Manchester, Eng. (Univ. Manchester)
Turner, Ruth, Cambridge, Mass. (Harvard Univ.)
Ubaghs, Georges, Liège, Belgium (Univ. Liège)
Utgaard, John, Carbondale, Ill. (Southern Illinois Univ.)
Van Couvering, John, Boulder, Colo. (Univ. Colorado)
Wagner, C. D., Fairbanks, Alaska (Univ. Alaska)
Wanner, Johannes (deceased)
Weir, John, Tayport, Fife, Scotland (Univ. Glasgow)
Weller, J. M. (deceased)
Wells, J. W., Ithaca, N.Y. (Cornell Univ.)
Whittington, H. B., Cambridge, Eng. (Univ. Cambridge)
Williams, Alwyn, Glasgow, Scot. (Univ. Glasgow)
Withers, T. H. (deceased)
Wood, T. S., Dayton, Ohio (Wright State Univ.)
Wright, A. D., Belfast, N. Ireland (Queen's Univ. Belfast)
Wright, C. W., Dorset, Eng. (unattached)
Wrigley, Arthur (deceased)
Yochelson, E. L., Washington, D.C. (U.S. Geol. Survey)
Yonge, Sir Maurice, Edinburgh, Scotland (unattached)
Zeller, D. E. Nodine, Lawrence, Kans. (Kansas Geol. Survey)
Ziegler, Willi, Marburg, Germany (Philipps Univ.)
Zullo, V. A., Wilmington, N.C. (Univ. North Carolina)

EDITORIAL PREFACE

The aim of the *Treatise on Invertebrate Paleontology,* as originally conceived and consistently pursued, is to present the most comprehensive and authoritative, yet compact statement of knowledge concerning invertebrate fossil groups that can be formulated by collaboration of competent specialists.

The *Treatise* is divided into parts that bear index letters, each except the initial and concluding ones being defined to include designated groups of invertebrates. This arrangement provides for independence of the several parts as regards date of publication of both original editions and revisions (updatings of entire parts) and supplements (updatings of portions of lettered parts). Pages in each part bear the assigned index letter joined with numbers beginning with 1 and running consecutively to the end of the part. Several parts are of such length that they are published in two or more volumes with continuous pagination through successive volumes.

The subjects treated in connection with each large group of invertebrates include: (1) morphological features, with special reference to hard parts, (2) ontogeny, (3) classification, (4) geologic and geographic distribution, (5) evolutionary trends and phylogeny, (6) paleoecology, and (7) systematic description of genera, subgenera, and higher taxonomic units. Selected lists of references only were furnished in earlier parts of the *Treatise,* but since the mid-1960's, the tendency has been to make these lists as comprehensive as possible, and in particular, to supply reliable bibliographical documentation for all taxonomic names dealt with in the text.

The *Treatise* project has received substantial financial support from several sources. From 1948 to 1980, the Geological Society of America contributed a total of $181,200 through the bequest of R. A. F. Penrose, Jr. From 1959 to 1977 the National Science Foundation of the United States made several grants totaling $791,700. From 1976 to 1978 contributions of $40,000 were received from the Burton McCollum Fund, which is administered by the Kansas University Endowment Association. Since 1976, significant support has been provided through the bequest of Raymond C. and Lilian Moore, administered by the Kansas University Endowment Association. These collective funds have been used primarily to maintain editorial operations at the University of Kansas, exclusive of any stipend for the editor, and to provide assistance to authors in preparation of manuscripts and illustrations. Grateful acknowledgment is expressed on behalf of the societies sponsoring the *Treatise,* the University of Kansas, and innumerable persons benefited by the *Treatise* project.

Conodonts were first covered in the *Treatise* in 1962 in Part W, Miscellanea, a volume that in 250-odd pages also reviewed conoidal shells, worms, trace fossils, and problematica. Since then, increase in knowledge of the conodonts has been far more rapid than during any comparable period in their history of study. By at least 1970, Curt Teichert, Editor, had begun discussion with D. L. Clark concerning the need for a *Treatise* revision of the conodonts. The first detailed plans for the review were made in 1971, when several authors of this volume met in Marburg, Germany. Coordination of the task was assigned to D. L. Clark, and a deadline of September 1974 was set for authors.

As with virtually all *Treatise* volumes with many authors, some authors of the conodont revision were unable to meet the assigned deadline. A nearly complete manuscript was received at the *Treatise* editorial office in December 1976. At about the same time, major editorial responsibility for the conodont volume was transferred to R. A. Robison. Since 1974, however, manuscript had been completed and was in physical production for six other *Treatise* volumes: the three volumes of original Part T (1978), original Part A (1979), and the two volumes of Part F Supplement 1, Rugosa and Tabulata (1981). These having already received priority in processing, publication of the conodont revision was regrettably delayed. During late 1979, just before editorial processing of Part W Supplement 2 began, authors were extended the opportunity for minor updating of the conodont manuscript but not illustrations. An addendum, which lists the names, type species, and family affiliations for conodont genera described between about 1975 and mid-1980, was added.

Authors of this volume are commended for their continued cooperation and general patience. Partial funding for the color frontispiece was provided by Exxon Corporation through a grant to the University of Wisconsin.

ZOOLOGICAL NAMES

Many questions arise in connection with zoological names, especially including those related to acceptability and to alterations of some that may be allowed or demanded. Procedure in obtaining answers to these questions is guided and to a large extent governed by regulations published (1961) in the *International Code of Zoological Nomenclature*[1] (hereinafter cited simply as the *Code*). The prime object of the *Code* is to promote stability and universality in the use of the scientific names of animals, ensuring also that each name is distinct and unique while avoiding restrictions on freedom of taxonomic thought or action. Priority is a basic principle, but under specified conditions its application can be modified. This is all well and good, yet nomenclatural tasks confronting the zoological taxonomist are formidable. They warrant the complaint of some that zoology, including paleozoology, should be the study

[1] N. R. Stoll and others (ed. comm.), *International Code of Zoological Nomenclature, adopted by the XV International Congress of Zoology,* xvii + 176 p. (International Trust for Zoological Nomenclature, London, 1961; 2d edit., xx + 176 p., 1964).

of animals rather than of names applied to them.

Several ensuing pages are devoted to aspects of zoological nomenclature that are judged to have chief importance in relation to procedures adopted in the *Treatise*. Terminology is explained, and examples of style employed in the nomenclatural parts of systematic descriptions are given.

A draft of a revised edition of the *Code* was submitted to the meeting of the International Union of Biological Sciences at Helsinki, Finland, in August 1979. It is expected that this revised edition will not come into force before some time in 1982 (R. V. Melville, written commun., July, 1981) and the existing *Code* of 1961 is, therefore, strictly followed in the present volume.

TAXA GROUPS

Each taxonomic unit (taxon, pl., taxa) belongs to a rank in the adopted hierarchy of classificatory divisions. In part, this hierarchy is defined by the *Code* to include a species-group of taxa, a genus-group, and a family-group. Units of lower rank than subspecies are excluded from zoological nomenclature and those higher than superfamily of the family-group are not regulated by the *Code*. It is natural and convenient to discuss nomenclatural matters in general terms first and then to consider each of the taxa groups separately. Especially important is the provision that within each taxa group, classificatory units are coordinate (equal in rank), whereas units of different taxa groups are not coordinate.

FORMS OF NAMES

All zoological names are divisible into groups based on their form (spelling). The first-published form (or forms) of a name is defined as original spelling (*Code,* Art. 32) and any later-published form (or forms) of the same name is designated as subsequent spelling (Art. 33). Obviously, original and subsequent spellings of a given name may or may not be identical and this affects consideration of their correctness. Further, examination of original spellings of names shows that by no means all can be distinguished as correct. Some are incorrect, and the same is true of subsequent spellings.

Original Spellings

If the first-published form of a name is consistent and unambiguous, the original spelling is defined as correct unless it contravenes some stipulation of the *Code* (Arts. 26-31), or the original publication contains clear evidence of an inadvertent error, in the sense of the *Code,* or, among names belonging to the family-group, unless correction of the termination or the stem of the type genus is required. An original spelling that fails to meet these requirements is defined as incorrect.

If a name is spelled in more than one way in the original publication, the form adopted by the first reviser is accepted as the correct original spelling, provided that it complies with mandatory stipulations of the *Code* (Arts. 26-31).

Incorrect original spellings are any that fail to satisfy requirements of the *Code,* represent an inadvertent error, or are one of multiple original spellings not adopted by a first reviser. These have no separate status in zoological nomenclature and therefore cannot enter into homonymy or be used as replacement names and they call for correction. For example, a name originally published with a diacritic mark, apostrophe, diaeresis, or hyphen requires correction by deleting such features and uniting parts of the name originally separated by them, except that deletion of an umlaut from a vowel in a name derived from a German word or personal name requires the insertion of "e" after the vowel.

Subsequent Spellings

If a name classed as a subsequent spelling is identical with an original spelling, it is distinguishable as correct or incorrect on the same criteria that apply to the original spelling. This means that a subsequent spelling identical with a correct original spelling is also correct, and one identical with an incorrect original spelling is also incorrect. In the latter case, both original and subsequent spellings require correction (authorship and date of the original incorrect spelling being retained).

If a subsequent spelling differs from an

original spelling in any way, even by the omission, addition, or alteration of a single letter, the subsequent spelling must be defined as a different name (except that such changes as altered terminations of adjectival specific names to obtain agreement in gender with associated generic names, of family-group names to denote assigned taxonomic rank, and corrections for originally used diacritic marks, hyphens, and the like are excluded from spelling changes conceived to produce a different name). In certain cases species-group names having variable spellings are regarded as homonyms as specified in Art. 58 of the *Code.*

Altered subsequent spellings other than the exceptions noted may be either intentional or unintentional. If demonstrably intentional, the change is designated as an emendation. Emendations may be either justifiable or unjustifiable. Justifiable emendations are corrections of incorrect original spellings, and these take the authorship and date of the original spellings. Unjustifiable emendations are names having their own status in nomenclature, with author and date of their publication; they are junior objective synonyms of the name in its original form.

Subsequent spellings that differ in any way from the original spellings, other than previously noted exceptions, and that are not classifiable as emendations are defined as incorrect subsequent spellings. They have no status in nomenclature, do not enter into homonymy, and cannot be used as replacement names. It is the purpose of the following chapters to explain in some detail the implications of various kinds of subsequent spellings and how these are dealt with in the *Treatise.*

AVAILABLE AND UNAVAILABLE NAMES

Available Names

An available zoological name is any that conforms to all mandatory provisions of the *Code.* Such names are classifiable in groups which are recognized in the *Treatise,* though not explicitly differentiated in the *Code.* They are as follows:

1) So-called *"inviolate names"* include all available names that are not subject to alteration from their originally published form. They comprise correct original spellings and commonly include correct subsequent spellings, but include no names classed as emendations. Here belong most genus-group names (including those for collective groups), some of which differ in spelling from others by only a single letter or by the sequential order of their letters.

2) Names may be termed *"perfect names"* if, as originally published, they meet all mandatory requirements, needing no correction of any kind, but nevertheless are legally alterable in such ways as changing the termination (e.g., many species-group names, family-group names). This group does not include emended incorrect original spellings (e.g., *Boucekites,* replacement of *Boučekites*).

3) *"Imperfect names"* are available names that as originally published contain mandatorily emendable defects. Incorrect original spellings are imperfect names. Examples of emended imperfect names are: among species-group names, *guerini* (not *Guérini*), *obrienae* (not *O'Brienae*), *terranovae* (not *terra-novae*), *nunezi* (not *Nuñezi*), *Spironema rectum* (not *Spironema recta,* because generic name is neuter, not feminine); among genus-group names, *Broeggeria* (not *Bröggeria*), *Obrienia* (not *O'Brienia*), *Maccookites* (not *McCookites*); among family-group names Guembellotriinae (not Gümbellotriinae), Spironematidae (not Spironemidae, incorrect stem), Athyrididae (not Athyridae, incorrect stem). The use of "variety" for named divisions of fossil species, according to common practice of some paleontologists, gives rise to imperfect names, which generally are emendable (*Code,* Art. 45e) by omitting this term so as to indicate the status of this taxon as a subspecies. The name of a variety is always of feminine gender. If the variety is converted into a species or subspecies, the name takes on the gender of the associated genus.

4) *"Vain names"* are available names consisting of unjustified intentional emendations of previously published names. The emendations are unjustified because they are not demonstrable as corrections of incorrect original spellings as defined by the *Code* (Art. 32c). Vain names have status in nomenclature under their own author-

ship and date. They constitute junior objective synonyms of names in their original form. Examples are: among species-group names, *geneae* (published as replacement of original unexplained masculine, *geni,* which now is not alterable), *ohioae* (invalid change from original *ohioensis*); among genus-group names, *Graphiodactylus* (invalid change from original *Graphiadactyllis*); among family-group names, Graphiodactylidae (based on junior objective synonym having invalid vain name).

5) An important group of available zoological names can be distinguished as *"transferred names."* These comprise authorized sorts of altered names in which the change depends on transfer from one taxonomic rank to another, or possibly on transfers in taxonomic assignment of subgenera, species, or subspecies. Most commonly the transfer calls for a change in termination of the name so as to comply with stipulations of the *Code* on endings of family-group taxa and agreement in gender of specific names with associated generic names. Transferred names may be derived from any of the preceding groups except the first. Examples are: among species-group names, *Spirifer ambiguus* (masc.) to *Composita ambigua* (fem.), *Neochonetes transversalis* to *N. granulifer transversalis* or vice versa; among genus-group names, *Schizoculina* to *Oculina (Schizoculina)* or vice versa; among family-group names, Orthidae to Orthinae or vice versa, or superfamily Orthacea derived from Orthidae or Orthinae; among suprafamilial taxa (not governed by the *Code*), order Orthida to suborder Orthina or vice versa. The authorship and date of transferred names are not affected by the transfer, but the author responsible for the transfer and the date of his action are recorded in the *Treatise.*

6) Improved or *"corrected names"* include both mandatory and allowable emendations of imperfect names and of suprafamilial names, which are not subject to regulation as to name form. Examples of corrected imperfect names are given with the discussion of group 3. Change from the originally published ordinal name Endoceroidea (Teichert, 1933) to the presently recognized Endocerida illustrates a "corrected" suprafamilial name. Group 6 names differ from those in group 5 in not being dependent on transfers in taxonomic rank or assignment, but some names are classifiable in both groups.

7) *"Substitute names"* are available names expressly proposed as replacements for invalid zoological names, such as junior homonyms. These may be classifiable also as belonging in groups 1, 2, or 3. The glossary appended to the *Code* refers to these as "new names" *(nomina nova)* but they are better designated as substitute names, since their newness is temporary and relative. The first-published substitute name that complies with the definition here given takes precedence over any other. An example is *Marieita* Loeblich & Tappan, 1964, as substitute for *Reichelina* Marie, 1955, *non* Erk, 1942.

8) *"Conserved names"* include a relatively small number of species-group, genus-group, and family-group names which have come to be classed as available and valid by action of the International Commission on Zoological Nomenclature exercising its plenary powers to this end or ruling to conserve a junior synonym in place of a rejected "forgotten" name *(nomen oblitum)* (Art. 23b). Currently, such names are entered on appropriate "Official Lists," which are published from time to time.

It is useful for convenience and brevity of distinction in recording these groups of available zoological names to employ Latin designations in the pattern of *nomen nudum* (abbr., *nom. nud.*) and others. Thus we recognize the preceding numbered groups as follows: 1) *nomina inviolata* (sing., *nomen inviolatum,* abbr., *nom. inviol.*), 2) *nomina perfecta (nomen perfectum, nom. perf.),* 3) *nomina imperfecta (nomen imperfectum, nom. imperf.)* 4) *nomina vana (nomen vanum, nom. van.)* 5) *nomina translata (nomen translatum, nom. transl.),* 6) *nomina correcta (nomen correctum, nom. correct.),* 7) *nomina substituta (nomen substitutum, nom. subst.),* 8) *nomina conservata (nomen conservatum, nom. conserv.).* It should be noted that the *Code* does not differentiate between different kinds of subsequent intentional changes of spelling, all of which are grouped as "emendations" (see below).

Additional to the groups differentiated

above, the *Code* (Art. 17) specifies that a zoological name is not prevented from availability a) by becoming a junior synonym, for under various conditions this may be reemployed, b) for a species-group name by finding that original description of the taxon relates to more than a single taxonomic entity or to parts of animals belonging to two or more such entities, c) for species-group names by determining that it first was combined with an invalid or unavailable genus-group name, d) by being based only on part of an animal, sex of a species, ontogenetic stage, or one form of a polymorphic species, e) by being originally proposed for an organism not considered to be an animal but now so regarded, f) by incorrect original spelling which is correctable under the *Code,* g) by anonymous publication before 1951, h) by conditional proposal before 1961, i) by designation as a variety or form before 1961, j) by concluding that a name is inappropriate (Art. 18), or k) for a specific name by observing that it is tautonymous (Art. 18).

Unavailable Names

All zoological names which fail to comply with mandatory provisions of the *Code* are unavailable names and have no status in zoological nomenclature. None can be used under authorship and date of original publication as a replacement name *(nom. subst.)* and none preoccupies for purposes of the Law of Homonymy. Names identical in spelling with some, but not all, unavailable names can be classed as available if and when they are published in conformance to stipulations of the *Code,* and they are then assigned authorship and take date of the accepted publication. Different groups of unavailable names can be discriminated as follows.

9) *"Naked names"* include all those that fail to satisfy provisions stipulated in Article 11 of the *Code,* which states general requirements of availability. In addition they include names that, if published before 1931, were unaccompanied by a description, definition, or indication (Arts. 12, 16), as well as names published after 1930 that lacked accompanying statement of characters purporting to serve for differentiation of the taxon, or definite bibliographic reference to such a statement, or that were not proposed expressly as replacement *(nom. subst.)* of a preexisting available name (Art. 13a), or that were unaccompanied by definite fixation of a type species by original designation or indication (Art. 13b). Examples of "naked names" are: among species-group taxa, *Valvulina mixta* PARKER & JONES, 1865 (=*Cribrobulimina mixta* CUSHMAN, 1927, available and valid); among genus-group taxa, *Orbitolinopsis* SILVESTRI, 1932 (=*Orbitolinopsis* HENSON, 1948, available but classed as invalid junior synonym of *Orbitolina* D'ORBIGNY, 1850); among family-group taxa, Aequilateralidae D'ORBIGNY, 1846 (lacking type-genus), Hélicostègues D'ORBIGNY, 1826 (vernacular not latinized by later authors, Art. 11e(iii)), Poteriocrinidae AUSTIN & AUSTIN, 1843, =family Poteriocrinoidea AUSTIN & AUSTIN, 1842 (neither 1843 or 1842 names complying with Art. 11e, which states that "a family-group name must, when first published, be based on the name then valid for a contained genus," such valid name in the case of this family being *Poteriocrinites* MILLER, 1821).

10) "Denied names" include all those that are defined by the *Code* (Art. 32c) as incorrect original spellings. Examples are: specific names, *nova-zelandica, mülleri, 10-brachiatus*; generic names, *M'Coyia, Størmerella, Römerina, Westergårdia*; family name, Růžičkinidae. Uncorrected "imperfect names" are "denied names" and unavailable, whereas corrected "imperfect names" are available.

11) *"Impermissible names"* include all those employed for alleged genus-group taxa other than genus and subgenus (Art. 42a) (e.g., supraspecific divisions of subgenera), and all those published after 1930 that are unaccompanied by definite fixation of a type species (Art. 13b). Examples of impermissible names are: *Martellispirifer* GATINAUD, 1949, and *Mirtellispirifer* GAUTINAUD, 1949, indicated respectively as a section and subsection of the subgenus *Cyrtospirifer; Fusarchaias* REICHEL, 1949, without definitely fixed type species (=*Fusarchaias* REICHEL, 1952, with *F. bermudezi* designated as type species).

12) *"Null names"* include all those that are defined by the *Code* (Art. 33b) as in-

correct subsequent spellings, which are any changes of original spelling not demonstrably intentional. Such names are found in all ranks of taxa. It is not always evident from the original publication whether an incorrect subsequent spelling is intentional, resulting in a "vain name" which is invalid but available (category 4 above), or unintentional, resulting in a "null name" which is invalid and unavailable. In such cases, the decision of a subsequent author will sometimes have to be arbitrary according to his best judgment.

13) *"Forgotten names"* are defined (Art. 23b) as senior synonyms that have remained unused in primary zoological literature for more than 50 years. Such names are not to be used unless so directed by ICZN.

Latin designations for the discussed groups of unavailable zoological names are as follows: 9) *nomina nuda* (sing., *nomen nudum,* abbr. *nom. nud.*), 10) *nomina negata (nomen negatum, nom. neg.),* 11) *nomina vetita (nomen vetitum, nom. vet.),* 12) *nomina nulla (nomen nullum, nom. null.),* 13) *nomina oblita (nomen oblitum, nom. oblit.).*

VALID AND INVALID NAMES

Important distinctions relate to valid and available names, on one hand, and to invalid and unavailable names, on the other. Whereas determination of availability is based entirely on objective considerations guided by Articles of the *Code,* conclusions as to validity of zoological names may be partly subjective. A valid name is the correct one for a given taxon, which may have two or more available names but only a single correct name, generally the oldest. Obviously, no valid name can also be an unavailable name, but invalid names may include both available and unavailable names. Any name for a given taxon other than the valid name is an invalid name.

A sort of nomenclatorial no-man's-land is encountered in considering the status of some zoological names, such as *"doubtful names," "names under inquiry,"* and *"forgotten names."* Latin designations of these are *nomina dubia, nomina inquirenda,* and *nomina oblita,* respectively. Each of these groups may include both available and unavailable names, but the latter can well be ignored. Names considered to possess availability conduce to uncertainty and instability, which ordinarily can be removed only by appealed action of ICZN. Because few zoologists care to bother in seeking such remedy, the "wastebasket" names persist.

SUMMARY OF NAME GROUPS

Partly because only in such publications as the *Treatise* is special attention to groups of zoological names called for and partly because new designations are here introduced as means of recording distinctions explicitly as well as compactly, a summary may be useful. In the following tabulation valid groups of names are indicated in boldface type, whereas invalid ones are printed in italic.

Definitions of Name Groups

nomen conservatum (nom. conserv.). Name unacceptable under regulations of the *Code* which is made valid, either with original or altered spelling, through procedures specified by the *Code* or by action of ICZN exercising its plenary powers.

nomen correctum (nom. correct.). Name with intentionally altered spelling of sort required or allowable by the *Code* but not dependent on transfer from one taxonomic rank to another ("improved name"). (See *Code,* Arts. 26b, 27, 29, 30a(i)(3), 31, 32c(i), 33a; in addition, change of endings for suprafamilial taxa not regulated by the *Code.*)

nomen imperfectum (nom. imperf.). Name that as originally published meets all mandatory requirements of the *Code* but contains defect needing correction ("imperfect name"). (See *Code,* Arts. 26b, 27, 29, 32c, 33a.)

nomen inviolatum (nom. inviol.). Name that as originally published meets all mandatory requirements of the *Code* and also is not correctable or alterable in any way ("inviolate name").

nomen negatum (nom. neg.). Name that as originally published constitutes invalid original spelling, and although possibly meeting all other mandatory requirements of the *Code,* cannot be used and has no separate status in nomenclature ("denied name"). It is to be corrected wherever found.

nomen nudum (nom. nud.). Name that as originally published fails to meet mandatory requirements of the *Code* and, having no status in nomenclature, is not correctable to establish original authorship and date ("naked name").

nomen nullum (nom. null.). Name consisting of an unintentional alteration in form (spelling) of a previously published name (either available name, as *nom. inviol., nom. perf., nom. imperf., nom. transl.*; or unavailable name, as *nom. neg., nom. nud., nom. van.,* or another *nom. null.*) ("null name").

nomen oblitum (nom. oblit.). Name of senior synonym unused in primary zoological literature in more than 50 years, not to be used unless so directed by ICZN ("forgotten name").

nomen perfectum (nom. perf.). Name that as originally published meets all mandatory requirements of the *Code* and needs no correction of any kind but which nevertheless is validly alterable by change of ending ("perfect name").

nomen substitutum (nom. subst.). Replacement name published as substitute for an invalid name, such as junior homonym (equivalent to "new name").

nomen translatum (nom. transl.). Name that is derived by valid emendation of a previously published name as result of transfer from one taxonomic rank to another within the group to which it belongs ("transferred name").

nomen vanum (nom. van.). Name consisting of an invalid intentional change in form (spelling) from a previously published name, such invalid emendation having status in nomenclature as a junior objective synonym ("vain name").

nomen vetitum (nom. vet.). Name of genus-group taxon not authorized by the *Code* or, if first published after 1930, without definitely fixed type species ("impermissible name").

Except as specified otherwise, zoological names accepted in the *Treatise* may be understood to be classifiable either as *nomina inviolata* or *nomina perfecta* (omitting from notice *nomina correcta* among specific names) and these are not discriminated. Names which are not accepted for one reason or another include junior homonyms, senior synonyms classifiable as *nomina negata* or *nomina nuda,* and numerous junior synonyms which include both objective *(nomina vana)* and subjective types; rejected names are classified as completely as possible.

NAME CHANGES IN RELATION TO TAXA GROUPS

Species-group Names

Detailed consideration of valid emendation of specific and subspecific names is unnecessary here because it is well understood and relatively inconsequential. When the form of adjectival specific names is changed to obtain agreement with the gender of a generic name in transferring a species from one genus to another, it is never needful to label the changed name as a *nom. correct.* Likewise, transliteration of a letter accompanied by a diacritical mark in manner now called for by the *Code* (as in changing originally published *bröggeri* to *broeggeri*) or elimination of a hyphen (as in changing originally published *cornu-oryx* to *cornuoryx*) does not require "*nom. correct.*" with it.

Genus-group Names

So rare are conditions warranting change of the originally published valid form of generic and subgeneric names that lengthy discussion may be omitted. Only elimination of diacritical marks of some names in this category seems to furnish basis for valid emendation. It is true that many changes of generic and subgeneric names have been published, but virtually all of these are either *nomina vana* or *nomina nulla.* Various names which formerly were classed as homonyms now are not, for two names that differ only by a single letter (or in original publication by presence or absence of a diacritical mark) are construed to be entirely distinct.

Examples in use of classificatory designations for genus-group names as previously given are the following, which also illustrate designation of type species as explained later.

Paleomeandron PERUZZI, 1881, p. 8 [**P. elegans*; SD HÄNTZSCHEL, 1975, p. *W*91] [=*Palaeomeandron* FUCHS, 1885, p. 395, *nom. van.*].

Stichophyma POMEL, 1872 [**Manon turbinatum* RÖMER, 1841; SD RAUFF, 1893] [=*Stychophyma* VOSMAER, 1885, *nom null.; Sticophyma* MORET, 1924, *nom. null.*].

Vacuocyathus OKULITCH, 1950, p. 392 [**Coelocyathus kidrjassovensis* VOLOGDIN, 1937, p. 478, *nom. nud.*; 1939, p. 237; OD] [=*Coelocyathus* VOLOGDIN, 1934, p. 502, *nom. nud.*; 1937, p. 472, *nom. nud.*].

Cyrtograptus CARRUTHERS, 1867, p. 540, *nom. correct.* LAPWORTH, 1873, *pro Crytograpsus* CARRUTHERS, 1867, ICZN Op. 650, 1963 [**Cyrtograpsus murchisoni*; OD].

As has been pointed out above, it is in many cases difficult to decide whether a change in spelling of a name by a subsequent author was intentional or unintentional, that is, whether it should be classified as *nomen vanum* or *nomen nullum,* and the decision will often have to be arbitrary.

Family-group Names: Use of "nom. transl."

The *Code* specifies the endings only for subfamily (-inae) and family (-idae) but all family-group taxa are defined as coordinate, signifying that for purposes of priority a name published for a taxon in any category and based on a particular type genus shall date from its original publication for a taxon in any category, retaining this priority (and authorship) when the taxon is treated as belonging to a lower or higher category. By exclusion of -inae and -idae, respectively reserved for subfamily and family, the endings of names used for tribes and superfamilies must be unspecified different letter combinations. These, if introduced subsequent to designation of a subfamily or family based on the same nominate genus, are *nomina translata,* as is also a subfamily that is elevated to family rank or a family reduced to subfamily rank. In the *Treatise* it is desirable to distinguish the valid alteration comprised in the changed ending of each transferred family-group name by the abbreviation *"nom. transl."* and record of the author and date belonging to this alteration. This is particularly important in the case of superfamilies, for it is the author who introduced this taxon that one wishes to know about rather than the author of the superfamily as defined by the *Code,* for the latter is merely the individual who first defined some lower-rank family-group taxon that contains the nominate genus of the superfamily. The publication containing introduction of the superfamily *nomen translatum* is likely to furnish the information on taxonomic considerations that support definition of the unit.

Examples of the use of *"nom. transl."* are the following.

Subfamily STYLININAE d'Orbigny, 1851

[*nom. transl.* VERRILL, 1864, *ex* Stylinidae D'ORBIGNY, 1851]

Superfamily ARCHAEOCTONOIDEA Petrunkevitch, 1949

[*nom. transl.* PETRUNKEVITCH, 1955, *ex* Archaeoctonidae PETRUNKEVITCH, 1949]

Superfamily ANCYLOCERATACEAE Meek, 1876

[*nom. transl.* WRIGHT, 1957, *ex* Ancyloceratidae MEEK, 1876]

Family-group Names: Use of "nom. correct."

Valid name changes classed as *nomina correcta* do not depend on transfer from one category of family-group units to another but most commonly involve correction of the stem of the nominate genus; in addition, they include somewhat arbitrarily chosen modifications of endings for names of tribes or superfamilies. Examples of the use of *"nom. correct."* are the following.

Family STREPTELASMATIDAE Nicholson, 1889

[*nom. correct.* WEDEKIND, 1927, *pro* Streptelasmidae NICHOLSON, 1889]

Family PALAEOSCORPIIDAE Lehmann, 1944

[*nom correct.* PETRUNKEVITCH, 1955, *pro* Palaeoscorpionidae LEHMANN, 1944]

Family AGLASPIDIDAE Miller, 1877

[*nom. correct.* STØRMER, 1959, *pro* Aglaspidae MILLER, 1877]

Superfamily AGARICIICAE Gray, 1847

[*nom. correct.* WELLS, 1956 (*pro* Agaricioidae VAUGHAN & WELLS, 1943, *nom. transl.* WELLS, 1956, *ex* Agaricidae GRAY, 1847)]

Family-group Names: Replacements

Family-group names are formed by adding letter combinations (prescribed for family and subfamily) to the stem of the name belonging to the genus (nominate genus) first chosen as type of the assemblage. The type genus need not be the oldest in terms of receiving its name and definition, but it must be the first-published as name-giver to a family-group taxon among all those included. Once fixed, the family-group name remains tied to the nominate genus even if its name is changed by reason of status as a junior homonym or junior synonym, either objective or subjective. Seemingly, the *Code* requires replacement of a family-group name only in the event that the nominate genus is found to have been invalid when it was proposed (Arts. 11e, 39), and then a substitute family-group name is accepted if it is formed from the oldest available substitute name for the nominate genus. Authorship and date attributed to the replacement family-group name are determined by first publication of the changed family-group name, but for purposes of the Law of Priority, they take the date of the replaced name. Numerous long-used family-group names are incorrect in being *nomina nuda,* since they fail to satisfy criteria of availability (Art. 11e). These also demand replacement by valid names.

The aim of family-group nomenclature is greatest possible stability and uniformity, just as in other zoological names. Experience indicates the wisdom of sustaining family-group names based on junior subjective synonyms if they have priority of publication, for opinions of different workers as to the synonymy of generic names founded on different type species may not agree and opinions of the same worker may alter from time to time. The retention similarly of first-published family-group names which are found to be based on junior objective synonyms is less clearly desirable, especially if a replacement name derived from the senior objective synonym has been recognized very long and widely. To displace a much-used family-group name based on the senior objective synonym by disinterring a forgotten and virtually unused family-group name based on a junior objective synonym because the latter happens to have priority of publication is unsettling.

Replacement of a family-group name may be needed if the former nominate genus is transferred to another family group. Then the first-published name-giver of the family-group assemblage in the remnant taxon is to be recognized in forming a replacement name.

Family-group Names: Authorship and Date

All family-group taxa having names based on the same type genus are attributed to the author who first published the name for any of these assemblages, whether tribe, subfamily, or family (superfamily being almost inevitably a later-conceived taxon). Accordingly, if a family is divided into subfamilies or a subfamily into tribes, the name of no such subfamily or tribe can antedate the family name. Also, every family containing differentiated subfamilies must have a nominate *(sensu stricto)* subfamily, which is based on the same type genus as that for the family, and the author and date set down for the nominate subfamily invariably are identical with those of the family, without reference to whether the author of the family or some subsequent author introduced subdivisions.

Changes in the form of family-group names of the sort constituting *nomina correcta,* as previously discussed, do not affect authorship and date of the taxon concerned, but in the *Treatise* it is desirable to record the authorship and date of the correction.

Suprafamilial Taxa

International rules of zoological nomenclature as given in the *Code* are limited to stipulations affecting lower-rank categories (subspecies to superfamily). Suprafamilial categories (suborder to phylum) are either unmentioned or explicitly placed outside of the application of zoological rules. The *Copenhagen Decisions on Zoological Nomenclature*[1] (1953, Arts. 59-69) proposed to adopt rules for naming suborders and higher taxonomic divisions up to and including phylum, with provision for designating a type genus for each, hopefully in such manner as not to interfere with the taxonomic freedom of workers. Procedures for applying the Law of Priority and Law of Homonymy to suprafamilial taxa were outlined and for dealing with the names for such units and their authorship, with assigned dates, when they should be transferred on taxonomic grounds from one rank to another. The adoption of terminations of names, different for each category but uniform within each, was recommended.

The Colloquium on Zoological Nomenclature which met in London during the week just before the XVth International Congress of Zoology convened in 1958 thoroughly discussed the proposals for regulating suprafamilial nomenclature, as well as many others advocated for inclusion in the new *Code* or recommended for exclusion from it. A decision which was supported by a wide majority of the participants in the Colloquium was against the establishment of rules for naming taxa above

[1] Francis Hemming, ed., *Copenhagen Decisions on Zoological Nomenclature. Additions to, and modifications of, the Règles Internationales de la Nomenclature Zoologique,* xxix + 135 p. (International Trust for Zoological Nomenclature, London, 1953).

family-group rank, mainly because it was judged that such regulation would unwisely tie the hands of taxonomists. For example, a class or order defined by an author at a given date, using chosen morphologic characters (e.g., gills of bivalves), should not be allowed to freeze nomenclature, taking precedence over another, later-proposed class or order distinguished by different characters (e.g., hinge-teeth of bivalves). Even the fixing of type genera for suprafamilial taxa might have small value, if any, hindering taxonomic work rather than aiding it. At all events, no legal basis for establishing such types and for naming these taxa has yet been provided.

The considerations just stated do not prevent the editors of the *Treatise* from making "rules" for dealing with suprafamilial groups of animals described and illustrated in this publication. At least a degree of uniform policy is thought to be needed, especially for the guidance of *Treatise* authors. This policy should accord with recognized general practice among zoologists; but where general practice is indeterminate or nonexistent, our own procedure in suprafamilial nomenclature needs to be specified as clearly as possible. This pertains especially to decisions about names themselves, about citation of authors and dates, and about treatment of suprafamilial taxa which on taxonomic grounds are changed from their originally assigned rank. Accordingly, a few "rules" expressing *Treatise* policy are given here, some with examples of their application.

1) The name of any suprafamilial taxon must be a Latin or latinized uninominal noun of plural form, or treated as such, a) with a capital initial letter, b) without diacritical mark, apostrophe, diaeresis, or hyphen, and c) if a component consisting of a numeral, numerical adjective, or adverb is used, this must be written in full (e.g., Stethostomata, Trionychi, Septemchitonina, Scorpiones, Subselliflorae). No uniformity in choice of ending for taxa of a given rank is demanded (e.g., orders named Gorgon*acea,* Millepor*ina,* Rugo*sa,* Scleractin*ia,* Stromatopor*oidea,* Phalang*ida*).

2) Names of suprafamilial taxa may be constructed in almost any way, a) intended to indicate morphological attributes (e.g., Lamellibranchiata, Cyclostomata, Toxoglossa), b) based on the stem of an included genus (e.g., Bellerophontina, Nautilida, Fungiina), or c) arbitrary combinations of letters (e.g., Yuania), but none of these can be allowed to end in -idae or -inae, reserved for family-group taxa. No suprafamilial name identical in form to that of a genus or to another published suprafamilial name should be employed (e.g., order Decapoda LATREILLE, 1803, crustaceans, and order Decapoda LEACH, 1818, cephalopods; suborder Chonetoidea MUIR-WOOD, 1955, and genus *Chonetoidea* JONES, 1928). Worthy of notice is the classificatory and nomenclatural distinction between suprafamilial and family-group taxa which respectively are named from the same type genus, since one is not considered to be transferable to the other (e.g., suborder Bellerophontina ULRICH & SCOFIELD, 1897; superfamily Bellerophontacea M'COY, 1851; family Bellerophontidae M'COY, 1851). Family-group names and suprafamilial names are not coordinate.

3) The Laws of Priority and Homonymy lack any force of international agreement as applied to suprafamilial names, yet in the interest of nomenclatural stability and the avoidance of confusion these laws are widely applied by zoologists to taxa above the family-group level wherever they do not infringe on taxonomic freedom and long-established usage.

4) Authors who accept priority as a determinant in nomenclature of a suprafamilial taxon may change its assigned rank at will, with or without modifying the terminal letters of the name, but such change(s) cannot rationally be judged to alter the authorship and date of the taxon as published originally. a) A name revised from its previously published rank is a "transferred name" *(nom. transl.),* as illustrated in the following.

Order CORYNEXOCHIDA Kobayashi, 1935

[*nom. transl.* MOORE, 1959, *ex* suborder Corynexochida KOBAYASHI, 1935]

b) A name revised from its previously published form merely by adoption of a different termination, without changing taxonomic rank is an "altered name" *(nom. correct.).* Examples follow.

Order DISPARIDA Moore & Laudon, 1943

[*nom. correct.* MOORE in MOORE, LALICKER, & FISCHER, 1952, *pro* order Disparata MOORE & LAUDON, 1943]

Suborder AGNOSTINA Salter, 1864

[*nom. correct.* HARRINGTON & LEANZA, 1957, *pro* suborder Agnostini SALTER, 1864]

c) A suprafamilial name revised from its previously published rank with accompanying change of termination (which may or may not be intended to signalize the change of rank) is recorded as *nom. transl. et correct.*

Order HYBOCRINIDA Jaekel, 1918

[*nom. transl. et correct.* MOORE in MOORE, LALICKER, & FISCHER, 1952, p. 613, *ex* suborder Hybocrinites JAEKEL, 1918, p. 90]

5) The authorship and date of nominate subordinate and superordinate taxa among suprafamilial taxa are considered in the *Treatise* to be identical since each actually or potentially has the same type. Examples are given below.

Subclass ENDOCERATOIDEA Teichert, 1933

[*nom. transl.* TEICHERT in TEICHERT *et al.*, 1964, p. *K*128 (*ex* superorder Endoceratoidea SHIMANSKIY & ZHURAVLEVA, 1961, *nom. transl.* TEICHERT in TEICHERT *et al.*, 1964, p. *K*128, *ex* order Endoceroidea TEICHERT, 1933)]

Order ENDOCERIDA Teichert, 1933

[*nom. correct.* TEICHERT in TEICHERT *et al.*, 1964, p. *K*165, *pro* order Endoceroidea TEICHERT, 1933]

Suborder ENDOCERINA Teichert, 1933

[*nom. correct.*, herein, *ex* Endoceratina SWEET, 1958, suborder]

TAXONOMIC EMENDATION

Emendation has two distinct meanings as regards zoological nomenclature. These are: 1) alteration of a name itself in various ways for various reasons, as has been reviewed, and 2) alteration of taxonomic scope or concept in application of a given zoological name. The *Code* (Art. 33a and Glossary p. 148) concerns itself with only the first type of emendation, applying the term to either justified or unjustified changes, both intentional, of the original spelling of a name. These categories are identified in the *Treatise* as *nomina correcta* and *nomina vana*, respectively. The second type of emendation primarily concerns classification and inherently is not associated with change of name. Little attention generally has been paid to this distinction in spite of its significance.

Most zoologists, including paleozoologists, who have signified emendation of zoological names refer to what they consider a material change in application of the name such as may be expressed by an importantly altered diagnosis of the assemblage covered by the name. The abbreviation *"emend."* then may accompany the name, with statement of the author and date of the emendation. On the other hand, many workers concerned with systematic zoology think that publication of *"emend."* with a zoological name is valueless, because more or less alteration of taxonomic sort is introduced whenever a subspecies, species, genus, or other assemblage of animals is incorporated under or removed from the coverage of a given zoological name. Inevitably associated with such classificatory expansions and restrictions is some degree of emendation affecting diagnosis. Granting this, still it is true that now and then somewhat radical revisions are put forward, generally with published statement of reasons for changing the application of a name. To erect a signpost at such points of most significant change is worthwhile, both as aid to subsequent workers in taking account of the altered nomenclatural usage and as indication that not-to-be-overlooked discussion may be found at a particular place in the literature. Authors of contributions to the *Treatise* are encouraged to include records of all specially noteworthy emendations of this nature, using the abbreviation *"emend."* with the name to which it refers and citing the author and date of the emendation.

Examples from *Treatise* volumes follow.

Order ORTHIDA Schuchert & Cooper, 1932

[*nom. transl. et correct.* MOORE in MOORE, LALICKER, & FISCHER, 1952, p. 220, *ex* suborder Orthoidea SCHUCHERT & COOPER, 1932, p. 43] [*emend.* WILLIAMS & WRIGHT, 1965]

Subfamily ROVEACRININAE Peck, 1943

[Roveacrininae PECK, 1943, p. 465; *emend.* PECK in MOORE & TEICHERT, eds., 1978, p. *T*921]

STYLE IN GENERIC DESCRIPTIONS

Citation of Type Species

The name of the type species of each genus and subgenus is given next following the generic name with its accompanying author, date, and page reference or after entries needed for definition of the name if it is involved in homonymy. The originally published combination of generic and trivial names for this species is cited, accompanied by an asterisk (*), with notation of the author and date of original publication. An exception in this procedure is made, however, if the species was first published in the same paper and by the same author as that containing definition of the genus that it serves as type; in such case, the initial letter of the generic name followed by the trivial name is given without repeating the name of the author and date. Examples of these two sorts of citations are as follows:

Diplotrypa NICHOLSON, 1879 [**Favosites petropolitanus* PANDER, 1830].
Chainodictyon FOERSTE, 1887 [**C. laxum*].

If the cited type species is a junior synonym of some other species, the name of this latter also is given, as follows:

Acervularia SCHWEIGGER, 1819 [**A. baltica*; M; =**Madrepora ananas* LINNÉ, 1758].

In the *Treatise,* the name of the type species is always given in the exact form it had in the original publication; in cases where mandatory changes are required, these are introduced later in the text, mostly in a figure caption. Examples are:

Ceratostreon BAYLE, 1978, pl. 133-134 explanations [**Exogira spinosa* MATHERON, 1843, p. 192]. Misspelling of *Exogyra.*
Obinautilus KOBAYASHI, 1954 [**O. pulchra*]. Wrong gender for species name *(recte pulcher).*

It is desirable to record the manner of establishing the type species, whether by original designation or by subsequent designation.

Fixation of type species originally. The type species of a genus or subgenus, according to provisions of the *Code,* may be fixed in various ways in the original publication or it may be fixed in specified ways subsequent to the original publication as stipulated by the *Code* (Art. 68) in order of precedence as 1) *original designation* (in the *Treatise* indicated as "OD") when the type species is explicitly stated or (before 1931) indicated by "n. gen., n. sp." (or its equivalent) applied to a single species included in a new genus, 2) defined by use of *typus* or *typicus* for one of the species included in a new genus (adequately indicated in the *Treatise* by the specific name), 3) established by *monotypy* if a new genus or subgenus has only one originally included species (in the *Treatise* indicated as "M"), and 4) fixed by *tautonymy* if the genus-group name is identical to an included species name not indicated as type belonging to one of the three preceding categories.

Fixation of type species subsequently. The type species of many genera are not determinable from the publication in which the generic name was introduced and therefore such genera can acquire a type species only by some manner of subsequent designation. Most commonly this is established by publishing a statement naming as type species one of the species originally included in the genus, and in the *Treatise* fixation of the type species in this manner is indicated by the letters "SD" accompanied by the name of the subsequent author (who may be the same person as the original author) and the date of publishing the subsequent designation. Some genera, as first described and named, included no mentioned species and these necessarily lack a type species until a date subsequent to that of the original publication when one or more species are assigned to such a genus. If only a single species is thus assigned, it automatically becomes the type species and in the *Treatise* this subsequent monotypy is indicated by the letters "SM." Of course, the first publication containing assignment of species to the genus which originally lacked any included species is the one concerned in fixation of the type species, and if this named two or more species as belonging to the genus but did not designate a type species, then a later "SD" designation is necessary. Examples of the use of "SD" and "SM" as employed in the *Treatise* follow.

Hexagonaria GÜRICH, 1896 [**Cyathophyllum hexa-*

gonum GOLDFUSS, 1826; SD LANG, SMITH, & THOMAS, 1940].

Muriceides STUDER, 1887 [**M. fragilis* WRIGHT & STUDER, 1889; SM WRIGHT & STUDER, 1889].

Another mode of fixing the type species of a genus is action of the International Commission on Zoological Nomenclature using its plenary powers. Definition in this way may set aside application of the *Code* so as to arrive at a decision considered to be in the best interest of continuity and stability of zoological nomenclature. When made, it is binding and commonly is cited in the *Treatise* by the letters "ICZN," accompanied by the date of announced decision and reference to the appropriate numbered Opinion.

It should be noted that *subsequent designation* of a type species is admissible only for genera established prior to 1931. A new genus-group name established after 1930, and not accompanied by fixation of a type species through original designation or original indication, is invalid (*Code,* Art. 13b). Effort of a subsequent author to "validate" such a name by subsequent designation of a type species constitutes an original publication making the name available under authorship and date of the subsequent author. This provision of the *Code* has not been consistently applied in all earlier *Treatise* volumes, but is rigidly adhered to in the present volume.

Type species of synonyms. In about 1969 a decision was made by the editors to include the names of type species of genera that were placed in subjective synonymy. Such species are simply identified as "type." An example is:

Trachycardium MORCH, 1853 [**Cardium isocardia* LINNÉ, 1758; SD VON MARTENS, 1870] [=*Kathocardia* TUCKER & WILSON, 1932 (type, *Cardium (K.) aclinense,* OD)].

Fixation of types of type species. The present Supplement to Part F introduces an important innovation in that documentation on the type specimen (or specimens) of type species is supplied, which has not been included in any previous *Treatise* volume. Citation of type species and indication of the manner of its designation is followed by a symbol (†) that precedes the museum designation, usually a number of the type specimen or specimens and the name and location of the repository. The status of the types is indicated next. When no information follows the repository citation, it is understood that a holotype was satisfactorily designated in the original publication (including designation by monotypy in species established on the basis of only one specimen). When syntypes only are available, this has been indicated. If no holotype was selected by the original author, subsequent choice of a lectotype, if any, is indicated by author and date, and where the original holotype has been lost, the same procedure is followed for the neotype if one has been selected. Holotype, paratypes and syntypes, lectotype, and neotype are the only categories of types recognized in this Supplement.

The procedure described and followed here does not establish a precedent to be followed necessarily in future *Treatise* parts and supplements.

Homonyms

Most generic names are distinct from all others and are indicated without ambiguity by citing their originally published spelling accompanied by name of the author and date of first publication. If the same generic name has been applied to two or more distinct taxonomic units, however, it is necessary to differentiate such homonyms, and this calls for distinction between junior homonyms and senior homonyms. Because a junior homonym is invalid, it must be replaced by some other name. For example, *Callopora* HALL, 1851, introduced for Paleozoic trepostome bryozoans, is invalid because GRAY in 1848 published the same name for Cretaceous-to-Holocene cheilostome bryozoans, and BASSLER in 1911 introduced the new name *Hallopora* to replace Hall's homonym. The *Treatise* style of entry is:

Hallopora BASSLER, 1911, *nom. subst. pro Callopora* HALL, 1851 *non* GRAY, 1848.

In like manner, a needed replacement generic name may be introduced in the *Treatise* (even though first publication of generic names otherwise in this work is generally avoided). The requirement that an exact bibliographic reference must be given for the replaced name commonly can be met in the *Treatise* by citing a publication recorded in the list of references, as shown in the following example.

Mysterium DE LAUBENFELS, herein, *nom. subst. pro Mystrium* SCHRAMMEN, 1936, p. 60, *non* ROGER, 1862 [**Mystrium porosum* SCHRAMMEN, 1936].

Otherwise, no mention of the existence of a junior homonym generally is made.

Synonymous homonyms. An author sometimes publishes a generic name in two or more papers of different date, each of which indicates that the name is new. This is a bothersome source of errors for later workers who are unaware that a supposed first publication that they have in hand is not actually the original one. Although the names were separately published, they are identical and therefore definable as homonyms; at the same time they are absolute synonyms. For the guidance of all concerned, it seems desirable to record such names as synonymous homonyms, and in the *Treatise* the junior one of these is indicated by the abbreviation "jr. syn. hom."

Identical family-group names not infrequently are published as new names by different authors, the author of the later-introduced name being ignorant of previous publication(s) by one or more other workers. In spite of differences in taxonomic concepts as indicated by diagnoses and grouping of genera and possibly in assigned rank, these family-group taxa are nomenclatural homonyms, based on the same type genus, and they are also synonyms. Wherever encountered, such synonymous homonyms are distinguished in the *Treatise* as in dealing with generic names.

A special, though rare, case of synonymy exists when identical family names are formed from generic names having the same stem but differing in their endings. An example is the family name Scutellidae R. & E. RICHTER, 1925, based on *Scutellum* PUSCH, 1833, a trilobite. This name is a junior synonym of Scutellidae GRAY, 1825, based on *Scutella* LAMARCK, 1816, an echinoid. The name of the trilobite family was later changed to Scutelluidae (ICZN, Op. 1004, 1974).

Synonyms

Citation of synonyms is given next following record of the type species and if two or more synonyms of differing date are recognized, these are arranged in chronological order. Objective synonyms are indicated by accompanying designation "obj.," others being understood to constitute subjective synonyms, of which the types are also indicated. Examples showing *Treatise* style in listing synonyms follow.

Modiomorpha HALL & WHITFIELD, 1869, p. 72 [**Pterinea concentrica* CONRAD, 1838; SD HALL, 1885] [=*Palanatina* HALL & WHITFIELD, 1870 (type, *P. typa,* OD)].

Staurocyclia HAECKEL, 1882 [**S. cruciata* HAECKEL, 1887] [=*Coccostaurus* HAECKEL, 1882 (obj.); *Phacostaurus* HAECKEL, 1887 (obj.)].

Graphiocrinus DE KONINCK & LE HON, 1854, p. 115 [**G. encrinoides*; M] [=*Scaphiocrinus* HALL, 1858b, p. 550 (type, *S. simplex,* OD)].

Some junior synonyms of either objective or subjective sort may take precedence desirably over senior synonyms wherever uniformity and continuity of nomenclature are served by retaining a widely used but technically rejectable name for a generic assemblage. This requires action of ICZN using its plenary powers to set aside the unwanted name and validate the wanted one, with placement of the concerned names on appropriate official lists.

ABBREVIATIONS

Abbreviations used in this part of the *Treatise* are explained in the following alphabetically arranged list. Standard abbreviations or those found only in the references are not included here.

Afr., Africa
Ala., Alabama
ant., anterior
approx., approximately
Arenig., Arenigian
Arg., Argentina
Ark., Arkansas
Ashgill., Ashgillian
Asia M., Asia Minor
Aus., Austria

Belg., Belgium

CAI, color alteration index
Cam., Cambrian
Can., Canada
Carb., Carboniferous
Chazy., Chazyan
Chester., Chesterian
Colo., Colorado
cosmop., cosmopolitan

Desmoines., Desmoinesian
Dev., Devonian

diagr., diagram
Dinant., Dinantian
Distr., District
Dol., Dolomite

Eifel., Eifelian
Ems., Emsian
Eng., England
Est., Estonia
Eu., Europe

F., Formation
Famenn., Famennian
Frasn., Frasnian

G.Brit., Great Britain
Ger., Germany
Givet., Givetian
Gr., Group
Greenl., Greenland

Ill., Illinois
Ind., Indiana
Ire., Ireland

Kans., Kansas
Kazakh., Kazakhstan
Kinderhook., Kinderhookian
Ky., Kentucky

L., Lower
lat., lateral
Llandov., Llandoverian
low., lower
Ls., Limestone
Ludlov., Ludlovian

M, monotypy
M., Middle
Manit., Manitoba
Mbr., Member
Mich., Michigan
Minn., Minnesota
Miss., Mississippi, Mississippian
Missour., Missourian
Mo., Missouri
Mont., Montana
Mts., Mountains
m.y., million years

n, new
N.Am., North America, -n
Namur., Namurian
NE., Northeast
Nev., Nevada
Newf., Newfoundland
nom. correct., *nomen correctum*, corrected or intentionally altered name
nom. neg., *nomen negatum*, denied name
nom. nud., *nomen nudum*, naked name
nom. subst., *nomen substitutum*, substitute name
nom. transl., *nomen translatum*, transferred name
Nor., Norway
n. sp., new species
NW., Northwest
N.Y., New York
N.Z., New Zealand

obj., objective [synonym]
OD, original designation
Okla., Oklahoma
Ont., Ontario
Ord., Ordovician
Osag., Osagian

Pa., Pennsylvania
Pak., Pakistan
Penn., Pennsylvanian
Perm., Permian
pers. commun., personal communication
Pol., Poland
post., posterior
Precam., Precambrian
Pridol., Pridolian

Que., Quebec
Queensl., Queensland

S.Am., South America
Scand., Scandinavia
Scot., Scotland
SD, subsequent designation
S.Dak., South Dakota
sec., section
SEM, scanning electron microscope
Sh., Shale
Sib., Siberia
Sil., Silurian
sp., species (spp., plural)
Ss., Sandstone
SW., Southwest

Tenn., Tennessee
Tommot., Tommotian
Tournais., Tournaisian
transv., transverse
Trempeal., Trempealeauan
Trias., Triassic

U., Upper
up., upper
USA, United States (America)
USSR, Union of Soviet Socialist Republics

Vict., Victoria

W., West
Wenlock., Wenlockian
Westphal., Westphalian
Wis., Wisconsin
Wyo., Wyoming

Yudom., Yudomian

Z., Zone

REFERENCES TO LITERATURE

The titles of serials cited in the references are abbreviated as recommended in the *Bibliographical Guide for Editors and Authors* (1974, The American Chemical Society, Washington, D.C.); titles of serials not covered in the *Guide* have been abbreviated according to the standard established in International Standards Organization (ISO) recommendation 833-1974. The names of authors and titles of works in Cyrillic have been transliterated for the most part according to the method suggested by the Geographical Society of London and the U.S. Board on Geographic Names. A translation of each Cyrillic title is given in brackets at the end of the reference. Full citations of references containing senior homonyms applied to organisms other than members of the Conodonta are not included, but may be found in contracted form in S. A. Neave, *Nomenclator Zoologicus* (1939-1975, 7 v., Zoological Society, London).

SOURCES OF ILLUSTRATIONS

At the end of each figure caption, the name of the author of the illustration and the date of publication are given, full citation of the publication being provided in the References. Although original sources do not always produce the best illustrations, they are, historically speaking, definitive and are commonly selected by *Treatise* authors. Previously unpublished illustrations are indicated by the name of the author and the letter n ("new").

STRATIGRAPHIC DIVISIONS

As commonly cited in the *Treatise,* classification of rocks forming the geologic column is reasonably uniform and firm throughout most of the world as regards major divisions (e.g., series, systems, and rocks representing eras), but it may be variable and unfirm as regards minor division (e.g., substages, stages, and subseries), which tend to be provincial in application. Users of the *Treatise* have suggested the desirability of publishing reference lists showing the stratigraphic arrangement of at least the most commonly cited divisions. Accordingly, a tabulation of European and North American units, which generally follows usage by authors of this volume, is given here. Divisions of post-Triassic systems are not listed because of the extinction of Conodonta during the Triassic.

Generally Recognized Divisions of Geologic Column

EUROPE	NORTH AMERICA
CENOZOIC ERATHEM	**CENOZOIC ERATHEM**
QUATERNARY SYSTEM	**QUATERNARY SYSTEM**
TERTIARY SYSTEM	**TERTIARY SYSTEM**
MESOZOIC ERATHEM	**MESOZOIC ERATHEM**
CRETACEOUS SYSTEM	**CRETACEOUS SYSTEM**
JURASSIC SYSTEM	**JURASSIC SYSTEM**
TRIASSIC SYSTEM	**TRIASSIC SYSTEM**
Upper Triassic Series	**Upper Triassic Series**
Rhaetian Stage	Rhaetian Stage
Norian Stage	Norian Stage
Carnian Stage	Carnian Stage
Middle Triassic Series	**Middle Triassic Series**
Ladinian Stage	Ladinian Stage
Anisian Stage	Anisian Stage
Lower Triassic Series	**Lower Triassic Series (Scythian)**
	Spathian Stage
	Smithian Stage
Scythian Stage	Dienerian Stage
	Griesbachian Stage
PALEOZOIC ERATHEM	**PALEOZOIC ERATHEM**
PERMIAN SYSTEM	**PERMIAN SYSTEM**
Upper Permian Series	**Upper Permian Series**
Tartarian Stage	Ochoan Stage
Kazanian Stage	Guadalupian Stage
Lower Permian Series	**Lower Permian Series**
Artinskian Stage	Leonardian Stage
Sakmarian Stage	Wolfcampian Stage
Asselian Stage	

CARBONIFEROUS SYSTEM

- Silesian Subsystem
  - Stephanian Series
  - Westphalian Series
  - Namurian Series
- Dinantian Subsystem
  - Visean Series
  - Tournaisian Series

DEVONIAN SYSTEM

- Upper Devonian Series
  - Famennian Stage
  - Frasnian Stage
- Middle Devonian Series
  - Givetian Stage
  - Eifelian Stage
- Lower Devonian Series
  - Emsian Stage
  - Siegenian Stage
  - Gedinnian Stage

SILURIAN SYSTEM

- Pridolian Series
- Ludlovian Series
- Wenlockian Series
- Llandoverian Series

ORDOVICIAN SYSTEM

- Ashgillian Series
- Caradocian Series
- Llandeilian Series
- Llanvirnian Series
- Arenigian Series
- Tremadocian Series[1]

CAMBRIAN SYSTEM

- Upper Cambrian Series
- Middle Cambrian Series
- Lower Cambrian Series

ROCKS OF PRECAMBRIAN ERAS

PENNSYLVANIAN SYSTEM

- Virgilian Series
- Missourian Series
- Desmoinesian Series
- Atokan Series
- Morrowan Series

MISSISSIPPIAN SYSTEM

- Chesterian Series
- Meramecian Series
- Osagian Series
- Kinderhookian Series

DEVONIAN SYSTEM

- Upper Devonian Series
  - Famennian Stage
  - Frasnian Stage
- Middle Devonian Series
  - Givetian Stage
  - Eifelian Stage
- Lower Devonian Series
  - Emsian Stage
  - Siegenian Stage
  - Gedinnian Stage

SILURIAN SYSTEM

- Pridolian Series
- Ludlovian Series
- Wenlockian Series
- Llandoverian Series

ORDOVICIAN SYSTEM

- Cincinnatian Series (Upper Ordovician)
  - Richmondian Stage
  - Maysvillian Stage
  - Edenian Stage
- Champlainian Series (Middle Ordovician)
  - Mohawkian Stage
  - Chazyan Stage
  - Whiterockian Stage
- Canadian Series (Lower Ordovician)

CAMBRIAN SYSTEM

- Upper Cambrian Series
  - Trempealeauan Stage
  - Franconian Stage
  - Dresbachian Stage
- Middle Cambrian Series
- Lower Cambrian Series

ROCKS OF PRECAMBRIAN ERAS

[1] Tremadocian is placed in Cambrian by some authors.

PART W SUPPLEMENT 2
CONODONTA

By David L. Clark, Walter C. Sweet, Stig M. Bergström, Gilbert Klapper, Ronald L. Austin, Frank H. T. Rhodes, Klaus J. Müller, Willi Ziegler, Maurits Lindström, James F. Miller, and Anita G. Harris

CONTENTS

INTRODUCTION TO THE CONODONTA

SYSTEMATIC DESCRIPTIONS

INTRODUCTION TO THE CONODONTA

GENERAL ASPECTS

By David L. Clark

[University of Wisconsin]

Conodonts are an extinct group of marine animals whose most commonly preserved parts are microscopic elements, some of which superficially resemble small fish teeth and worm jaws. This superficial similarity is responsible for their name. These elements have a range of Early Cambrian and possibly Late Precambrian to Late Triassic.

Conodont elements were first described in 1856 from Ordovician strata in Estonia but since have been reported from all parts of the world in considerable abundance. The greatest diversity of conodont genera is found in the present northern hemisphere and in Ordovician rocks. Almost half of all conodont genera were in existence during the Ordovician. Conodonts were a declining group of organisms from the Devonian until their extinction at the close of the Triassic.

Their global distribution in widely different marine rocks has convinced students that most conodonts were pelagic organisms. Some probably attained a benthic mode.

The exact biology of conodonts has eluded workers, but this has not been a serious factor affecting their value in biostratigraphy and paleoecology. Conodonts have been demonstrated to be superb tools in worldwide biostratigraphy, and their value in Cambrian through Triassic rocks is not exceeded by any other group of fossils.

Most conodonts were bilaterally symmetrical animals probably only a few centimeters in greatest dimension. Their hard parts consist of microscopic elements ranging up to about three millimeters in maximum dimension and are composed of carbonate apatite. These elements were embedded in a fleshy medium and probably functioned as supports for such physiological activities as respiration or ingestion. The elements are the only parts commonly preserved, and laboratory separation of rock matrix and conodonts usually produces isolated elements rather than whole assemblages representing complete hard-part anatomy.

Pander (1856) first described conodonts and coined their name. He believed that the isolated elements in his Ordovician collection were teeth of some extinct group of fish. His publication included description of 56 conodont "species" and 14 "genera"; the taxonomy was based on morphology of the isolated elements. Hinde (1879) agreed with Pander that conodonts are an extinct group of fish. But his material included an assemblage of different morphologic types on Upper Devonian shale surfaces, the occurrence of which convinced him that all the morphologically different elements belonged to the same animal. It was not until 1934 that Schmidt and Scott, working separately, found morphologically different elements occurring together in such a way that the assemblage nature of elements was really proven. Their work showed that different kinds of elements occurred in a single conodont. Because unmistakable assemblages similar to those described by Schmidt (1934) in Europe and Scott (1934) in North America are rare, most elements described in the following 30 years were treated as distinct taxa. A complex system of taxonomy based on gross morphology of each element developed in spite of the fact that most workers recognized that more than one kind of element occurred in most conodonts.

Much important taxonomic work with conodonts was accomplished during the period 1930 to 1966. Most of this work reflected the form-taxa system used by Pander in 1856, and each element was given a different name.

This time-honored scheme of form taxa was challenged in 1966, in separate publications by Bergström and Sweet, Webers, and Schopf. These students demonstrated that individual Ordovician conodonts were multielement-bearing and that as many as nine different names were in use for as many elements representing no more than a single biologic species. This concept, although dating from Hinde (1879) and proven by Schmidt (1934) and Scott (1934), has now revolutionized conodont taxonomy. The student of conodonts has had to make judgments on the original elemental composition of conodont species and, in most cases, base his judgment on isolated elements occurring in acid residues. Numerical and morphologic studies have aided in this monumental effort, and this volume, the text of which was largely completed in 1975, represents a first attempt at a unified multielement taxonomy for most conodont genera.

When *Treatise* Part W was published in 1962, advances in conodont research already had outdated that volume. Since publication of the original volume, a more or less complete biostratigraphy for conodonts has been formulated, the geologic range of conodonts (still in doubt in 1962) is well established, a natural taxonomy is recognized, and numerous problems of microstructure, paleoecology, and biology, unknown in 1962, can be discussed with some degree of confidence.

Field collection and sample preparation.—Conodonts commonly occur in considerable abundance: hundreds and even thousands of conodont elements can be recovered from a few grams of rock. In addition, conodonts were rapidly changing organisms whose elements reflect this evolution with considerable morphologic variety. For these reasons, special care is taken in the field to collect closely spaced but discrete samples. Channel sampling, even over a few centimeters of rock, is likely to result in a considerable mixture of taxa that otherwise are discrete, morphologically and stratigraphically.

Geosynclinal rock, which usually reflects more rapid sedimentation rates than does rock deposited on cratonal areas, poses special problems because of the great thickness of rock that is available. Field marking of precisely sampled intervals often is helpful, especially if additional collecting is planned. It is particularly necessary when precise biostratigraphic or paleoecologic research is the objective.

The microscopic size of conodont hard parts has resulted in the development of special preparation techniques. Their composition of carbonate apatite permits treatment of host rock in a variety of acids. In practice, at least one kilogram of rock is collected and crushed to 2- to 3-centimeter pieces. The crushed sample is treated with acetic or formic acid (10 to 15% solution) if it is calcareous, or with Quaternary O or kerosene if it is noncalcareous. After digestion of the matrix is complete, the residue is caught on a 100- to 200-mesh sieve, dried, and the residue examined for conodont elements. Picking of large residues is facilitated if the residue is reduced, using standard heavy-liquid treatment or treatment in a magnetic separator. All of these techniques and procedures have been described in detail by Collinson (1963, 1965) and Dow (1965).

MORPHOLOGY AND COMPOSITION OF ELEMENTS

MACROMORPHOLOGY OF ELEMENTS AND APPARATUSES

By Walter C. Sweet

[Ohio State University]

It is now recognized that the skeletal apparatus of conodonts consisted of a number of independent, mineralized elements of different shape, which were mostly paired and served collectively as internal supports for an organ or organs of unknown anatomy and function. Commonly, it is the disassociated skeletal elements that are collected and studied, and all but a few conodont taxa are based on such **discrete elements;** however, a few **natural assemblages** provide valuable evidence about the organization of discrete elements in individuals of the species they represent. Reconstruction of skeletal apparatuses from collections of discrete elements has been effected lately through use of a variety of empirical and quantitative techniques. Clearly, a discussion of conodont macromorphology must give attention not only to the shapes and features of discrete elements, but also to the organizational plan of the more elaborate skeletal apparatuses of which those discrete elements were once parts. Furthermore, morphologic terminology, shape classification, and the names for shape categories must be separated plainly from Linnean nomenclature and should be designed to facilitate the description and comparison of multielement apparatuses as they are recognized and named. Thus, the shapes, shape categories, and terminology of discrete elements are discussed first, and this discussion is followed by a consideration of the organization and terminology of multielement conodont skeletal apparatuses.

SHAPES OF DISCRETE ELEMENTS

During the first century of conodont research, the shapes of discrete elements served as the principal basis for specific, generic, and familial concepts. There was little need in such a form-taxonomy for a set of formally designated shape categories with non-Linnean names; *Cordylodus,* for example, stood not only for a biologic concept, but also for a shape category. When the need arose to discuss a collective shape category of suprageneric rank, colloquial contractions of familial names (e.g., cordylodid, distacodontid, prioniodid) were employed, or more general terms such as simple cone, bar, blade, plate, or platform were used informally. Most such terms lack descriptive precision, however, and they have been used by different authors with various meanings.

As interest in multielement taxonomy has grown, so has terminological confusion, and this confusion has added considerably to the nomenclatural problems that are inevitable in a period of taxonomic transition. As an illustration, it may be noted that the skeletal apparatus of the Ordovician genus *Plectodina* was apparently composed of discrete elements that can be referred in form-taxonomy to the genera *Trichonodella, Zygognathus, Cordylodus, Eoligonodina, Plectodina, Cyrtoniodus, Prioniodina, Dichognathus,* and *Ozarkodina.* Because each of these form-genera represents a distinctive shape category, Bergström and Sweet (1966) chose to describe the skeletal components of multielement *Plectodina* as

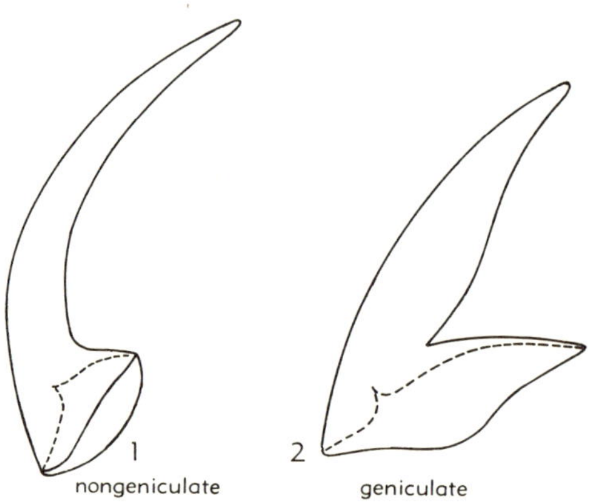

Fig. 1. Shapes of coniform elements (after Lindström, 1964).

cordylodus-like, trichonodella-like, or, in later descriptions (e.g., Sweet & Bergström, 1972), as cordylodiform or trichonodelliform. Others have followed different procedures. Sweet (1970), Klapper and Philip (1971), and Kozur and Mostler (1971) assigned capital letters, with or without numerical subscripts, to positions in the multielement apparatuses they described; Scott (1934) referred to discrete components of natural assemblages as "hindeodells" and "cavusgnaths" because in a form-taxonomic system such elements would have been assigned to the genera *Hindeodella* and *Cavusgnathus*; and Jeppsson (1971, 1972) and Sweet and Bergström (1972) employed such letter symbols as tr, hi, and others for the locations of elements in various multielement apparatuses. None of these procedures is dismissible on objective grounds, but all are susceptible of subjective criticism. That is, form and apparatus location are not clearly distinguished, and it is not clearly apparent from the notation or terminology employed that, for example, the trichonodelliform (or trichonodella-like) elements of *Plectodina,* the hibbardelliform elements of *Prioniodus* and *Amorphognathus,* the A_3 or B_3 elements of Devonian species described by Klapper and Philip (1971), and the U-elements of Triassic species discussed by Sweet (1970) are homologous structures that probably had comparable positions and similar or identical functions in the various apparatuses of which they were once parts. Further, to state that the skeletal apparatus of multielement *Phragmodus,* for example, was composed of distacodontid, prioniodid, and ligonodinid elements would convey the absurd, if unintentional, suggestion that components of the skeletal apparatus of a single species represent three different families.

In the descriptions of conodont genera in this volume, major positions in a common type of skeletal apparatus are designated by letters, as explained in a subsequent section of this chapter, and elements occupying these positions in various apparatuses are described in accord with the system outlined in Table 1 and discussed in greater detail on the next several pages. Linnean names and contracted familial names are reserved for discussions of taxa; they are not used for components of the skeletal apparatuses of individuals or for larger units assigned to those taxa.

MAJOR SHAPE CATEGORIES

CONIFORM ELEMENTS

Major types of coniform elements are illustrated in Figure 1, and the orientation and terminology of elements of this shape category are diagrammed in Figure 2.

From figures 1 and 2 it is clear that coniform elements are basically conical units that can be divided into two major parts: a more or less expanded **base,** which encloses a subconical **basal cavity;** and a **cusp,** which is solid, has a straight or curved long axis, and tapers to the **apex,** or **tip,** of the unit. Coniform elements are oriented for descriptive purposes so that the tip of the cusp is up and the upper margin of the base is approximately horizontal. The sides, or **lateral faces,** are those parts of the element on either side of the **midplane,** which is an imaginary surface that includes the apexes of the cusp and basal cavity and the uppermost and lowermost points on

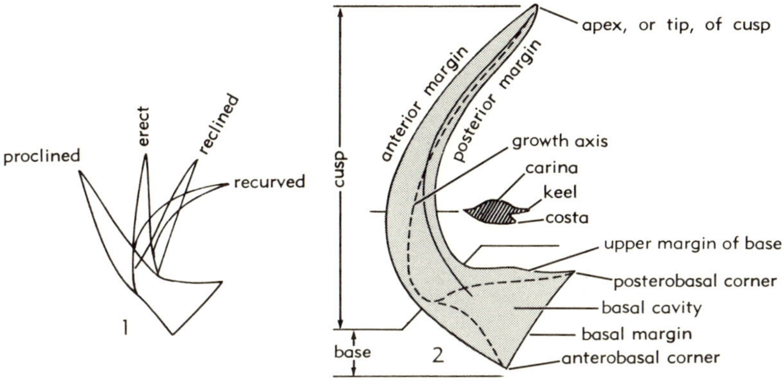

FIG. 2. Orientation and morphology of coniform elements (after Lindström, 1955).

the basal margin (the **anterobasal** and **posterobasal corners** of Fig. 2,*2*). The **anterior** and **posterior margins** of a unit so oriented are the traces in the midplane of the sides of the element that are respectively farthest from, and nearest to, the uppermost point on the basal margin. In most coniform elements, the posterior margin is concave in lateral view, the anterior margin is convex, and orientation may safely be determined on this basis; however, in a few coniform elements (and in a number of ramiform ones) this informal orientation rule does not apply. It is noted that in such elements, and in most others, one margin of the basal cavity is concave near its tip, and this side is taken as posterior.

The base of coniform elements is that part of the unit that includes the basal cavity. In most elements of this category, the base is rather distinctly set off from the cusp by its greater lateral expansion, and, at least in thermally unaltered specimens, the outlines of the basal cavity are readily visible through the thin walls of the base. The basal cavity varies greatly in size and shape. In most coniform elements it is of modified conical shape, and its apex is commonly directed toward the anterior margin. Commonly the basal cavity (and thus the base) makes up less than half of the entire unit. In a few coniform elements, however, virtually the entire element is hollow and all but a very short apical portion must then be regarded as base.

The cusp of coniform elements is the solid portion of the unit above the apex of the basal cavity. Depending on its relationship with the base, the cusp may be described as **proclined, erect, reclined,** or **recurved** (Fig. 2,*1*). In coniform elements with little "white matter," a fine longitudinal line, termed a **growth axis,** is commonly visible in the interior of the cusp. This line begins at the apex of the basal cavity, passes through the apexes of each of the lamellae that compose the cusp, and terminates at the cusp apex. The growth axis is commonly less markedly curved than the anterior margin of the cusp.

The anterior, posterior, and lateral faces of coniform elements may be **smooth** and uniformly rounded, longitudinally **channeled,** incised by fine longitudinal **striae,** or characterized by coarser longitudinal markings, which are termed **carinae** (sing., **carina**) if they are broadly rounded elevations, **costae** (sing., **costa**) if they are more narrowly rounded or sharp-edged elevations, and **grooves** if they are longitudinal depressions below the general surface of the element. Sharp edges along either the anterior or posterior margin, which are produced in flangelike longitudinal structures in some coniform elements, have been termed **keels.**

In Table 1, coniform elements are di-

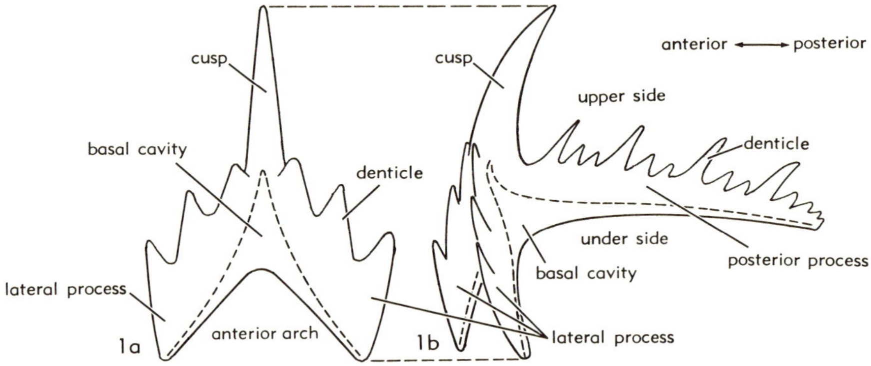

FIG. 3. Morphology of an alate ramiform element seen in anterior *(1a)* and lateral *(1b)* views (Sweet, n).

vided into two subordinate groups, termed **geniculate** and **nongeniculate** (Fig. 1). Geniculate coniform elements are those in which the posterior margin of the cusp joins the upper edge of the base to enclose an acute angle. Nongeniculate coniform elements are distinguished by a smooth transition from posterior cusp margin to upper basal edge, which takes the form in lateral view of a straight or smoothly arcuate line. Division of coniform elements into just these two categories is admittedly crude, for the variety of form in this shape category is at least as great as in the next one to be described, the ramiform elements. Names or descriptions are not supplied for subordinate groups of coniform elements, other than the two introduced here, because homologies with components of apparatuses with ramiform and pectiniform elements are obvious in some multimembrate apparatuses composed of coniform elements, and this condition may turn out to be general when we know more about such apparatuses. Consequently, proposal of a special terminology for subordinate shape categories of coniform elements might serve to obscure relationships that would be made more obvious through use of the same (or a closely similar) terminology for both coniform and ramiform categories. In multimembrate *Panderodus* and *Staufferella,* for example, there are bilaterally symmetrical coniform elements that clearly are homologues of alate elements in the apparatuses of other genera, which are composed of ramiform and pectiniform types.

RAMIFORM ELEMENTS

Ramiform conodont elements are structures in which at least one of the sides or edges of the base is drawn out laterally, anteriorly, or posteriorly from the cusp into a process that is serrate on its upper edge. Elements of this shape category and its principal subdivisions are illustrated in figures 3 to 8. Names applied to the subdivisions are listed in Table 1.

Each ramiform element consists of two fundamental parts, base and cusp. As in coniform elements, the **base** is defined as the part of the unit that includes the basal cavity, and the **cusp** is the conical structure that is developed above the apex of the basal cavity. Well-defined ridges, carinae, costae, or keels that extend across, and are parts of, the base are **processes.** For an element to be termed ramiform, at least one of these processes must be serrate on its upper edge, but all processes need not be serrate. Individual elements of the serrate edge of a process are **denticles,** which may be of minor size, or may rival or exceed the cusp in length and width. Processes with denticles are described as **denticulate;** those without denticles may be described as **adenticulate.**

For descriptive and comparative purposes, ramiform elements are oriented in the same manner as coniform elements. That is, the

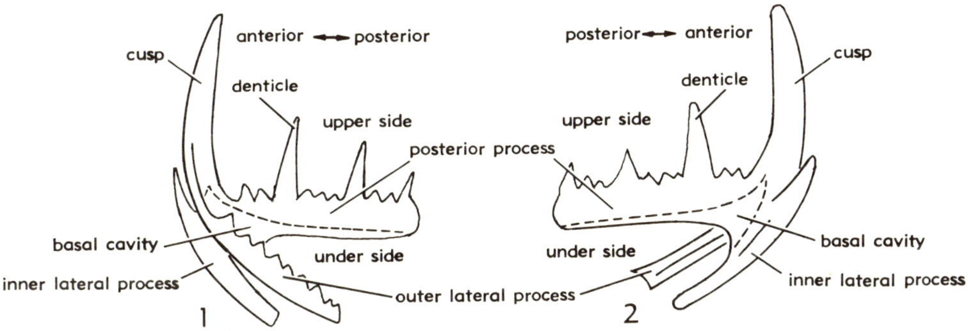

FIG. 4. Morphology of tertiopedate ramiform elements viewed from outer *(1)* and inner *(2)* sides (Sweet, n).

apex of the cusp is up and the posterior side of the unit is on the concave side of the cusp, or is the side of the element toward which the cusp apex points, or is the side of the element congruent with the margin of the basal cavity that shows a concavity just below its apex. A basal process that is continuous with the posterior side or margin of the cusp is a **posterior process.** By convention, as well as by analogy with coniform elements, the long axis of the posterior process is oriented horizontally and the position of all other processes is thereby determined. An **anterior process** is one that is continuous proximally with the anterior face or margin of the cusp; and a **lateral process** is one that is continuous with one of the lateral faces of the cusp.

The base of ramiform elements is comparable with that of coniform elements, but is commonly more complex. The **basal margin** coincides with the lower edge of the element; the upper limit of the base is the irregular surface defined by the upper margins of all processes and the outer surface of any intraprocess areas above the basal cavity. The basal cavity, which varies considerably in size and extent, is essentially a subconical space, with its apex beneath the proximal part of the cusp and with groovelike extensions along the under sides of the processes. In many ramiform elements the basal cavity is capacious and its groovelike extensions are prominent and continue to the extremities of all processes. In others, however, the basal opening is restricted to a small **basal pit** beneath the cusp, and process under sides are flat, or are sharp edges that are bordered on immediately adjacent parts of the process by surfaces that are faintly striated parallel to the process axis. In the growth of such elements, the basal margins of successively younger lamellae extended no farther downward along processes than the basal margins of previous lamellae, or actually retreated upward along the lateral faces of processes as the elements grew larger. Portions of the base of either coniform or ramiform elements that document such a growth history have been termed **inverted basal cavities** (LINDSTRÖM, 1955), but the expression **zone of recessive basal margin** may have greater descriptive precision. The basal cavity of elements of this type consists, then, of basal pit surrounded by (or partially surrounded by) a zone of recessive basal margin.

The **proximal** and **distal ends** of a process are the ends closest to, and farthest from, the cusp, respectively. The **upper edge** or **upper margin** is the denticulate edge or its equivalent processes. **Height** is any measurement taken normal to the upper or lower margin; **width** is any measurement at right angles to height and length.

Denticles are said to be **discrete** (e.g., Fig. 3) if they do not touch adjacent denticles at any point above their junction with the undenticulated part of the base.

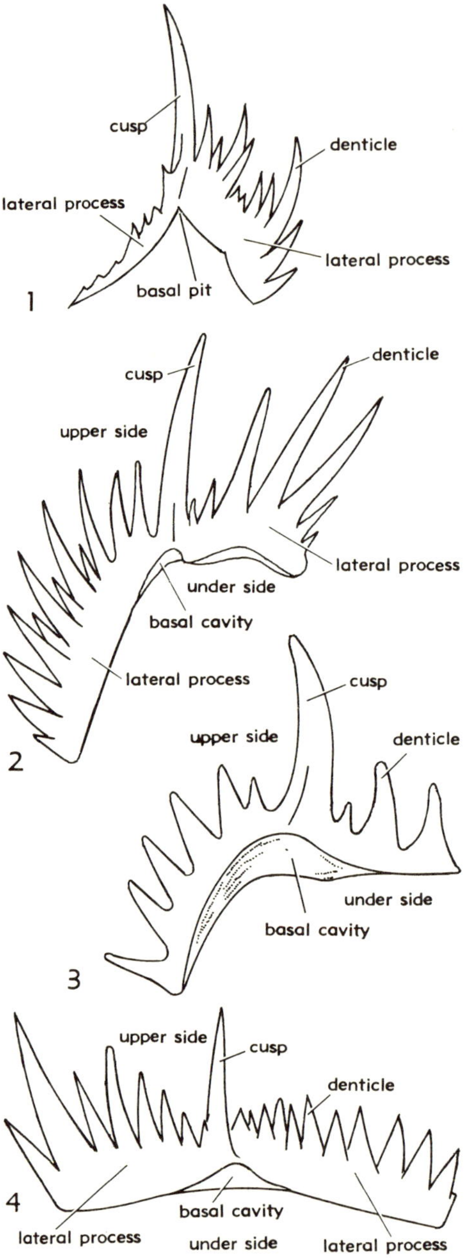

FIG. 5. Morphology of digyrate ramiform elements seen in anterior *(1)* and posterior *(2-4)* views (Sweet, n).

Denticles whose margins merge with those of adjacent denticles for a significant part of their height are described as **fused,** or **laterally confluent** (e.g., Fig. 6,*1*). **Overgrown** or **submerged denticles** do not appear as distinct serrations along the upper edge of a process, but their outlines are visible within the undenticulated body of the process base. Such denticles have been termed "germ" or "unerupted" denticles by some authors, but these terms are inappropriate because they suggest conditions of growth not known to occur in conodonts.

The cusp of ramiform elements is commonly longer than any of the process denticles, but this is not invariably the case. Terms used to describe the cusp of coniform elements are also appropriate for the cusp of ramiform elements.

TABLE 1. *Major Shape Categories of Conodont Elements.*

Element	Shape category	
Coniform ("simple cones")	Geniculate Nongeniculate	
Ramiform ("bars")	Alate Tertiopedate Digyrate Bipennate Dolabrate Quadriramate Multiramate	
Pectiniform ("blades," "plates," "platforms")	Stellate	Stelliplanate Stelliscaphate
	Pastinate	Pastiniplanate Pastiniscaphate
	Carminate	Carminiplanate Carminiscaphate
	Angulate	Anguliplanate Anguliscaphate
	Segminate	Segminiplanate Segminiscaphate

Major Types of Ramiform Elements

Ramiform elements are assigned to one or another of the major subdivisions of this category named in Table 1.

Bilaterally symmetrical ramiform elements that lack an anterior process, but have a posterior process and a lateral process on each side of the cusp, are termed **alate** ramiform elements. In alate ramiform elements (Fig. 3), the midplane is a plane of bilateral symmetry. That is, the lateral processes are identical, or essentially so, and are symmetrically disposed on either side of the cusp. Because lateral processes of alate ramiform elements are commonly directed downward and away from the cusp, they characteristically form a distinctive arcuate structure in anterior or posterior view that has been termed an **anterior arch,** with the cusp at its summit. The posterior process of some alate ramiform elements is long and denticulate; that of others is short and adenticulate; and, in a few, existence of a posterior process is indicated only by a slight swelling of the cusp base on its posterior side. Elements of the latter type differ strikingly in appearance from alate elements with a well-developed posterior process; however, they are clearly functional homologues of more typical alate elements and there is no reason to refer them to a separately named shape subcategory.

Tertiopedate ramiform elements (Fig. 4) have a posterior process and a lateral process on each side of the cusp. The latter,

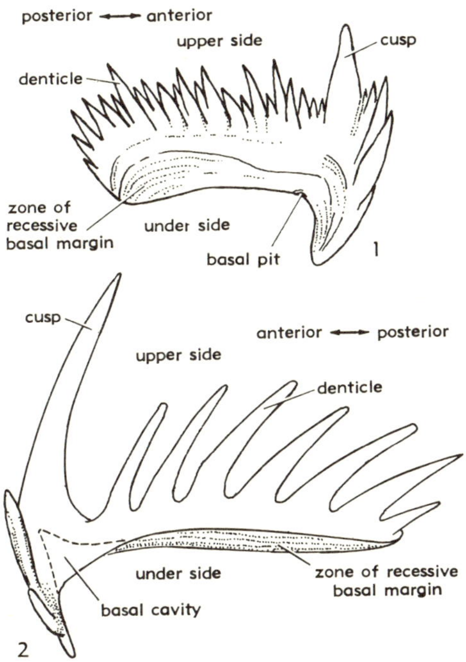

Fig. 6. Morphology of bipennate ramiform elements, both viewed from the inner side (Sweet, n).

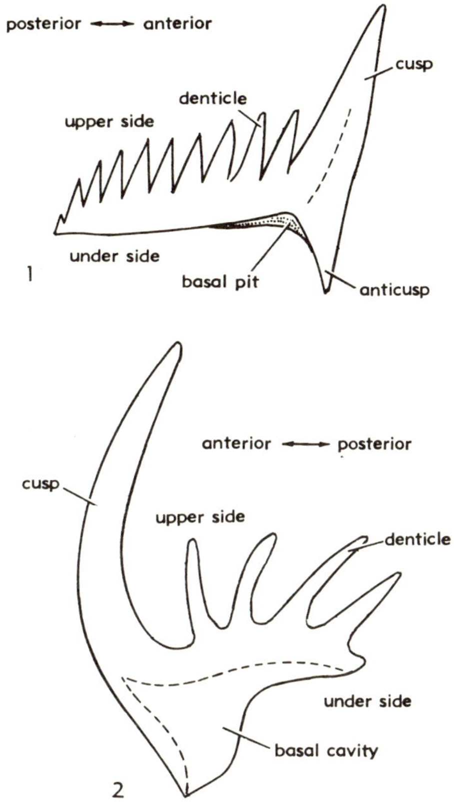

Fig. 7. Morphology of dolabrate ramiform elements (Sweet, n).

however, are not symmetrically disposed with respect to the midplane and the posterior process is commonly long and denticulate.

Digyrate ramiform elements (Fig. 5) are comparable to alate and tertiopedate elements in process number and orientation, but the midplane is not a plane of bilateral symmetry, the posterior process is short and adenticulate, and distal process extremities commonly twist in opposite directions.

Bipennate ramiform elements (Fig. 6) have only two processes, anterior and posterior. Elements of this shape category intergrade at one end of their variation spectrum with digyrate elements, and at the other with dolabrate types. Characteristically, the posterior process is longer than the anterior one, which in many bipennate elements is also curved, bent, or deflected

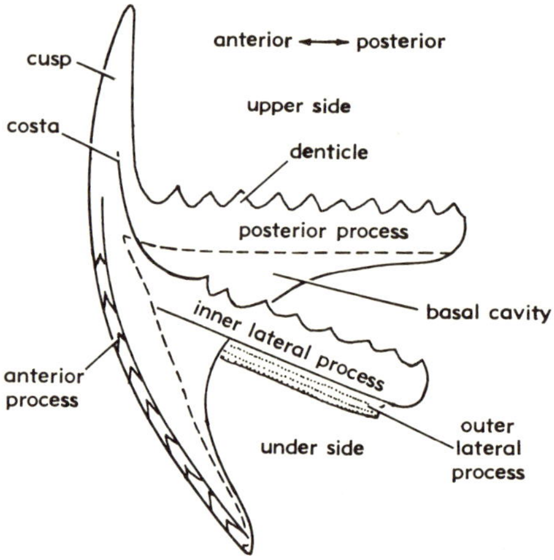

FIG. 8. Morphology of a quadriramate ramiform element (Sweet, n).

toward one side and may even be adenticulate.

Dolabrate ramiform elements (Fig. 7) have only a posterior process and are commonly pickshaped in lateral aspect. An elongate downward projection of the anterior part of the basal margin in some dolabrate elements forms a distinctive structure termed an **anticusp.**

Ramiform elements with more than three basic processes are rare; however, a few are known and others may be discovered. Subdivision of the ramiform shape category must then include a place for them. Thus, **quadriramate** elements (Fig. 8) are those with four processes (anterior, posterior, and a lateral process on each side of the cusp), and **multiramate** elements are ramiform elements with more than four basic processes.

PECTINIFORM ELEMENTS

Basically, pectiniform elements are comb-shaped units of the sort termed "pectinate teeth" by HINDE (1879), but most commonly described by later authors as "blades." Comb- or blade-shaped pectiniform elements are closely related in form, however, to more elaborate types that are expanded or produced laterally in various ways and have often been termed "plates"

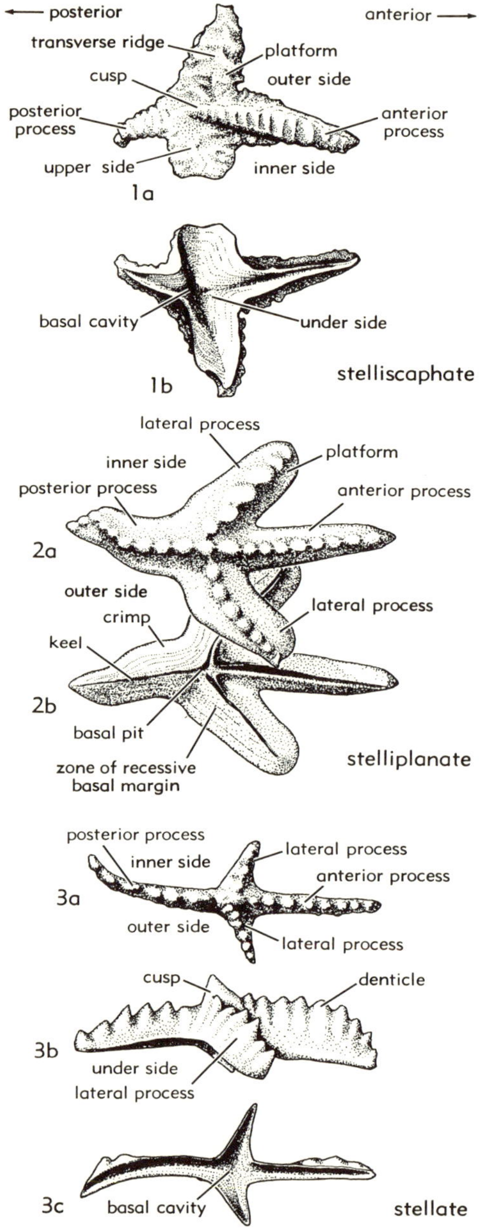

FIG. 9. Morphology of a stellate pectiniform element *(3)*, and its planate *(2)* and scaphate *(1)* counterparts (*1,2,* after Lindström, 1964; *3,* Sweet, n).

or "platforms." In view of substantial evidence that all these elements, simple or laterally elaborate, occupied similar positions in their respective skeletal apparatuses and probably had the same or similar functions, and recognizing the great variety of such elements, it seems overly artificial to continue the recognition of just two categories (i.e., blades and plates or platforms). Consequently, although all of them are regarded basically as pectiniform elements, they are assigned for descriptive purposes to the 15 named categories listed in Table 1.

Comb- or blade-shaped pectiniform elements of the sort most commonly termed "blades" in the literature are divided into five shape categories, termed "stellate," "pastinate," "carminate," "angulate," and "segminate." Elements of each of these types may develop more or less elaborate lateral extensions in two basic ways, resulting in categories of "platform" elements for which the general terms "planate" and "scaphate" may be employed; however, to indicate that a planate or scaphate pectiniform element is related either in shape or origin to elements in one or another of the named categories of bladelike pectiniform elements, terms such as "stelliplanate" (or "stelliscaphate") are used as names for major categories of platformed pectiniform elements. Thus, explanation of the 15 categories of pectiniform elements listed in Table 1 requires discussion of only seven terms: stellate, pastinate, carminate, angulate, segminate, planate, and scaphate.

Bladelike elements that lack significant lateral expansion, and belong in one or another of the first five categories of pectiniform elements listed in Table 1 have many features in common. Basically, they consist of a cusp, situated above the apex of a basal pit or basal cavity, and one or several processes that extend away from the cusp. These processes are regarded as **primary processes** if they project from the cusp and if their under surface or edge is occupied by an extension of the basal cavity or basal pit. **Secondary processes** are recognized as branches of primary ones by the fact that they join the latter at some point other than the latter's proximal end.

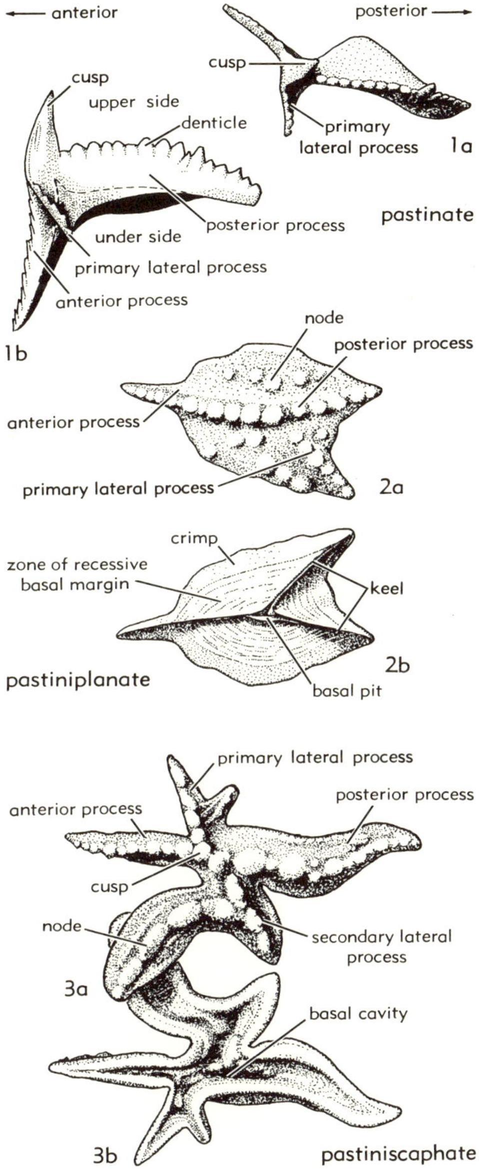

FIG. 10. Morphology of a pastinate pectiniform element *(1)*, and its planate *(2)* and scaphate *(3)* counterparts (*1a* after Lindström, 1964; others, Sweet, n).

Characteristically, processes of bladelike pectiniform elements are laterally compressed, higher than comparable structures of ramiform elements, and commonly divisible longitudinally into a lower, or basal part, which includes the **attachment surface,**

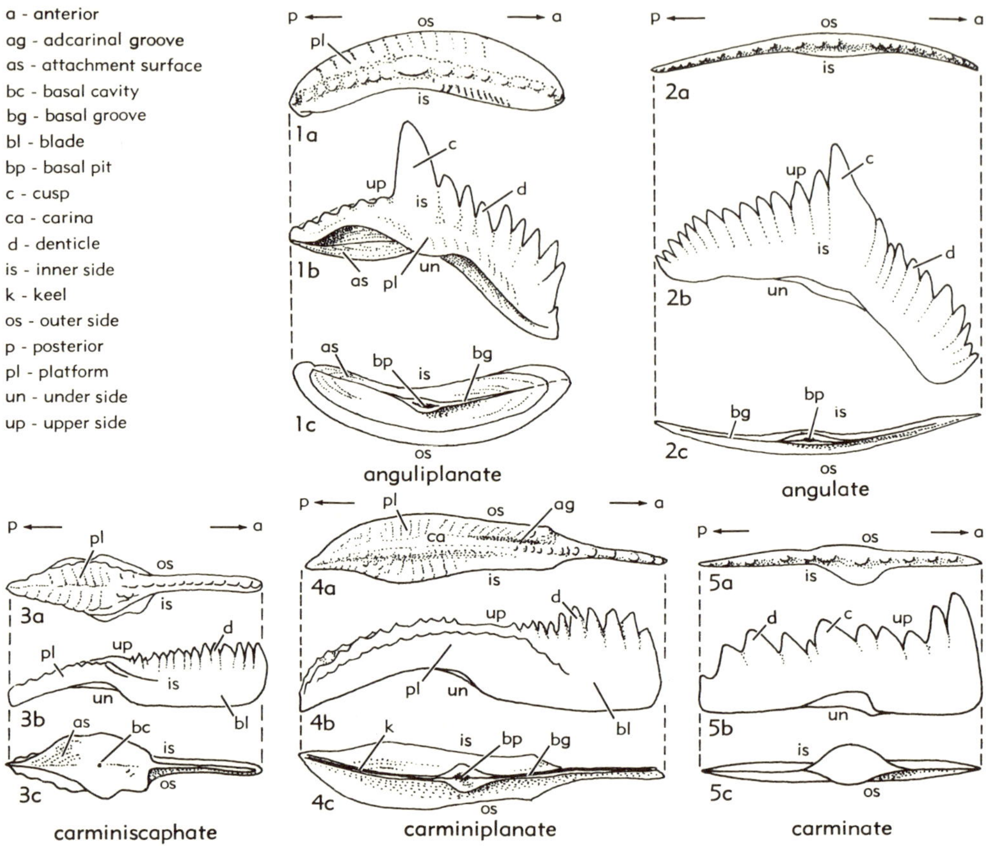

FIG. 11. Morphology of angulate and carminate pectiniform elements together with planate and scaphate counterparts (*1-3, 5,* Sweet, n; *4* after Lindström, 1964).

and an upper part that is denticulate in some types but adenticulate in others. Sides of processes may be smooth, or plane, or bear prominent longitudinal ribs or ledges at the level of the junction between basal and upper parts.

The under side of bladelike pectiniform elements is occupied by an attachment surface. This may have the form of a more or less capacious basal cavity with groovelike extensions beneath processes and an expanded portion with laterally flaring walls, or sheaths, beneath the cusp. Or the attachment surface may be composed of a relatively small basal pit beneath the cusp, which may or may not have narrow groovelike extensions beneath processes, and more or less extensive adjacent zones of recessive basal margin. The latter may form a flat surface on the underside of the element, or they may form scarlike areas of various shapes and disposition on the basal part of the inner and outer sides of processes.

Bladelike pectiniform elements are divided into five major categories in Table 1, primarily, but not exclusively, on the basis of the number and arrangement of primary processes. **Stellate** pectiniform elements (Fig. 9) have at least four primary processes, of which two are anterior and posterior. **Pastinate** pectiniform elements (Fig. 10) have three primary processes, anterior, posterior, and lateral. **Carminate** and **angulate** pectiniform elements (Fig. 11) have two primary processes, which are anterior and posterior, but differ in that the longitudinal axis of carminate elements is straight, or essentially so, in lateral view, whereas in angulate elements processes intersect beneath the cusp to form an angle and are said to be **arched.** **Segminate** pectiniform elements (Fig. 12) have just one process, which is anterior.

Laterally elaborate, or platformed, pectiniform elements are described as **planate** if they exhibit conspicuous lateral ledges, brims, platforms, or plates, and if the attachment surface on their under side is distinguished by a zone of recessive basal margin, which at least partially surrounds a basal pit that has groovelike extensions beneath equivalents of at least primary processes. Commonly, but not invariably, the basal pit and its subprocess extensions are situated within ridgelike structures, or **keels,** which project below the general level of the surrounding zones of recessive basal margin.

Laterally elaborate, or platformed, pectiniform elements whose under sides are marked by capacious basal cavities, rather than by the attachment-surface features distinctive of planate elements, are termed **scaphate.**

In both scaphate and planate pectiniform elements, an anterior portion of the element is commonly bladelike and is termed a **blade.** This is said to be a **free blade** (Fig. 12,*3b*) if it bears no ledge-, brim-, or platelike lateral extensions, although it may be marked by a midlateral rib of varying prominence. On the other hand, the blade may be described as a **fixed blade** if it is distinguished from the remainder of the element only by the fact that ledge-, brim-, or platelike projections along its sides are narrower than those of more posterior parts of the element.

Posterior to the blade, scaphate and planate elements develop prominent lateral extensions of various sorts on one or both sides of the longitudinal axis. It is difficult to draw a clear line between the prominent

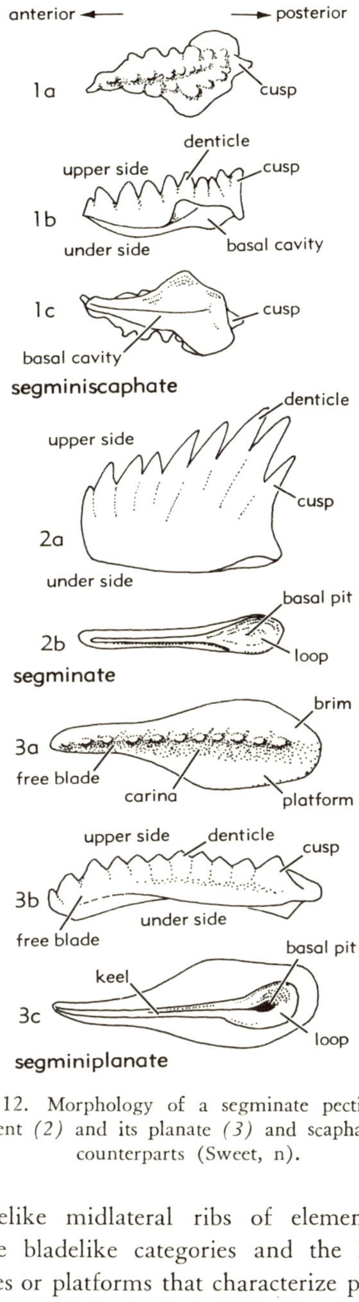

FIG. 12. Morphology of a segminate pectiniform element *(2)* and its planate *(3)* and scaphate *(1)* counterparts (Sweet, n).

ledgelike midlateral ribs of elements in some bladelike categories and the lateral plates or platforms that characterize planate elements. It is similarly difficult to provide mechanically applicable criteria that can be used to separate some fundamentally bladelike elements possessing laterally expanded basal cavities with unornamented upper surfaces from closely related forms that develop various patterns of ribs, ridges, and denticles above laterally expanded parts of their scoop-shaped basal cavities. No attempt is made to provide such criteria, for experience suggests that there was probably little basic difference in function between such borderline elements, and it thus makes little difference if they are described as carminate, for example, by one author, or as carminiscaphate by another.

A row of nodes or denticles that forms a continuation of the blade into the platformed segment of a planate or scaphate element is a **carina** (Fig. 12,*3a*), and laterally adjacent, smooth or less prominently nodose areas are **adcarinal grooves** (Fig. 11,*4a*). Distinct rows of nodes or denticles along the outer margins of lateral plates or platforms are **parapets** (as in *Cavusgnathus*), and rows of nodes transverse to the long axis of the element are **transverse ridges** (Fig. 9,*1a*). A pair of prominent longitudinal nodose ridges, one of which subdivides the adcarinal groove longitudinally on either side of the carina, is termed a **rostrum** (as in *Siphonodella*). Other terms used to describe various features of the upper and under sides of platformed elements are given in Figures 9 to 12, and are included in the glossary.

CLASSIFICATION AND TERMINOLOGY OF SKELETAL APPARATUSES

Discrete elements, described and classified by shape in preceding sections of this chapter, were components of more elaborate, multielement skeletal apparatuses, which are only rarely found in original association. A few more or less complete apparatuses, termed **natural assemblages** or **fused clusters,** have been discovered and described, but those of most species have been, or will have to be, reconstructed from collections of disassembled discrete elements.

Evidence from natural assemblages, or from skeletal apparatuses reconstructed from collections of disjunct elements, suggests that the number of major skeletal patterns may have been small, and several attempts have been made to classify apparatuses according to such patterns, or on the basis of distinctive features or structures shared by all the elements in apparatuses of given plan. For example, Rhodes (in Hass, 1962) recognized three skeletal plans among the Carboniferous natural assemblages with which he was familiar. He grouped the apparatuses of *Lochriea* Scott, *Lewistownella* Scott, *Westfalicus* Schmidt, and *Scottognathus* Rhodes, which are similar, as Class A apparatuses. These were contrasted collectively with those of *Illinella* Rhodes and *Duboisella* Rhodes, which differ from each other but were not assigned to lettered classes except by implication.

Klapper and Philip (1971) reconstructed four types of skeletal apparatuses from collections of discrete Devonian elements. In their terminology, apparatuses of types 1 and 2 include six morphologically distinct sorts of elements, but components of the two apparatus-types differ in style of denticulation and in the nature of the three element-types that compose a symmetry-transition series. Type 3 apparatuses include only five morphologically distinct elements, and Type 4 apparatuses include a scaphate pectiniform (or "icriodontan") element and two others that are coniform or modifications thereof.

Sweet (1970) recognized apparatuses of two types, but did not distinguish them formally by numbers, letters, or names. Multielement apparatuses, regarded as those of species of *Ellisonia* Müller, include five or six morphologically distinct types of elements; whereas others, thought then to represent species of *Anchignathodus, Neogondolella, Neospathodus,* and *Xaniognathus,* were interpreted to consist of elements of a single morphologic type.

Jeppsson (1971) recognized only two major apparatus types, which he termed Group 1 and Group 2. The first includes the Ordovician *Belodina* cluster described by

BARNES (1967), the Silurian *Panderodus* clusters described by POLLOCK (1969), and the Devonian *Belodella* and *Acodina* clusters discussed by LANGE (1968). Group 2 apparatuses include those with five or six types of elements like those assigned by JEPPSSON (1969) to *Hindeodella.* In 1972 JEPPSSON divided his Group 2 apparatuses into A- and B-type categories, based on denticulation of the elements and composition of the apparatus. JEPPSSON's A- and B-type apparatuses appear to correspond rather closely to the categories KLAPPER and PHILIP (1971) designated as Type 1 and Type 2–Type 3, respectively.

KOZUR and MOSTLER (1971) concluded that the skeletal apparatus of *Gladigondolella tethydis* (HUCKRIEDE) was multielement, and used shorthand designations to distinguish platform *(T),* paired *(PA, PB, PC, PD),* and symmetrical *(S)* elements. In various subsequent reports of other skeletal assemblies, KOZUR used this system to identify elements that he presumed were functional equivalents of those in the *G. tethydis* apparatus.

SWEET and BERGSTRÖM (1972) suggested division of known Ordovician multielement apparatuses into three major categories: simple-cone, ramiform, and ramiform-platform apparatuses.

LINDSTRÖM (1973) recognized four main types of conodont apparatuses, which he termed types 1, 2, 3, and 4. Type 1 apparatuses consist of geniculate and nongeniculate coniform elements and characterize conodonts that LINDSTRÖM (1970) included in his "Distacodontacea." Type 2 apparatuses have five or six element types and distinguish conodonts that LINDSTRÖM (1970) referred to his superfamilies "Prioniodontacea" and "Polygnathacea." Type 3 apparatuses, with five or six kinds of ramiform elements, characterize the "Prioniodinacea" and at least some "Chirognathacea," and Type 4 apparatuses, consisting solely of coniform elements of one general type, are distinctive of conodonts of the superfamily "Panderodontacea."

From this brief survey, it is apparent that there is potentially great value in a categorization of skeletal plans; however, it is also apparent that there is already a serious risk of confusion in terminology. Group 1 and 2 apparatuses in JEPPSSON's (1971) classification are not the same things as Type 1 and 2 apparatuses in the schemes proposed and used by KLAPPER and PHILIP (1971) or LINDSTRÖM (1973), and it is not apparent from the notation that JEPPSSON's A-type is based on considerations quite different from those that distinguish RHODES's Class A apparatuses. No criticism is made of any of these classificatory schemes, but it is suggested that confusion is inevitable in any scheme of classification that employs letters or numbers to designate major classes or categories. Further, it is probably premature to formalize anything more than a purely descriptive terminology, for it is by no means certain that all described apparatuses are complete, that all major skeletal plans are known, or that homologies between elements in apparatuses of different classes are completely understood.

Because it may be premature to suggest a classification of apparatuses or a nomenclature for them, a purely descriptive approach is advocated, which is genetically noncommittal but may perhaps be useful as a framework for description in the current period of taxonomic exploration. That is, it is suggested that conodont skeletal apparatuses, however reconstructed or interpreted, be described simply as **unimembrate** or **multimembrate,** with the latter category subdivided into **bimembrate, trimembrate, quadrimembrate, quinquimembrate, seximembrate,** or **septimembrate** apparatuses, depending on the number of morphologically distinct element types of which they are composed.

There is no intended implication that all apparatuses of a given category (e.g., quinquimembrate) are alike in construction of their component elements. The intention is to convey only the information that an apparatus is (or is thought to be) composed of a certain number of different element types (e.g., five in a quinquimembrate apparatus).

TABLE 2. *Comparison of Locational Notation Schemes for Conodont Skeletal Apparatuses.*

SWEET, herein; COOPER, 1975	SWEET & SCHÖN-LAUB, 1975	SWEET, 1970 1[a]	SWEET, 1970 2[b]	JEPPSSON, 1971	KLAPPER & PHILIP, 1971 Type 1	KLAPPER & PHILIP, 1971 Type 2	KLAPPER & PHILIP, 1971 Type 3	KLAPPER & PHILIP, 1971 Type 4[c]	KOZUR & MOSTLER, 1971
Pa	Pb	—	—	sp	P	P_1	—	I	PA
Pb	Pa	LA	LC	oz	O_1	O_2, B_3?	O_2	S	PD, T
M	M	LD	LA	ne	N	N	N	M	PB
Sc	Sc	LB	LB	hi	A_1	B_1	B_1	M	PC
Sb	Sb	LE	LB	pl	A_2	B_2	B_2	M	PC
Sa	Sa	U	U	tr	A_3	—	B_3	M	S

[a] Used for apparatus of *Ellisonia teicherti* SWEET [=*Hindeodus typicalis* (SWEET)].

[b] Used for apparatus of *Ellisonia gradata* SWEET [=*Xaniognathus gradatus* (SWEET)].

[c] Basic plan said by KLAPPER & PHILIP to be that of *Icriodella superba* RHODES, as reconstructed by BERGSTRÖM & SWEET (1966). In the apparatus of that species it seems likely that elements assigned to the form-species *I. superba* and *Sagittodontus robustus* occupied *P* positions; that those assigned to the form-species *S. dentatus* occupied the *M* position; and that elements assigned to form-species of *Rhynchognathodus* ETHINGTON occupied *S* positions, with *R. divaricatus* clearly being the *Sa* element. Apparatuses of the Silurian and Devonian species described by KLAPPER & PHILIP appear to have been somewhat simpler; hence, their notation for Type 4 apparatuses is difficult to compare exactly with the other schemes shown.

It should also be emphasized that the proposed terminology refers only to the number of morphologically distinct types of elements in a given apparatus, not to the total number of discrete elements of all categories that may originally have composed the apparatus. For example, the Class A apparatus of RHODES (in HASS, 1962) was interpreted to consist of four morphologically distinct types of elements, hence would be quadrimembrate in the scheme proposed here. RHODES indicated, however, that a complete Class A apparatus included at least 14 discrete elements, because those of three morphologic categories were represented by a pair, and the fourth by four pairs, of elements.

Perhaps more important than a scheme for describing the element-type composition of skeletal apparatuses is a means of identifying and naming homologous, or supposedly homologous, positions within the apparatus. Various methods of doing this have been advocated (JEPPSSON, 1969, 1971; KLAPPER & PHILIP, 1971; KOZUR & MOSTLER, 1971; SWEET & BERGSTRÖM, 1972). From Table 2 it is evident, however, that there is little uniformity in the notations suggested, and homologies are clear only in the writings of authors who have proposed the schemes. For these reasons, and because no one of the notational systems suggested has yet been widely adopted, a new scheme is proposed here, indicated diagrammatically in Figure 13, and explained more fully in the following paragraphs.

The skeletal apparatuses of some complex Ordovician species may have been septimembrate, and a number of supposedly uni-, bi-, tri-, quadri-, and quinquimembrate apparatuses have been described or postulated; however, a very large group of common species seems to have settled on a seximembrate plan, which varied little in its major aspects from the Ordovician through the Triassic periods. Because most known or postulated apparatuses can be easily compared with the seximembrate plan, it was used by SWEET and SCHÖNLAUB (1975) as the basis for developing a scheme of locational notation that can also be used to describe the postulated positions of homologous elements in apparatuses that contained fewer (or more) than six element-types.

Components of seximembrate apparatuses can be separated readily into three principal categories, which are designated *P, M,* and

S in Figure 13. *P* positions are occupied by pectiniform or specialized ramiform elements and, characteristically, there are two types, which are designated *Pa* and *Pb*. *M* positions are typically occupied by arched, pick-shaped dolabrate elements in one group of seximembrate apparatuses, but by bipennate, digyrate, or coniform elements in another large group. Elements occupying the three major positions in the *S* category form a symmetry-transition series of the sort first recognized by LINDSTRÖM (1964). Elements in the *Sa* position are characteristically alate, although a definite posterior process is lacking in some of them. *Sb* positions are filled by digyrate or tertiopedate elements, and the *Sc* position in the symmetry-transition series is occupied in seximembrate apparatuses by bipennate or dolabrate elements, commonly with a long posterior process and a laterally deflected or recurved anterior process. Also indicated in Figure 13 is an *Sd* position, which is occupied in septimembrate apparatuses by quadriramate elements that are bilaterally symmetrical in some complex Ordovician apparatuses. The *Sd* position, however, is not filled in typical seximembrate apparatuses.

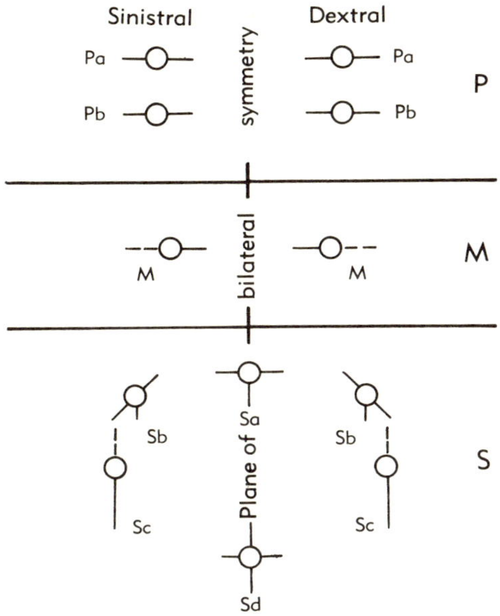

FIG. 13. Schematic arrangement and suggested notation for elements in a seximembrate apparatus. The position *Sd* is included for a few septimembrate apparatuses with quadriramate elements (Sweet, n).

As with any notational scheme, there are problems with the one just discussed. For example, it has been established beyond reasonable doubt (BERGSTRÖM, 1964; LANE, 1968) that elements in *P* positions in the apparatuses of some well-known Ordovician and Carboniferous species are paired, but that dextral (or right-handed) and sinistral (or left-handed) members of these pairs are not mirror images of one another as they are in a majority of known conodont apparatuses. Indeed, in *Amorphognathus* the two elements of one *P* pair are so strikingly different in morphology that they were originally referred to separate form-genera. Should the apparatuses of these species then be regarded as septimembrate (rather than seximembrate) and some special notation be devised for the asymmetric pairs in the *P* positions? Or should they be regarded as seximembrate apparatuses distinguished from those of other species with such apparatuses by the asymmetric pairs in *P* positions? The latter seems the more suitable choice. BERGSTRÖM (1971), after noting homologies that exist between the *Amorphognathus* and the seximembrate *Prioniodus* apparatuses, pointed out that those apparatuses are "basically the same in important respects." To describe that of *Amorphognathus,* then, as septimembrate because one of its *P*-pairs is asymmetric, and make elaborate formal distinction between components of the asymmetric *P*-pairs, would obscure the obvious homologies between the skeletal apparatuses of *Prioniodus* and *Amorphognathus* that suggest the two genera are really very closely related.

It should be emphasized that description of an element as the occupant of, for example, a *Pa* or *Pb* position does not (or should not) imply that it is necessarily homologous with elements described as occupants of *Pa* or *Pb* positions in other apparatuses. With time and increased understanding of the functions of conodont elements and the organ(s) of which they were parts, however, it is hoped that the connotation of homology can be attached to locational notation. Within certain major groups, this is probably the case now. For example, many conodont species assembled in this volume in the superfamily Polygnathacea seem to have had skeletal apparatuses built on the seximembrate plan of *Ozarkodina,* and it is logical from evidence available to assume, for example, that angulate pectiniform elements in the *Pb* positions of all these apparatuses are not only analogous structures, but that they are also truly homologous. Thus, in such a group of presumably closely related taxa, it is desirable that angulate pectiniform elements be consistently described as occupants of the *Pb* position. For such groups, conventions with regard to locational notation are established in the generic diagnoses elsewhere in this volume.

On the other hand, *Oulodus* and related genera of the Hibbardellacea apparently also formed seximembrate skeletal apparatuses, and it is presumed that two of the element-types recognized as components of these apparatuses occupied *P* positions. These elements, through relegation to *Pa* and *Pb* positions, are thus assumed to be analogous with those in *Pa* and *Pb* positions in the *Ozarkodina* apparatus, but they are probably not homologous with those elements. With the groups of genera typified by *Ozarkodina,* it is desirable to recognize supposed homologies between skeletal elements of different taxa within the group by assigning analogous elements to the same positions and using the same locational notation for them.

In brief, the scheme of locational notation proposed by Sweet and Schönlaub (1975), advocated in this chapter and used elsewhere in this volume, is designed to be a vehicle for expressing analogy. Homologies, however, must remain the basis for recognizing major taxonomic categories, and short of developing a separate notational scheme for apparatuses of each major group (e.g., Polygnathacea and Hibbardellacea) as Klapper and Philip (1971) seem to have been suggesting, no general system in which locational notation expresses both homology and analogy is apparent or suggested.

MICROMORPHOLOGY OF ELEMENTS

INTERNAL STRUCTURE

By Klaus J. Müller

[Rheinische Friedrich Wilhelms Universität, Bonn]

Conodont elements are laminated structures that have been built up by outer accretion around a more or less round nucleus. They have a highly complex histology, seemingly unique to the group, which is of relevance for comparison with other living or extinct animals. Such comparative studies eventually may lead to better understanding of the exact systematic position of conodonts. In addition, the changes in histology during evolution are useful for clarification of relationships between natural taxa. The mode of growth and its temporary changes may provide clues for the reconstruction of the paleoecology of conodonts.

MATERIAL AND PREPARATION TECHNIQUES

Because finer features of the infrastructure usually disappeared during the process of diagenesis, only well-preserved material can be used for detailed histological studies. Adequate preservation is rare and cannot

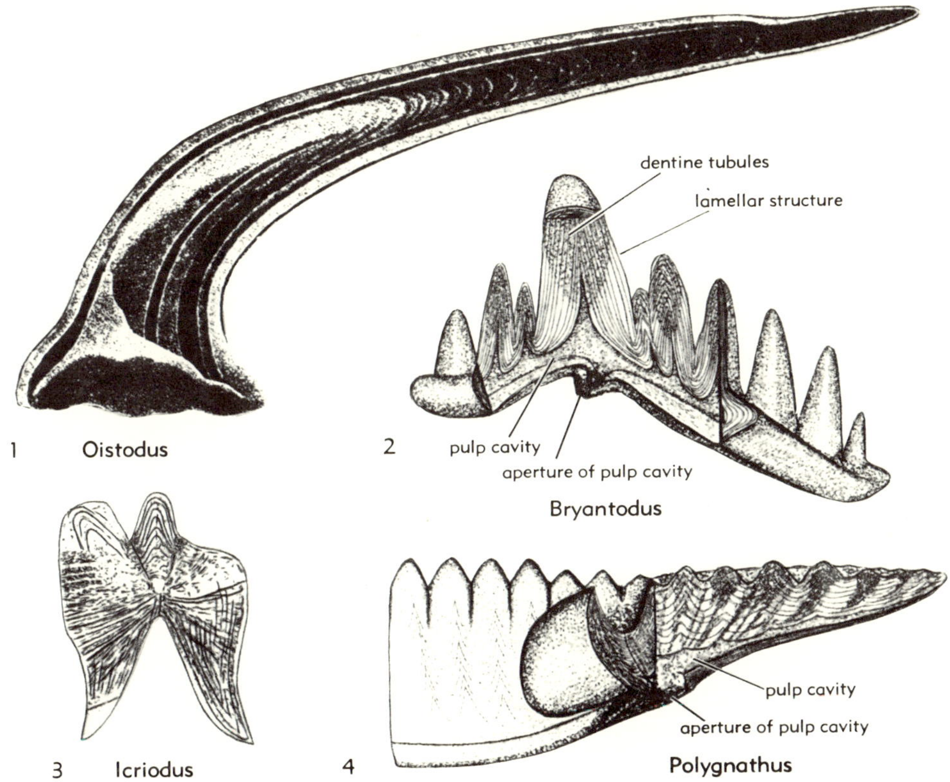

Fig. 14. Internal microstructure of conodont elements.——*1.* Coniform element of *Oistodus lanceolatus* Pander, longitudinal thick section showing lamellar mode of growth; first histological investigation recorded; L. Ord., ×90 (after Pander, 1856).——*2-4.* Morphology of elements according to Beckmann, based on material considerably altered through weathering; Dev., Ger. (after Beckmann, 1949). *2.* Ramiform element of *Bryantodus delicatus* Branson & Mehl, reconstruction with part of specimen cut away, ×60. *3.* Pectiniform element of *Icriodus symmetricus* Branson & Mehl, transverse section showing lamellar mode of growth and darker transverse bands, erroneously interpreted as dentine tubules, ×80. *4.* Pectiniform planate element of *Polygnathus pennatus* Hinde, reconstruction with part of specimen cut away, ×60.

be recognized through mere inspection of whole specimens, and "trial" thin sections are necessary. Material most suitable for such work can be obtained from slightly indurated sediments that contain a surplus of phosphatic matter, as in certain bone beds.

In order to obtain comprehensive information it is advisable not to rely on a single preparation technique but to apply all of the various techniques available in suitable combinations. These procedures are outlined below.

Transparent mounts.—Observations of the internal structures in clear, translucently preserved conodonts require little preparation. Quality of preservation can be checked simply by embedding thin fragments in Canada balsam or a similar resin, but layered structures cannot be easily studied by this technique alone. Hass (1941) partly used this technique.

Thin sections.—In order to reveal the finest structures (e.g., white matter), well-oriented preparations as thin as possible are needed. In many cases grinding down to a thickness of 10 to 15 microns is advisable. This sectioning technique, which Pander used as early as 1856, is still the best single one available and was extensively used by

Gross (1954, 1957, 1960) and Müller and Nogami (1971, 1972a).

Slicing with microtome.—Slicing of mineralized material leads to shatter effects that may obscure the true natural structure (Pietzner & others, 1968). Before satisfactory results can be obtained, the material must be fully demineralized. If well-preserved conodont elements are treated with the right solvent, the outer shape is fully preserved after demineralization. Little comparative work has been done as yet with microtome preparations of demineralized conodonts.

Observation of fractures.—Fractures may have developed in conodont elements before they were buried in sediment or during preparation. Most fractures are oriented at random, and it is difficult to break a specimen at a desired position and in a determined direction.

Some fractures have been naturally etched before deposition or during diagenesis in the rock. For study, however, specimens etched in the laboratory under controlled conditions seem to be preferable. This technique is useful for preliminary observations with the scanning electron microscope.

Oriented polished and etched sections.—These preparations are best suited for observation with the scanning electron microscope. The agent, concentration, and length of the etching process are critical factors and different preparations yield quite different visual results. Advantages and disadvantages of the technique have been discussed in detail by various authors (Lindström & Ziegler, 1971; Barnes, Rexroad, & Miller, 1973).

Particularly in earlier studies, inadequate preparation techniques or poorly preserved material led to misinterpretation of structures. For example, Pander (1856) investigated in transmitted light a single coniform element that had been ground on both sides; however, because the specimen was too thick to permit clear observation of the growth lines (Fig. 14,*1*), he misinterpreted it to have grown by internal apposition. Actually, most conodont elements were formed by outer apposition of mineral matter. Beckmann (1949) later based his study on material altered considerably through weathering and came to similar conclusions (Fig. 14,*2-4*). Because of this misunderstanding, he considered conodont elements to be closely related to vertebrate teeth. Based on further studies, this interpretation is no longer justified. Even today, many details of conodont element histology need more clarification before definite interpretation of the element nature and possible function can be attempted.

STRUCTURES OF PRIMITIVE ELEMENTS

The earliest conodonts may have had no phosphatic hard parts. From the beginning of the Middle Cambrian, a mineralogical evolution from organic matter with a small amount of calcium phosphate to predominantly phosphatic matter with little organic material has been noted (Clark & Miller, 1969). Cambrian conodont elements are composed of very fine phosphatic crystallites. In this respect they seem to be more similar to the basal plates of later elements than to their cusps. The material of the principal or cusp portion of later elements is formed by relatively coarser crystallites.

A basal plate is not differentiated in the oldest elements. They may be considered as equivalent to the basal plates of later elements. This condition, together with differences in internal structure (see section on white matter), may serve to subdivide conodont elements into two groups, an older one referred to as Paraconodontida (Müller, 1962c) and an advanced one, the Conodontophorida (Eichenberg, 1930), or true conodonts.

The oldest known conodont elements are from Precambrian-Cambrian boundary strata of the Siberian platform and Kazakhstan. These have been assigned to the genus *Protohertzina* by Missarzhevsky (1973) and they contain extensive solid phosphate.

Internal structures of Middle Cambrian elements of *Amphigeisina danica* (Poulsen) and *Hertzina? bisulcata* Müller have

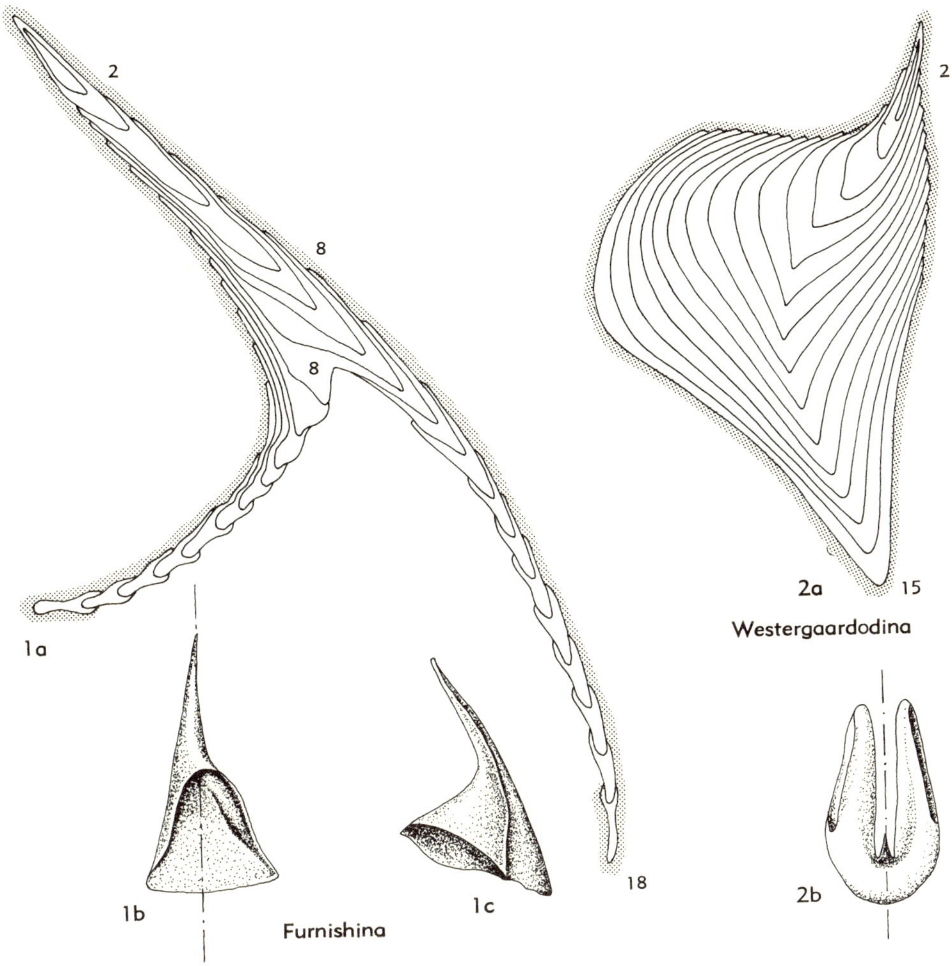

FIG. 15. Histology of early conodont elements (suborder Paraconodontida MÜLLER); longitudinal sections through growth center; all lamellae drawn, and organic cover finely dotted (after Müller & Nogami, 1972a).——*1. Furnishina furnishi* MÜLLER; first two lamellae closed on all sides, 3 to 8 interrupted at upper side, 9 to 18 disjunct at upper and lower sides; U. Cam. (Z. 3), Sweden, ×80.——*2. Westergaardodina bicuspidata* MÜLLER; all 15 growth lamellae uninterrupted on lower side; U. Cam. (Yencho F.), NE. China, ×120.

been investigated by BENGTSON (1976). These elements possess deep internal cavities and show growth lamellae that were added only at the inner surface. BENGTSON termed these forms "protoconodonts" and assumed that they represent a group ancestral to the Paraconodontida and later conodonts; however, the "protoconodonts" are not taxonomically differentiated from the Paraconodontida in this volume.

Growth lamellae develop differently in other representatives of the Paraconodontida. The later-formed lamellae in *Furnishina furnishi* MÜLLER terminate on both the upper and lower sides (Fig. 15,*1a*). In comparison, lamellae in all growth stages of *Westergaardodina bicuspidata* MÜLLER completely cover the lower side (Fig. 15,*2a*).

A thin organic layer seems to have been originally developed to cover the upper side

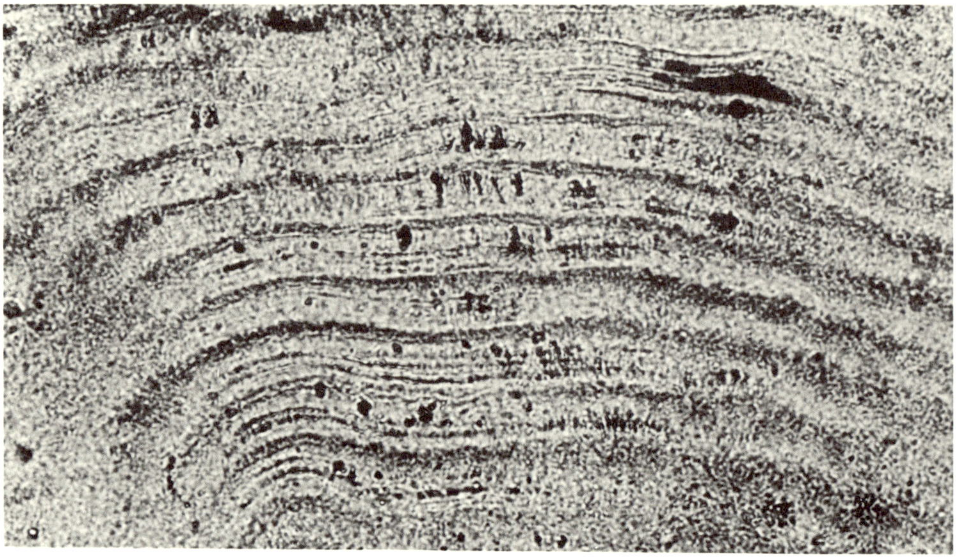

Fig. 16. Interlamellar striation and arrangement of crystallites in pectiniform element of *Polygnathus* sp.; very thin horizontal section across platform, more or less perpendicular to surface; U. Dev. (Maple Mill Sh.), USA (Iowa), ×400 (Müller & Nogami, 1971).

of paraconodontid elements. This can be observed in almost all specimens that are altered to the typical black color; however, it is not present on amber or whitish, translucent specimens. Such a cover is absent in advanced elements.

STRUCTURES OF ADVANCED ELEMENTS

The earliest advanced conodont elements are quite similar to the older coniform elements in outer morphology, but are clearly distinct from them in histology. In advanced conodont elements all growth lamellae are closed at the upper surface. Other important features are differentiation of the basal plate and the presence of white matter.

All advanced conodont elements are similar in general histological aspects but show differences in detail that are of taxonomic importance; however, because of technical difficulties in preparation, the amount of information available on the histology of some taxonomic units is still insufficient to obtain a full understanding of their relationships. Therefore, it seems advisable to describe and compare general features only.

Growth Lamellae

Growth lamellae are formed by outer apposition around a nucleus. The lamellae contain a considerable amount of organic matter. Embedded in this matter are crystallites of calcium phosphate. In places with isometric growth, such as the surface of a plate, they remain perpendicular to the lamellae, whereas in positions of accelerated

Fig. 17. Anisometric growth from a simple elongate nucleus into a planate pectiniform element of *Siphonodella quadriplicata* (Branson & Mehl); increase of width and development of sculpture attained by greater thickness of otherwise identically structured growth lamellae in directions of prevalent growth; drawn from thin section with every second lamella shown; Miss. (Chappel Ls.), USA (Texas), ×190 (Müller & Nogami, 1971).

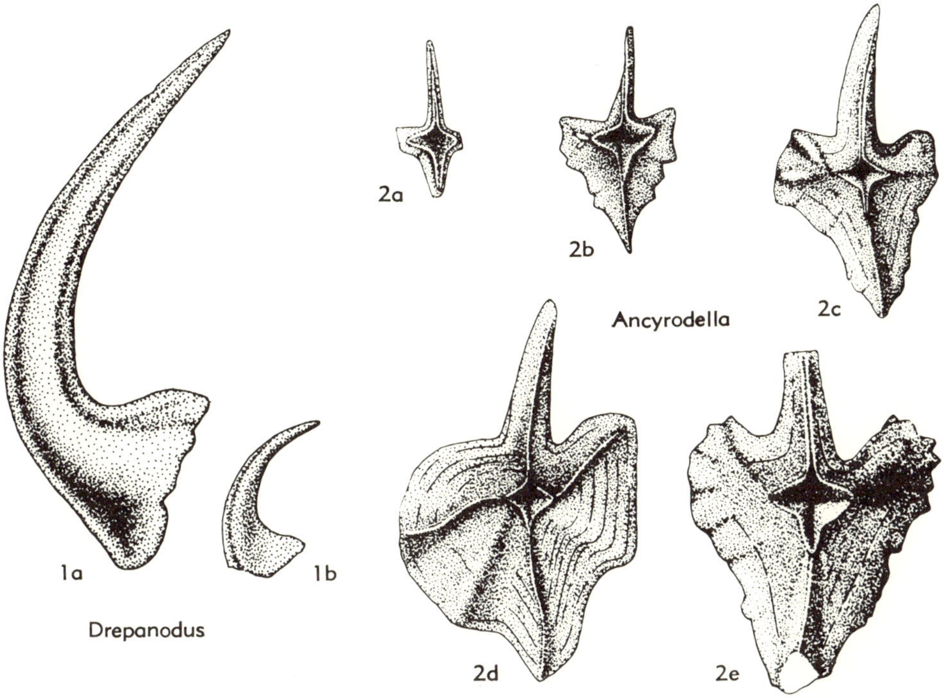

FIG. 18. Basal cavities and pits of conodont elements.——*1.* Example of elements having a basal cavity: two different growth stages of coniform element of *Drepanodus arcuatus* PANDER, 1856, demonstrating that entire smaller specimen *(1b)* would fit into basal cavity of larger one *(1a)*; both from same sample, L. Ord., Swed. (Öland), both ×45 (after Clark & Müller, 1968).——*2.* Example of elements having a pit: five pectiniform elements of *Ancyrodella rotundiloba* (BRYANT), showing size variation of pit, which is independent of growth stage; U. Dev. (Squaw Bay Ls.), USA (Mich.), all ×35 (after Müller & Clark, 1967).

growth their axes are pointed more or less in the direction of pronounced growth.

Interlamellar striation (Fig. 16) can be seen on excellently preserved material only (MÜLLER & NOGAMI, 1971). It demonstrates that a growth lamella was not deposited as a single event, but in three to four small layers. The striae most probably originated from intercalations of very thin organic layers within the growth lamella. Growth of the apatitic crystallites was not affected by these interlamellar structures, but they are set off at the much thicker organic layer that marks the boundary of a growth lamella.

The shape of an element, including nodes, pustules, ridges, or most other outer features, was formed by localized anisometric growth (Fig. 17). That is, individual lamellae are thicker in directions of prevalent growth and thinner in the areas between. Some features developed slowly and became more pronounced in subsequent lamellae, whereas others appeared spontaneously at a definite stage. Development of the basal excavations, with two distinct types, serves as an example (MÜLLER & CLARK, 1967, CLARK & MÜLLER, 1968).

Basal cavity.—A cavity is the more primitive type of basal structure, and the excavation increased in size as long as the element continued to grow (Fig. 18,*1*). Each subsequent growth lamella exceeds the lower limit of the preceding one and each diverges.

Basal pit.—In the earlier growth stages, a basal pit (Fig. 18,*2*) is similar to a basal cavity. Each succeeding growth lamella ex-

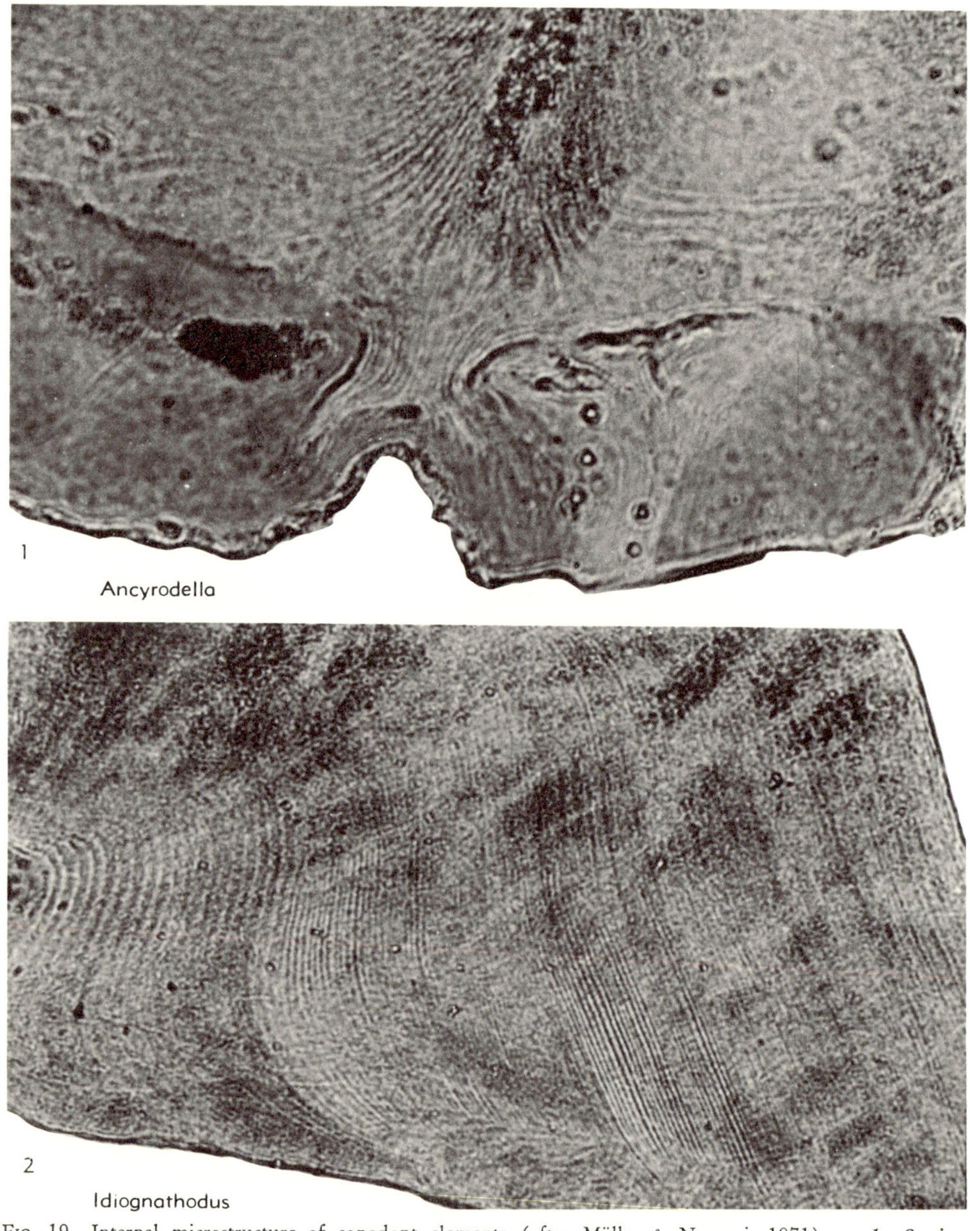

FIG. 19. Internal microstructure of conodont elements (after Müller & Nogami, 1971).——*1.* Section across initial stage of pit in pectiniform element of *Ancyrodella rotundiloba* (BRYANT), demonstrating distinct outer limit of pit after eighth growth lamella and continuous transition of growth lines from element into basal plate; U. Dev. (Independence F.), USA (Iowa), ×560.——*2.* Variation of width of growth lamellae in section across center of pectiniform element of *Idiognathodus*; early lamellae (left side) widely spaced and at maturity (right side) more closely spaced; Penn. (Desmoines.), USA (Ill.), ×480.

ceeds the one preceding it and diverges; however, after the development of a number of lamellae, which number is consistent for a specific element, subsequent lamellae

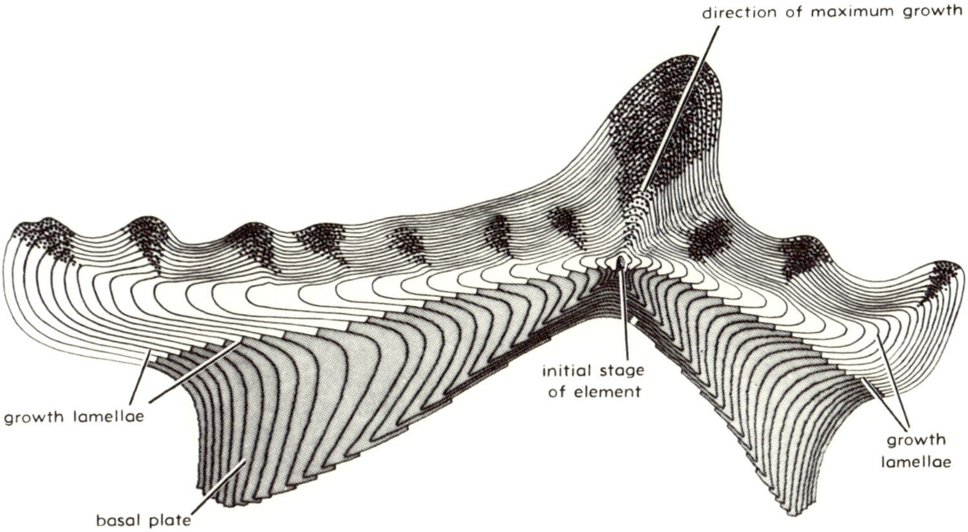

FIG. 20. Idealized section of the planate element of *Palmatolepis,* demonstrating the relationship between basal plate (gray pattern) and element; every second lamella shown; U. Dev., ×160 (after Müller & Nogami, 1971). The initial stage of element is not closed at its lower end, but is connected with the corresponding first lamella of basal plate. The growth lamellae of element and basal plate correspond, and do not alternate; there is no zone of resorption between the two. Even in the direction of maximum growth, there is no intercalation of additional lamellae, and the progression is caused by thickening of the lamellae only.

are shorter than preceding ones and, after forming a wall, they tend to converge. From the point at which lamellae cease to diverge, there is no change in size of the pit although the conodont element may increase many times in total size. Thus, the size of a basal pit is established fairly early in ontogeny, and may be a character useful for differentiation of specific elements. Although the number of growth lamellae forming the pit is more or less constant for an element, actual size may vary owing to variation in thickness of the lamellae (Fig. 19,*1*).

The local effects of anisometric growth aside, the general thickness of growth lamellae was quite variable during ontogeny. Earliest lamellae of an element are fairly thin, but succeeding lamellae increase in thickness, and in mature growth stages they are more closely spaced again (Fig. 19,*2*). Variation in thickness of growth lamellae may also have been controlled by external factors. In many individual elements, growth lamellae became thinner and closer together just before a dissolution took place, and in the earliest stages of redeposition they increased again in thickness.

Thickness of lamellae may also vary among elements of different taxa; however, this variation does not seem to be a character of superspecific stability. Within groups of individual elements, the thickness seems to be fairly constant, but may be different between closely related elements. For example, at a length of 1.0 mm, Upper Devonian pectiniform elements of *Palmatolepis helmsi* from widely separated localities in Europe bear 13 lamellae, whereas those of *Palmatolepis rugosa* display 16 to 17.

Basal Plate

All advanced conodont elements consist of two parts, the element proper, and a basal plate (Fig. 20). The first deposited growth lamella contains both element and basal plate. This lamella in early advanced elements was completely sclerotized, but in later stages of evolution may have been left

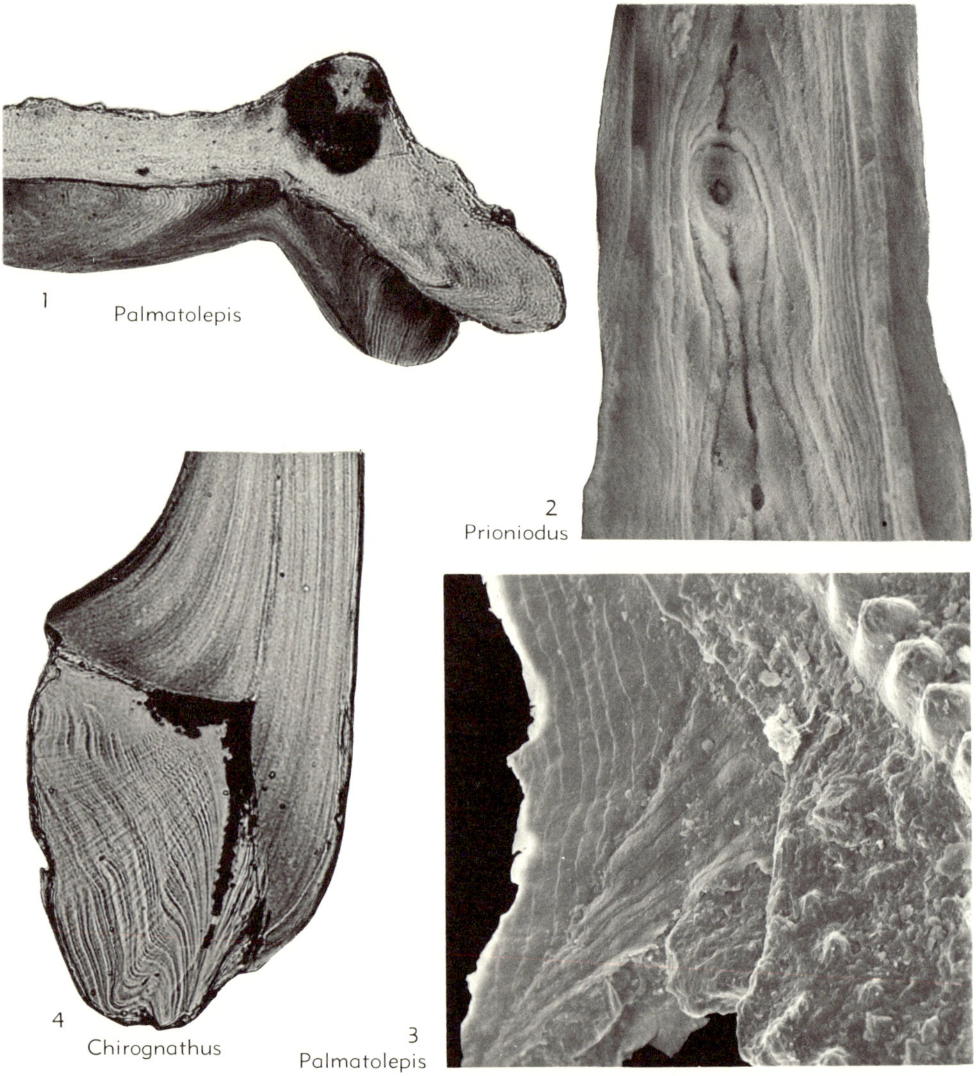

FIG. 21. Microstructure of conodont elements (*1, 3, 4,* Müller & Nogami, 1971; *2,* Müller, n).——*1.* Planate element of *Palmatolepis perlobata* ULRICH & BASSLER, demonstrating sharp angular recurvature of growth lamellae within basal plate; lamellae closed in earlier stages at lower limit, sclerotization limited to sides in later ones; U. Dev. (Maple Mill Sh.), USA (Iowa), ×150.——*2.* Lower side of pectiniform element of *Prioniodus* sp., showing the broken edges of growth lamellae that are transitional to basal plate; Ord. (drift), Ger., ×640.——*3.* Detail of a planate element of *Palmatolepis subrecta* MILLER & YOUNGQUIST; lobe of platform broken away, exposing upper side of basal plate showing growth lamellae; carina seen in upper right corner; U. Dev., Ger., ×250.——*4.* Section across main denticle of *Chirognathus duodactylus* BRANSON & MEHL showing distinct growth lamellae, which undulate in basal plate due to secondary shrinkage; M. Ord. (Harding Ss.), USA (Colo.), ×140.

FIG. 22. Development of growth lamellae in basal plates (gray pattern) of conodont elements from various ages (after Müller & Nogami, 1971).——*1.* Coniform *Furnishina furnishi* MÜLLER; differentiation into element and basal plate absent, and element possibly can be homologized with basal plate of more advanced elements; U. Cam., ×155.——*2.* Coniform *Oneotodus nakamurai* NOGAMI, showing distinct angular curvature of growth lamellae that are closed on lower side; uppermost Cam., ×130.——*3.* Coniform *Neocoleodus breviconus* BRANSON & MEHL; basal plate is area of most prominent growth, whereas in other genera it usually has a depression at the center; Ord., ×87.——*4.* Broad ramiform

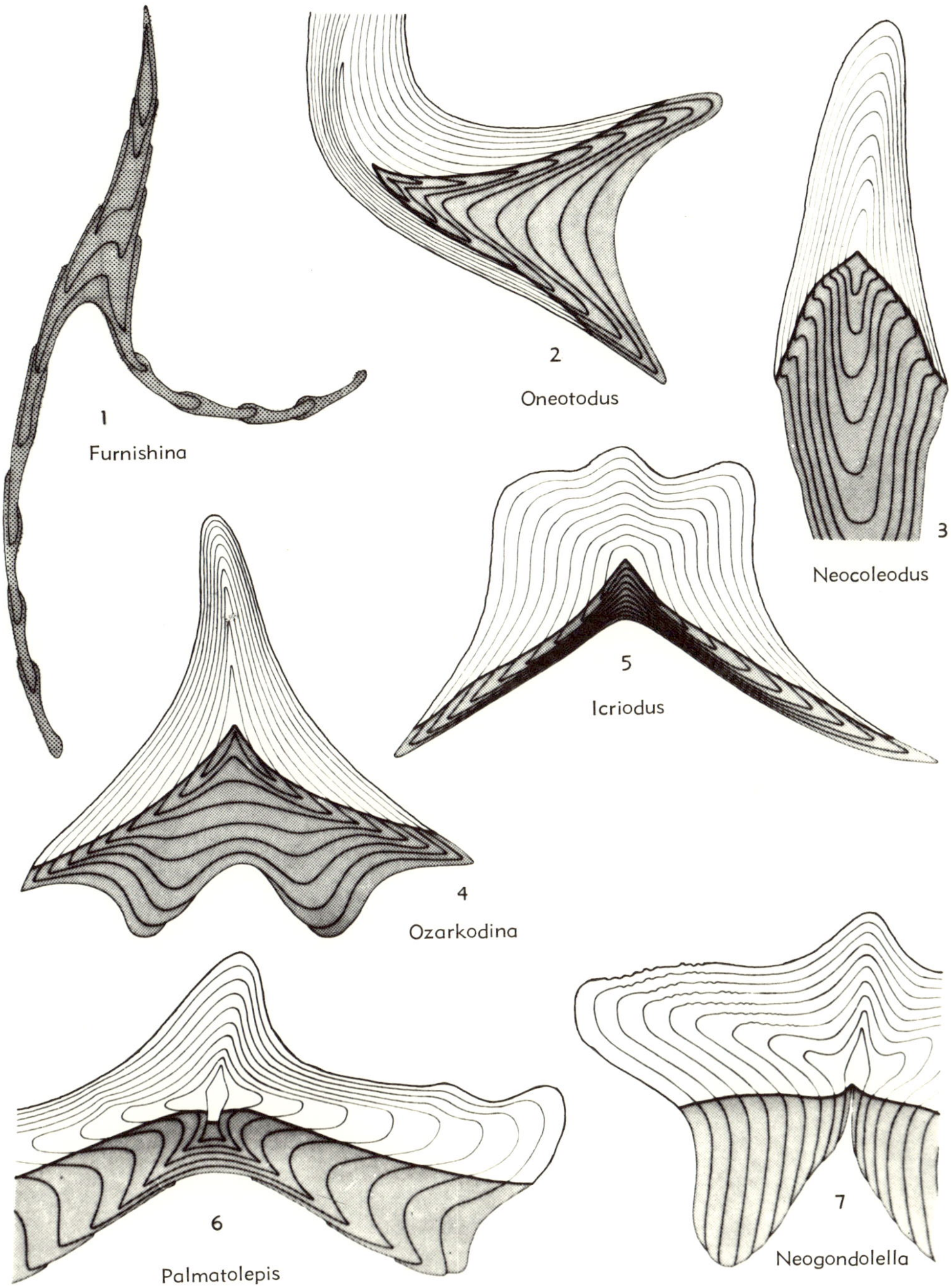

FIG. 22. *(Explanation continued from facing page.)*

element of *Ozarkodina prima* (BRANSON & MEHL), showing prominent growth in two directions; only last few lamellae not closed at middle; Sil., ×130.——*5*. Scaphate *Icriodus* sp., with large basal cavity and basal plate similar to that of the most primitive coniform elements; Dev., ×130.——*6*. Pectiniform element of *Palmatolepis perlobata* ULRICH & BASSLER, with first few growth lamellae closed in center but subsequent ones increasingly less sclerotized; Dev., ×172.——*7*. Pectiniform element of *Neogondolella navicula* (HUCKRIEDE), example of youngest elements, in which distinct recurvature below the transitional zone is lacking and all growth lamellae probably terminate below; Trias., ×172.

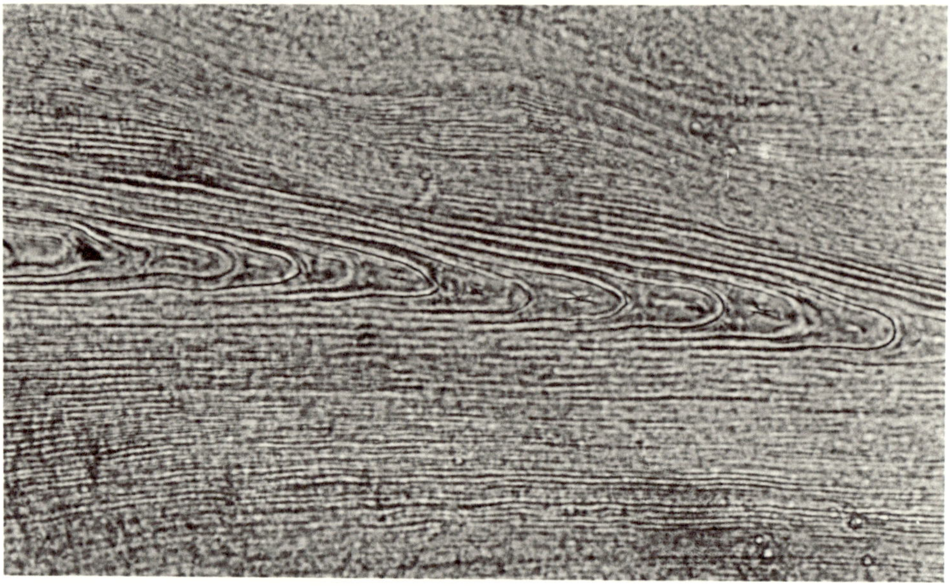

Fig. 23. Interlamellar spaces with white matter in pectiniform element of *Polygnathus* sp., U. Dev. (Maple Mill Sh.), USA (Iowa), ×450 (Müller & Nogami, 1971).

unmineralized in certain portions of the basal plate (Müller & Nogami, 1971).

Elements proper and basal plates are composed of the same type of carbonate apatite; however, in many specimens, the basal plate is not preserved and in others there are considerable differences in color and type of preservation between element proper and basal plate. This may be due to the higher organic content and the much finer apatite crystallites in the basal plates. Even well-preserved basal plates show undulations in growth lamellae that originated by shrinkage of the organic matter during diagenesis (Fig. 21,*4*).

Considered separately, element proper and basal plate have growth lamellae terminating with free edges at their lower or upper surfaces, respectively (Fig. 21,*2,3*). The plane of discontinuity occurs because the phosphatic matter deposited in the element proper is differently structured from that deposited in the basal plate. Its shape is of importance for the outer morphology of the under side of the element. In general terms, it may be concave (basal cavity or pit), even, or convex (forming a zone of recessive basal margin).

Between element and basal plate the growth lamellae are continuous (Fig. 19,*1*). In thin sections of well-preserved material, each growth lamella can be traced from the element into the basal plate without interruption; however, through inadequate preparation a grinding relief between the harder element and the softer basal plate may form. This may easily be misinterpreted as an offset of the growth lines in the transition.

The position of the plane of discontinuity within a unit does not seem to be a stable feature. Boersma (1973) observed that at least in some elements a portion of the original basal plate may be transformed during ontogeny. Prior to this transformation, a sharp recurvature of the basal growth lamellae occurred (Fig. 21,*1*).

A phylogenetic tendency toward gradual reduction of the basal plate has been noted (cf. Fig. 22). In the most primitive of the advanced elements, all growth lamellae of the basal plate are sclerotized completely at the lower rim. In later forms they are interrupted in the center, particularly in more

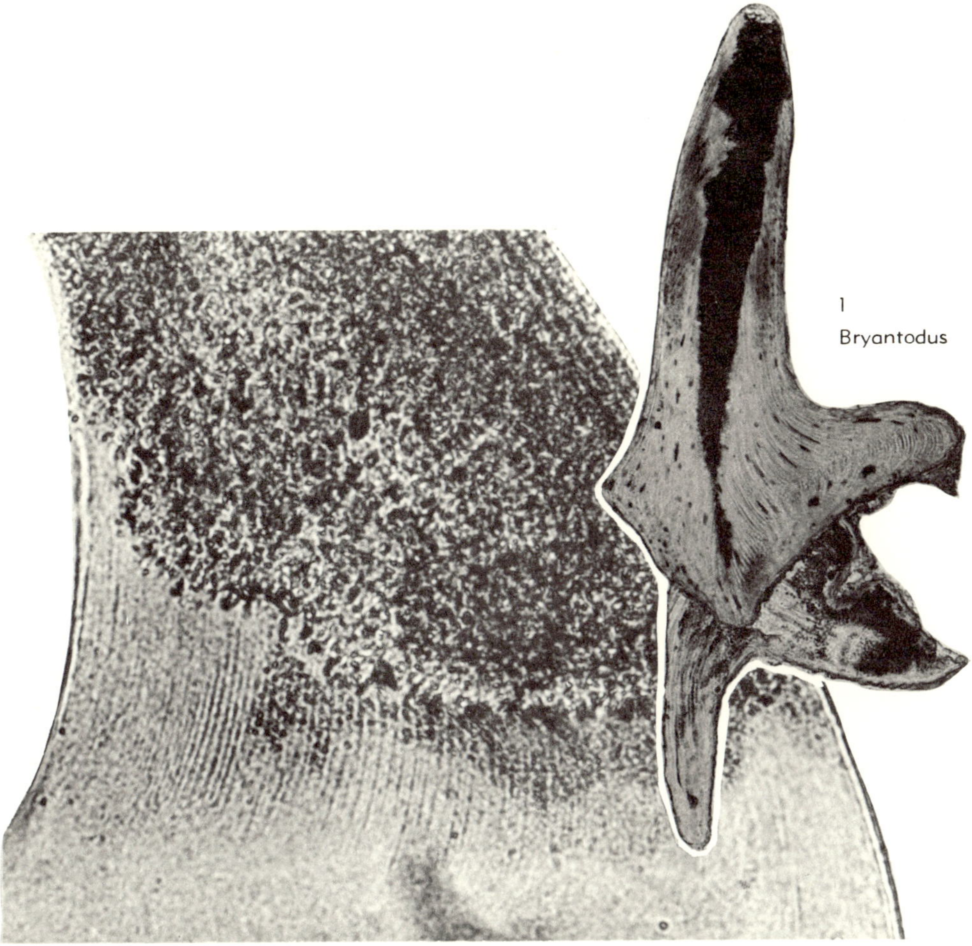

FIG. 24. Examples of white matter in conodont elements (after Müller & Nogami, 1971).——*1.* Element of *Bryantodus inaequalis* BRANSON & MEHL; section across secondary denticle with white matter visible in center, here appearing dark (see text); in upper left part, bubble zones cut through growth lines; in middle part, large pyrite particles are arranged parallel to growth lamellae; U. Dev. (Independence F.), USA (Iowa), ×140.——*2.* Ramiform element of *Plectodina inconstans* (WALLISER); longitudinal section through main denticle demonstrating transition between lamellar structure and white matter; U. Sil. (*Beyrichia* Ls.), Ger., ×600.

mature growth stages. The youngest forms, such as *Neogondolella navicula,* lack the distinct recurvature of growth lamellae below the plane of discontinuity and probably all growth lamellae in the basal plate are discontinuous at the lower rim (Fig. 22,7). The same tendency toward increased interruption of growth lamellae in basal plates can be observed in growth stages on a single specimen of more advanced genera (e.g., *Pa* elements of *Palmatolepis*).

White Matter

In reflected light the well-preserved conodont element is translucent above the rim of the base and light amber in color. Areas of prevalent growth, as in the center of denticles, appear opaque and light gray or pale brown. In thermally unaltered specimens these areas are lighter colored than

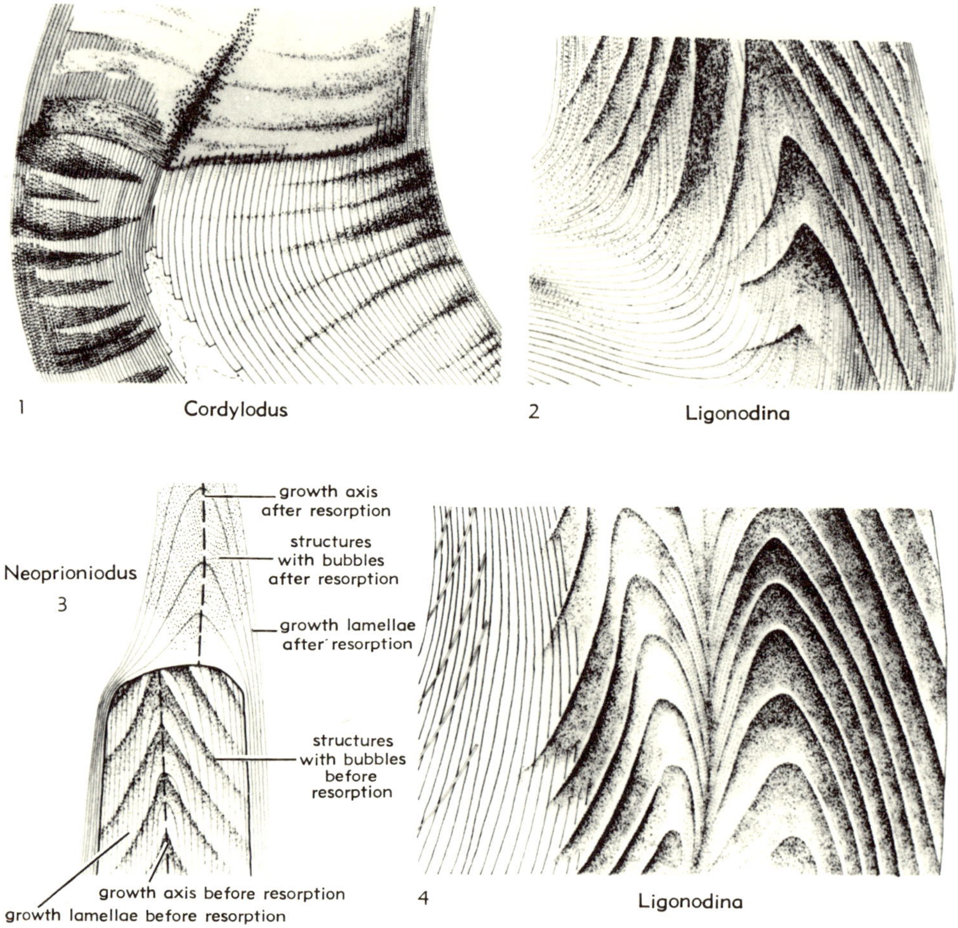

FIG. 25. Various types of white matter arranged in bubble layers perpendicular or oblique to growth lamellae; only every second lamella shown (after Müller & Nogami, 1971).——*1. Cordylodus proavus* MÜLLER; longitudinal section with stripes, which are broadest at the rims and taper toward the middle; L. Ord., USA (Wyo.), ×393.——*2.* Form-species *Ligonodina* sp., with conically arranged bubble structures that appear to be filled evenly with white matter; U. Dev. (Independence F.), USA (Iowa), ×220.——*3.* Longitudinal section through main denticle of form-species *Neoprioniodus* sp.; conelike structures with bubbles, similar to those in Figure 25,*2*, present in stub; however, direction reversed after regeneration; apex of cone in stub points toward apex of cone in regenerated zone; U. Dev. (Independence F.), USA (Iowa), ×263.——*4.* Longitudinal section through main denticle of bipennate *Sc* element of *Ligonodina* sp.; bubble-cones are inverted at tips, appearing as an M-shaped structure; Miss. (Wassonville F.), USA (Iowa), ×317.

FIG. 26. White matter arranged in structures perpendicular or oblique to growth lamellae (Müller & Nogami, 1971).——*1.* Longitudinal section across main denticle and proximal bar of bipennate element in *Ligonodina* sp.; in main denticle, cone-shaped bubble zones trespass in dark growth lamellae lying between light zones with almost no bubbles; above lower rim are many particles of pyrite (black); U. Dev. (Independence F.), USA (Iowa), ×150.——*2.* Longitudinal section of dolabriform *Cordylodus proavus* MÜLLER, with white matter more or less perpendicular to growth lamellae; L. Ord. (Deadwood F.), USA (Wyo.), ×150.——*3.* Longitudinal section across main denticle of bipennate element of

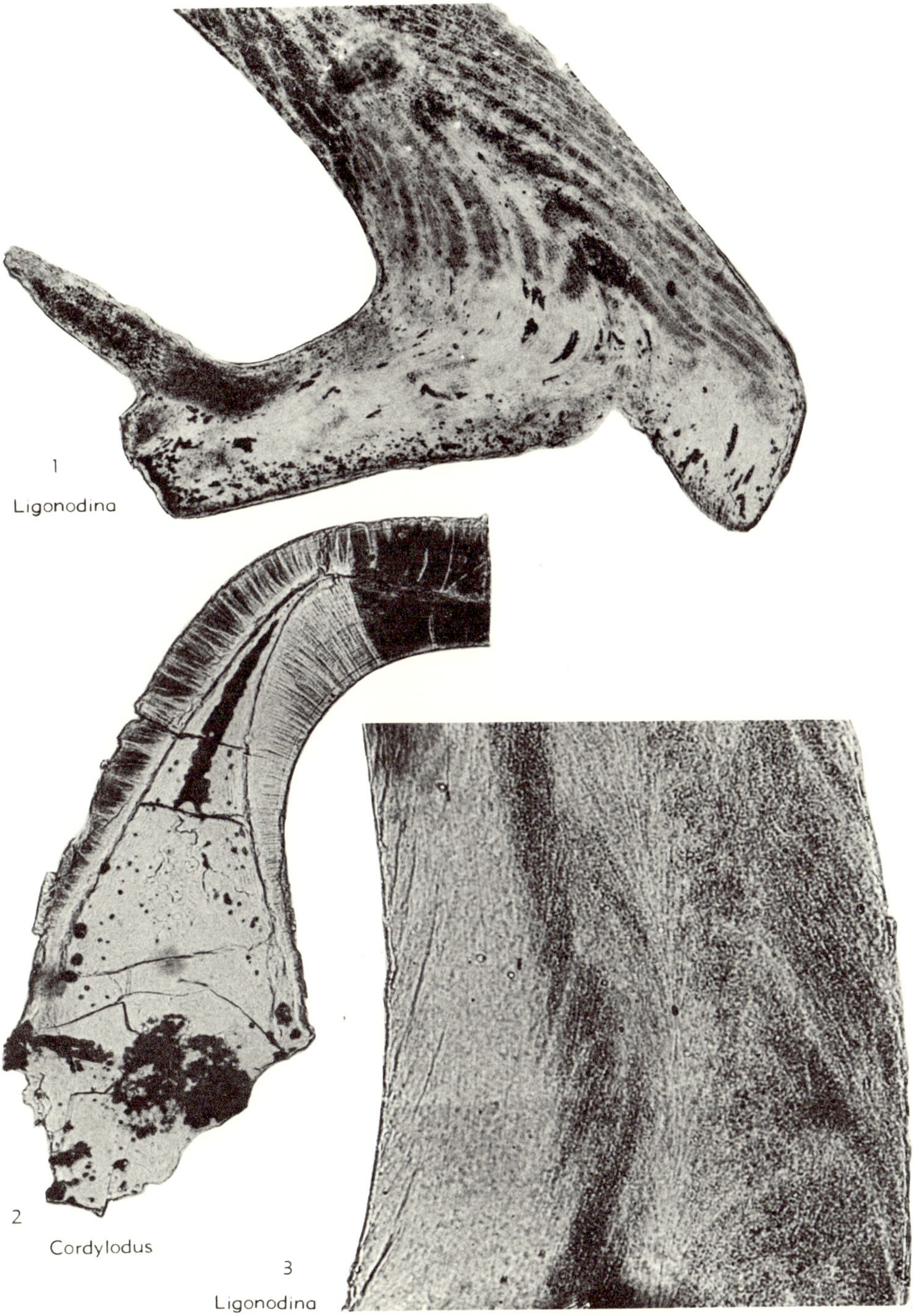

FIG. 26. *(Explanation continued from facing page.)*

Ligonodina sp., showing cone layers inverted in center to form M-shaped structures, which cross growth lamellae; Miss. (Wassonville F.), USA (Iowa), ×340.

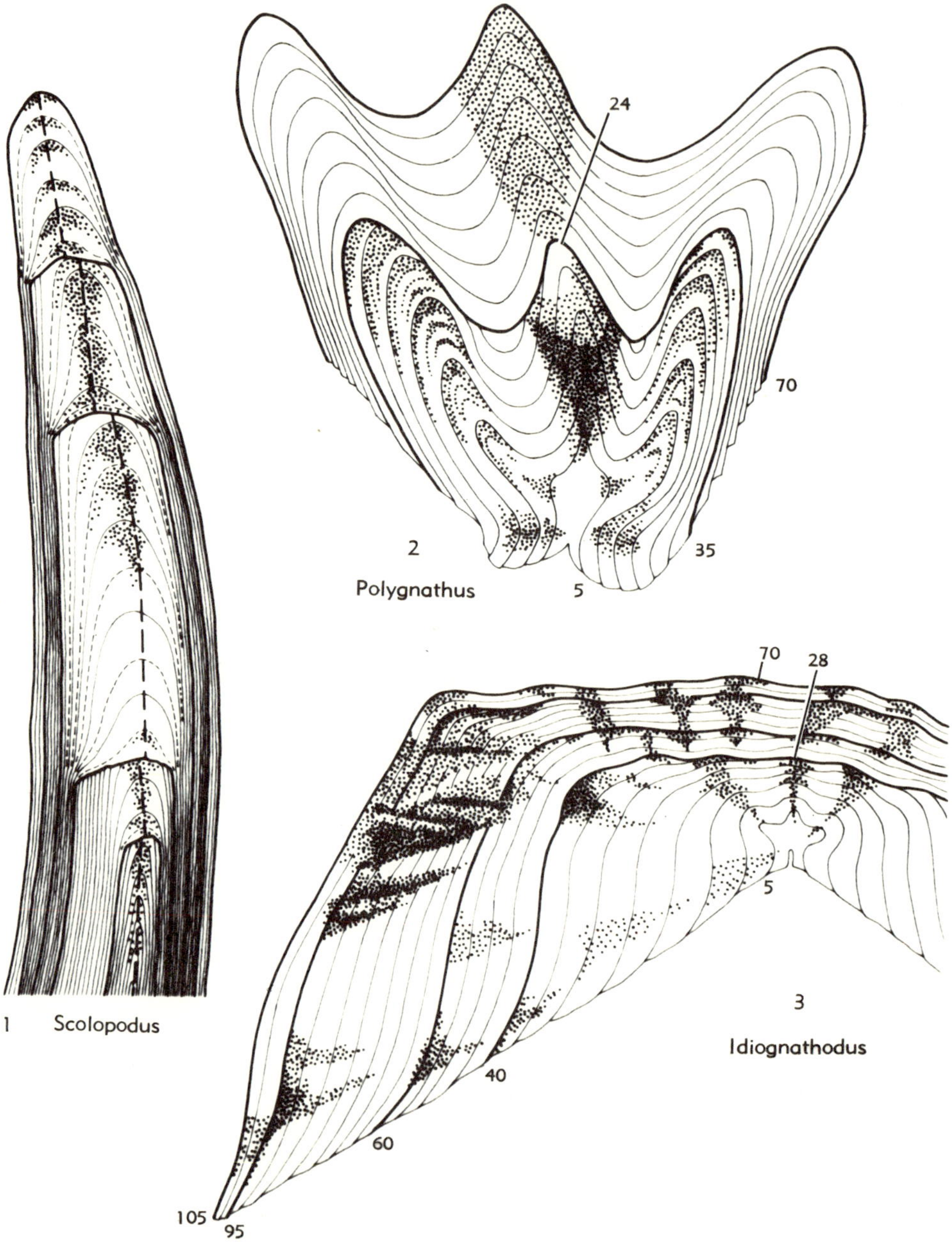

FIG. 27. Sections demonstrating resorption or fracture, and subsequent regeneration, a feature present in most conodont elements; as the zones of resorption do not coincide with those of accretion, a hiatus is developed between them (after Müller & Nogami, 1971).——*1.* Longitudinal section of coniform *Scolopodus rex* LINDSTRÖM, showing evidence of four periods of resorption or fracture, each followed by regeneration set off and in a slightly changed direction of growth (dashed line); L. Ord., Sweden, ×220.——*2.* Section through pit of pectiniform element of *Polygnathus angustidiscus* YOUNGQUIST, showing distinct zone of resorption with only 24 of 35 deposited growth lamellae still present in center; U. Dev. (Independence Sh.), USA (Iowa), ×275.——*3.* Pectiniform element of *Idiognathodus magnificus* STAUFFER & PLUMMER, show-

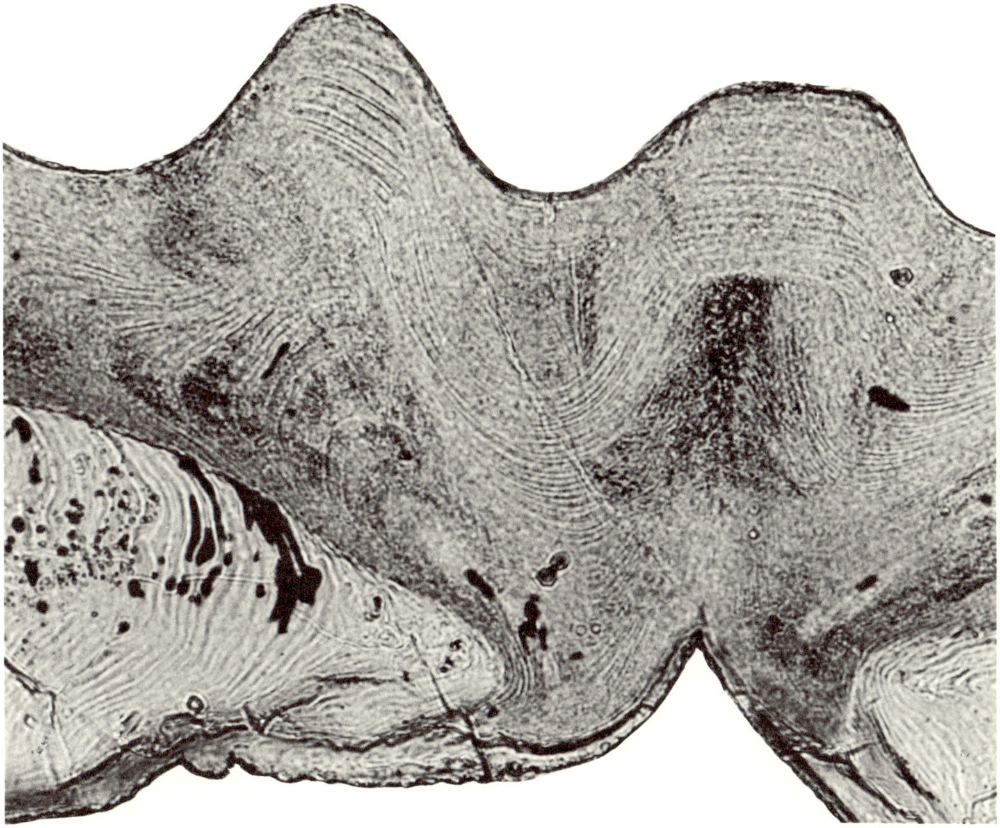

FIG. 28. Resorption and subsequent regeneration seen in section across pit and attached basal plate of pectiniform element of *Polygnathus tuberculatus* HINDE; in center of element, upper portion of a wide zone with white matter has been resorbed; growth lamellae continue from element into basal plate; U. Dev. (Genesee F.), USA (N.Y.), ×384 (Müller & Nogami, 1971).

surrounding areas and have been called **white matter.** In transmitted light, however, these areas appear dark. This feature can be observed in almost all advanced conodonts. Its position and structure is of systematic importance. Chemically, it seems to be identical with other parts, but it may contain a slightly higher content of organic material.

In thin sections, three main types of structures can be distinguished, and they commonly occur together in various combinations. Because they may develop differently in various regeneration stages within an element, all are considered to be primary structures that originated during the life of the conodont. They are:

1. "Interlamellar spaces" within the growth lamellae located in the middle of main growth axes where the lamellae are considerably broader (Fig. 23). They are probably caused by a deficiency of phosphatic matter at the places of fastest growth. These funnel-shaped cavities may be second-

---

FIG. 27. *(Explanation continued from facing page.)*

ing three zones of resorption; 35 of 105 growth lamellae preserved on the sides disappear at the top; Penn. (Desmoines.), USA (Iowa), ×220. Preferred resorption of the cuplike plate can be observed in many idiognathodontids.

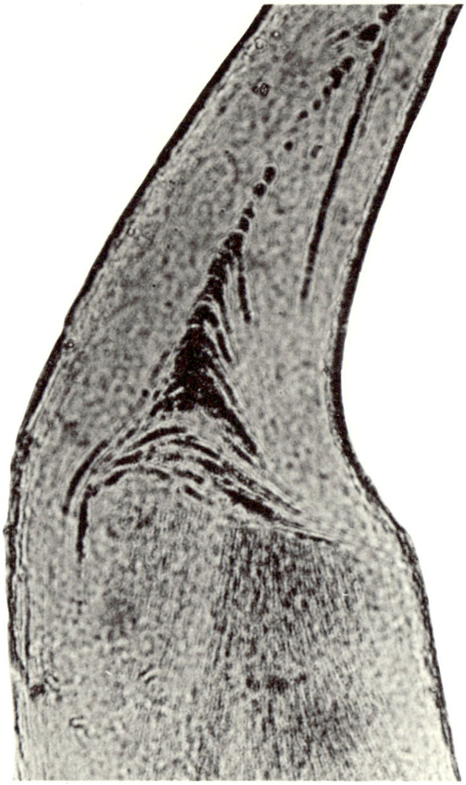

Fig. 29. Evidence of regeneration in longitudinal section of coniform element of *Drepanodus subarcuatus* Furnish; axis of growth slightly changed in regenerated portion, which is somewhat smaller than original element; Ord. (drift), Ger., ×335 (Müller & Nogami, 1971).

arily filled with bubble structures.

2. Peglike bubble structures formed in the center of denticles (Fig. 24,*1*). The so-called "germ denticles" in older conodont terminology (which are, in fact, suppressed denticles) are examples of these structures. An intergradation between lamellar and bubble structures can be observed. In very thin sections, round to elliptical bubbles can be observed to be inserted into growth lamellae at the outer limit of the white matter. Toward the center they become more abundant and the boundaries between lamellae disappear (Fig. 24,*2*).

3. Layers of bubbles cutting across the growth lamellae. The bubbles do not influence the shape of the growth lamellae, nor is the lamellar course influenced by them. This feature was observed by Pander (1856) and Staesche (1964), and both considered it to be of taxonomic significance. Nevertheless, hitherto it had little impact on conodont systematics, and its functional significance remained enigmatic. Because these structures can be studied only in thin section, little work has been done to elaborate their development within form-taxa or to use them for reconstruction of skeletal apparatuses. Four main types have been distinguished by Müller and Nogami (Fig. 25).

a. Bands perpendicular to growth axis (Fig. 25,*1*; 26,*2*). The earliest representatives with this type of white matter (e.g., Late Cambrian *Oneotodus nakamurai* Nogami and early species of *Cordylodus*) show somewhat irregular bubble bands, which are perpendicular to the carinae of the main denticle. Unlike other bubble layer types, these structures can be recognized in transmitted light through unsectioned elements.

b. Cone-shaped structures (Fig. 25,*2*; 26,*1*). Zones relatively rich in bubbles alternate with zones containing few bubbles. In thin sections these can be recognized as forming cone-in-cone structures. However, unsectioned elements of this type show only uniform, very broad bands of white matter when viewed in transmitted light because of the superposition of these zones. Differences may be observed in width and number of the bubble zones and in their more or less regular alternation.

c. Reversed cone-shaped structures (Fig. 25,*3*). The cones of type b point toward the tip of the unit. In regenerated parts, denticles may show similar but reversed cones

Fig. 30. Examples of secondary growth centers.——*1*. Section across center of platform and a protuberance on pectiniform element of *Siphonodella quadriplicata* (Branson & Mehl); L. Miss. (Wassonville F.), USA (Iowa), ×223 (Müller & Nogami, 1971). Resorption took place prior to formation of the secondary growth center.——*2*. Surface detail on element of *Pseudopolygnathus* sp.; the two protuber-

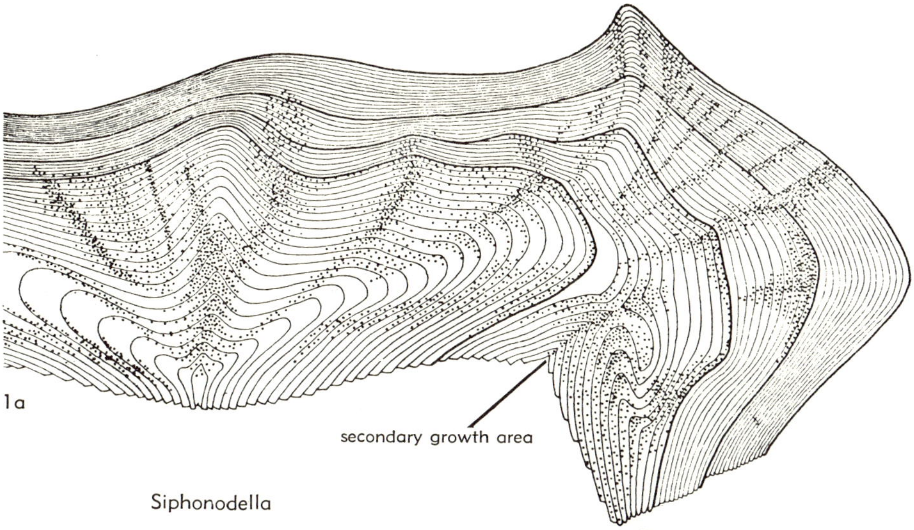

1a

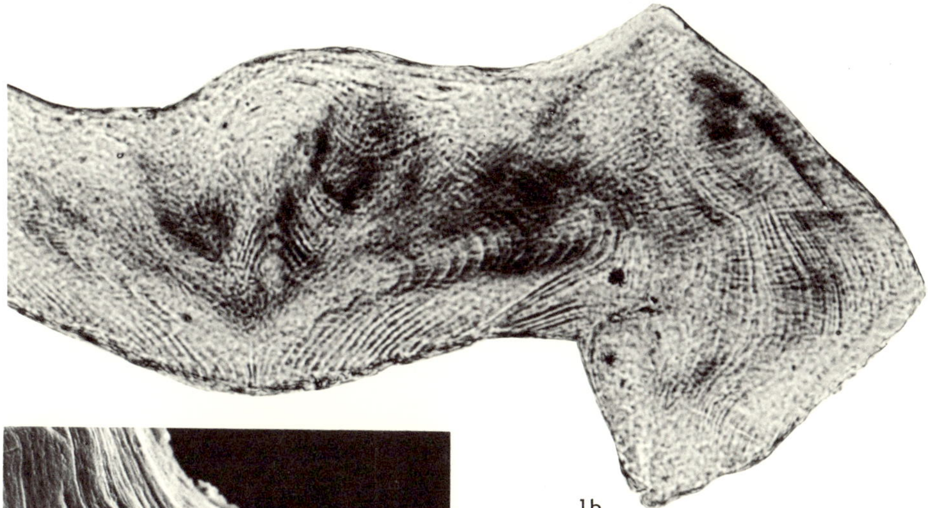

1b

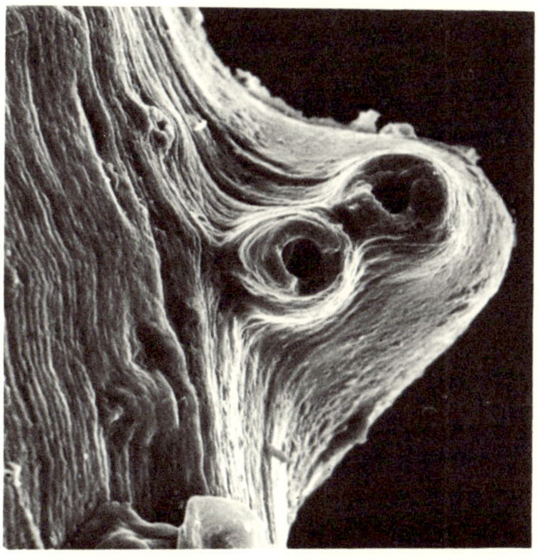

2b

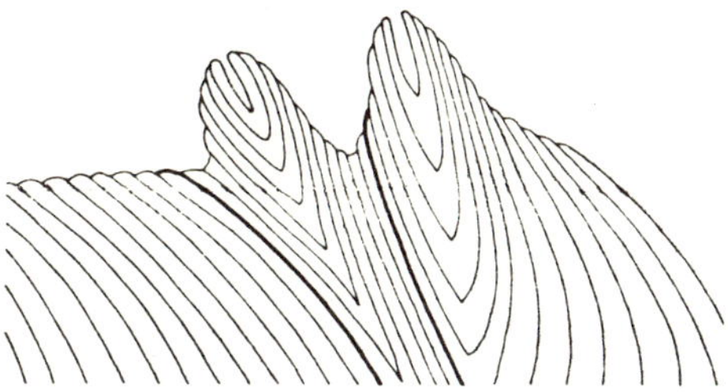

2a

Pseudopolygnathus

FIG. 30. *(Explanation continued from facing page.)*

ances cannot be interpreted to be formed "by a rather gentle disturbance in the course of lamellae," as suggested by ZIEGLER & LINDSTRÖM (1975); L. Miss. (Wassonville F.), USA (Iowa), ×306 (Müller, n).

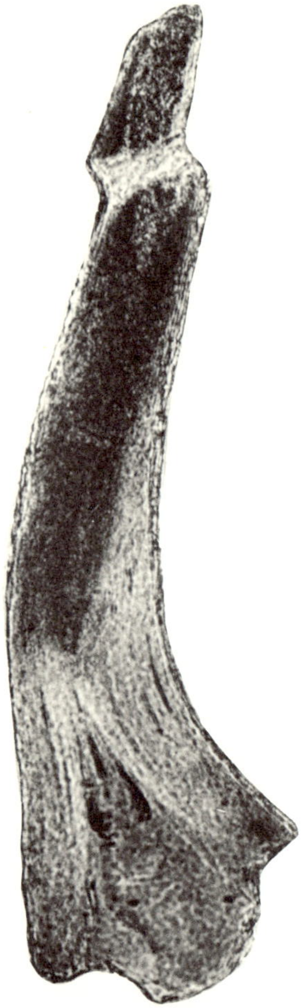

FIG. 31. Example of fracture and healing seen in longitudinal section of coniform element of *Icriodus* sp.; light portion below top consists of material newly formed in healing process; U. Dev. (Maple Mill Sh.), USA (Iowa), ×170 (Müller & Nogami, 1971).

that point toward the bottom of the element.

d. M-shaped structures (Fig. 25,*4*; 26,*3*). The cone-shaped structures, as in type b, may be further differentiated by an inversion of the tips of cones. These structures appear in the shape of an M with rounded points in longitudinal sections. This type can be explained as a combination of types b and c.

The superposition of lamellae and bubble bands occasionally results in a cancellate structure.

Structural Irregularities

Resorption and regeneration.—Tips of cusps and denticles commonly show regeneration (Fig. 27). As may be observed on complete translucent specimens in transmitted light, the white matter terminates abruptly and the regenerated portion contains little white matter and is somewhat more translucent.

The tips of elements either may have been resorbed or, if BENGTSON's (1976) theory is accepted, may have fallen off while the element was in a protracted position. Only rarely did the element remain in tissue where a fracture could be healed (Fig. 31). Commonly, the sharp edges of a broken denticle or cusp were smoothed before growth continued.

Regeneration of parts is a common feature. Large regeneration surfaces on pectiniform elements supersede relict structures of resorption. These discontinuities cannot be explained as having originated as fractures. Newly deposited matter attained approximately the original element shape (Fig. 28). Some large denticles are reduced to a smaller scale, producing a bevelled step about the "stump" portion. Lamellae deposited following a fracture or resorption period either covered the whole stump or filled in small depressions of the surface first and built up the denticle again before succeeding lamellae incorporated the whole stump (Fig. 29). In some elements, resorption and subsequent regeneration appear to be especially pronounced in certain areas. This could have a functional implication, such as protection or repair of portions exposed to particular stresses. An example is seen in *Ancyrodella rotundiloba* (BRYANT) (MÜLLER & CLARK, 1967, p. 905-906, fig. 2).

In some conodont collections, regeneration may be observed on nearly every specimen. Up to four resorption surfaces alternating with regeneration units have been observed in a single element (Fig. 27,*1*). The common occurrence of this phenomenon leads to the conclusion that resorption

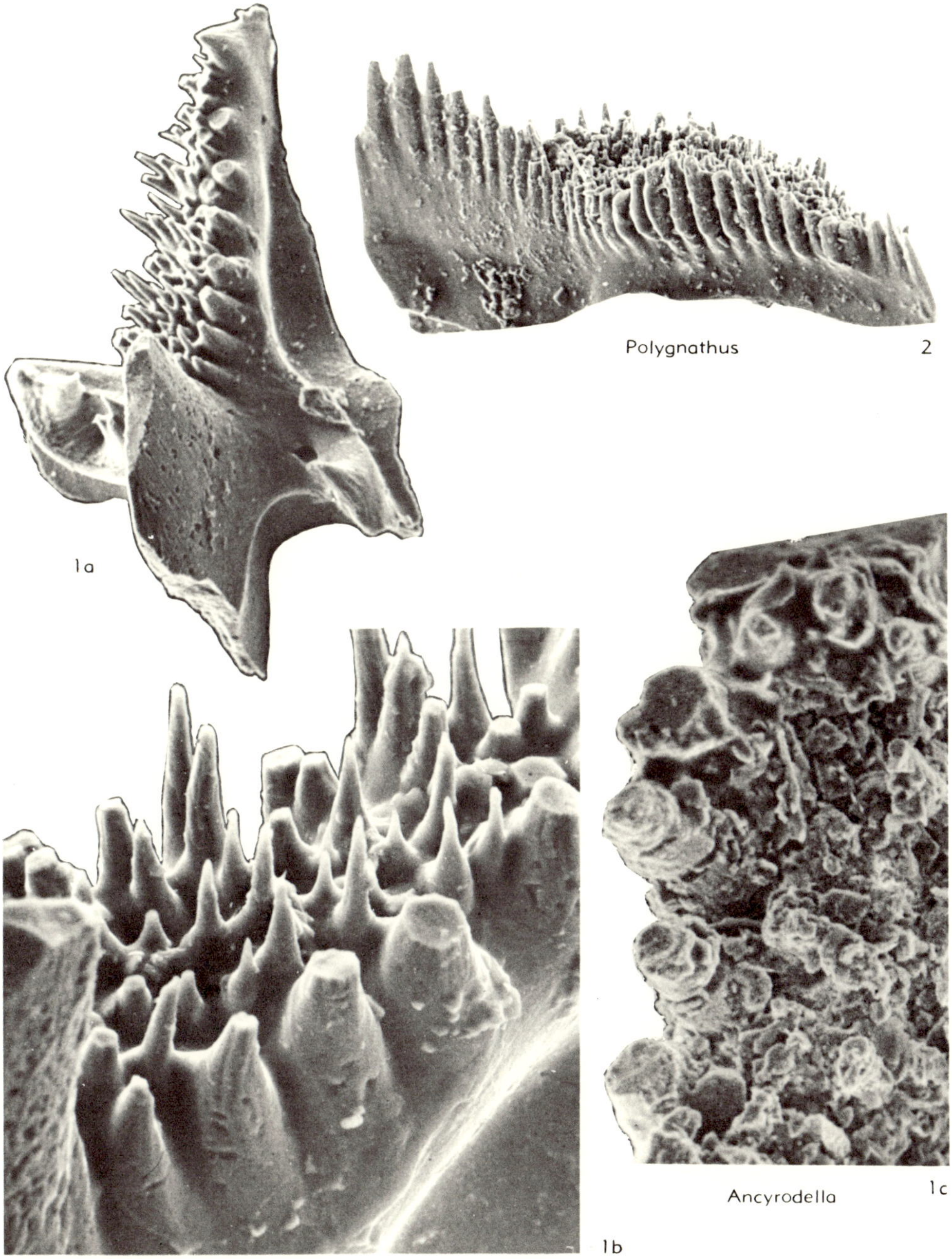

FIG. 32. Examples of unusual growth structures (after Müller, 1969).——*1. Ancyrodella* sp., U. Dev., Morocco; *1a,* Oblique view of platform surface with numerous needlelike structures, ×264; *1b,* Detail of specimen in *1a,* ×865; *1c,* Detail of similar specimen showing individual needles with structure like a stack of coins, ×865.——*2.* Side view of a brushlike pectiniform element of *Polygnathus* sp., U. Dev., Ger., ×144.

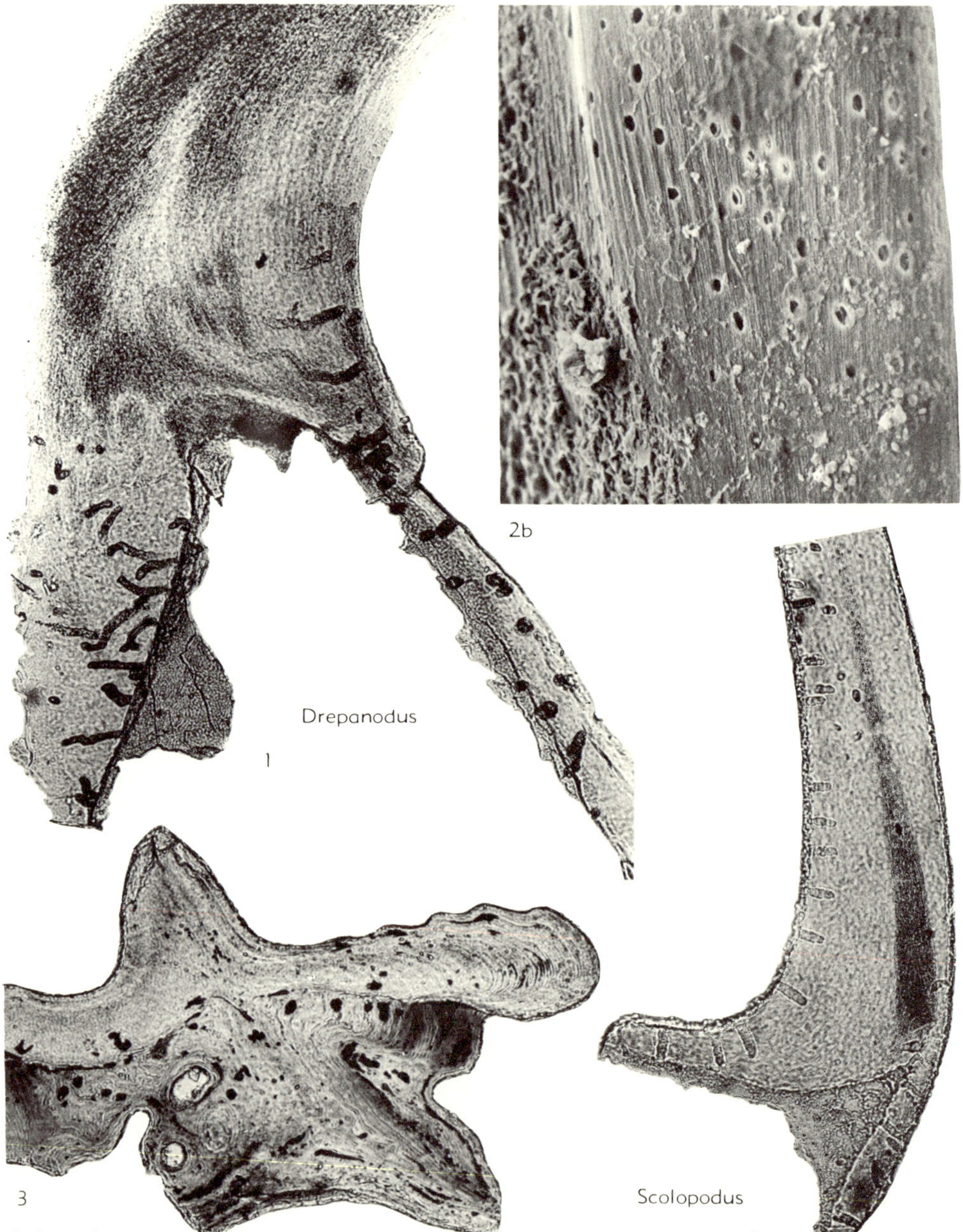

Fig. 33. Example of bored galleries (*1,2*, after Müller & Nogami, 1972b; *3*, after Müller & Nogami, 1971). ——*1*. Longitudinal section across basal cavity of coniform *Drepanodus* sp., showing irregular piled up and intersecting galleries that, observed from one side only, may be misinterpreted as being branched; L. Ord., Sweden, ×237.——*2*. *Scolopodus rex* Lindström, L. Ord. (drift), Ger.; *2a*, Longitudinal section across basal cavity; most bored galleries end blindly, are distributed almost equally on concave side, and were caused either by endobiont or by feeding on corpse; a few canules on convex side trespass into basal cavity; ×190; *2b*, Stereoscan photograph of detail on upper surface just above basal rim showing evidence of an endobiont, ×253.——*3*. Section across pectiniform element of *Palmatolepis perlobata* Ulrich & Bassler with attached basal plate, showing three tubules narrowed or even closed by apposition

may have served a specific purpose during the life of the conodont animal, and is not necessarily a pathological development. What use the conodont made of the resorbed material is open to speculation. Possible explanations include: (1) use as building material for additional elements as the animal became larger, (2) as a supplemental energy source (high-energy phosphate) in time of need, and (3) for osmoregulation by addition of recovered phosphate ions as a buffer.

Secondary growth centers.—Elements sometimes show not only resorbed surfaces, but new independent centers of growth. In later stages such a new center may be partly overgrown by succeeding lamellae, but it retains its identity as a small depression or funnel on the lower side (Fig. 30).

Healing of fractures.—The healing of fractures was comparatively rare. Generally, the fragments did not fit exactly into their original position, and the gap between was filled by a healing substance of the same mineralogical composition as the element. Before healing took place, the stumps may have been somewhat smoothed by resorption. Most specimens with healed fractures continued to grow, and the healed portion is covered by subsequent growth lamellae (Fig. 31).

Unusual growth structures.—On various elements, such as the *P* elements of *Ozarkodina, Polygnathus, Ancyrognathus,* and *Bryantodus,* an unusual type of growth has been observed. The upper surface of the specimen is differentiated in various minute spinelike processes and they appear brushlike (Müller, 1969). The edges of the growth lamellae sometimes can be observed on the sides of spine structures. Brushlike elements were formed when secretion tissue failed to deposit continuous phosphatic layers. Instead, phosphate was deposited as "islands." These elements are regarded as pathologic forms (Fig. 32).

Bored galleries.—Domicile burrows and bored galleries (Fig. 33) are common in conodont elements and have been found in an unexpected variety of forms (Rohon & Zittel, 1887; Müller & Nogami, 1972b; Eisenack, 1973). Obviously they originated through activity of different organisms. Pitting on the surface may be caused not only by the process of digestion in conodont-eating animals but also by acid used during preparation of specimens.

Tubes in basal plates were formed by vermiform organisms while the conodont was still alive, as indicated in some cases by secondary closure. Galleries also were caused either by epibionts or by organisms feeding on the dead animals, mainly thallophytes, probably fungi, and perhaps nematodes or annelids if they existed in this size range.

SURFACE MICRO-ORNAMENTATION AND OBSERVATIONS ON INTERNAL COMPOSITION

By Maurits Lindström and Willi Ziegler

[Philipps Universität, Marburg and Forschungsinstitut Senckenberg, Frankfurt]

Inclusions of pyrite within or between the growth lamellae could have been caused by sulphur bacteria.

ELEMENT SURFACES

Electron microscopy makes visible different kinds of micro-ornament much smaller than such well-known optic features as nodes, tubercles, ridges, and ribs. Such micro-ornamentation, first discussed by Ziegler (1970), has been found even on these optically observable ornaments. Most com-

Fig. 33. *(Explanation continued from facing page.)*

from inside; tubules surrounded by growth lamellae must have been present prior to formation of basal plate; U. Dev. (Maple Mill Sh.), USA (Iowa), ×134.

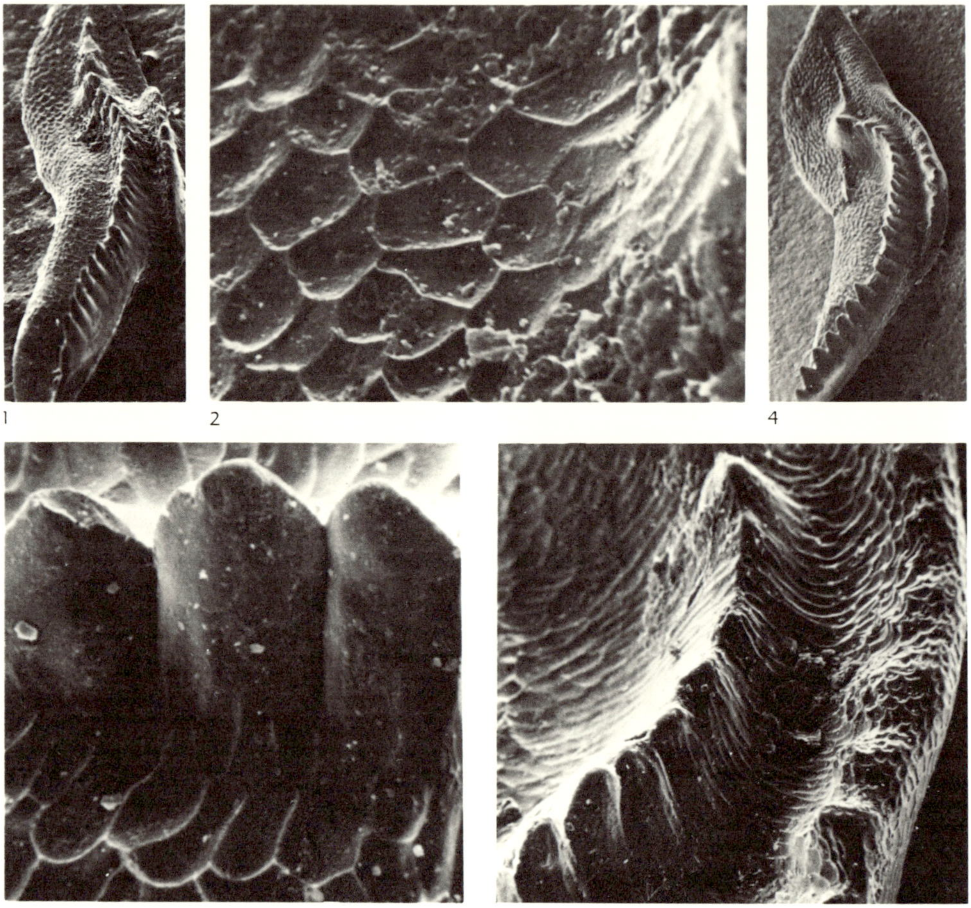

FIG. 34. Reticulation pattern of platforms; *Palmatolepis* species, U. Dev. (Holt's Summit F.), USA (Mo.) (Lindström & Ziegler, n).——*1-3. Palmatolepis distorta* BRANSON & MEHL. *1.* View of small juvenile specimen, ×141. *2.* Same specimen as *1*; outer half of platform next to central node where reticulate pattern extends to base of central node (bright area at right margin) as elongate loops, ×1,410. *3.* Different specimen; area posterior to central node showing that reticulation does not encroach on carina denticles, ×1,215.——*4-5. Palmatolepis marginifera* HELMS. *4.* View of adult specimen, ×60. *5.* Magnification of inner side of specimen in *4* showing honeycomb reticulation on nodes of parapet and on flat of platform; elongate loops present in furrow (between parapet and carina) and on carina denticles (note that carina denticles do bear reticulation, unlike those shown in *3*), ×382.

mon are primary micro-ornamentation striae that may occur on cusps, denticles, and other parts of many conodont elements. In most pectiniform elements a reticulate pattern may variously cover nodes, ridges, denticles, and platform surfaces. Commonly, it surrounds numerous small pits; in places it is a honeycomblike, polygonal pattern (Fig. 34,*2*). This reticulate pattern was derived from longitudinal striations of the Prioniodontacea (LINDSTRÖM & ZIEGLER, 1971; LINDSTRÖM, MCTAVISH, & ZIEGLER, 1972), where its development from anastomosing longitudinal striae has been observed on the anterolateral surface of the basal parts. In more advanced elements, the reticulate pattern commonly encroaches onto all horizontally expanded parts (bulges, platforms,

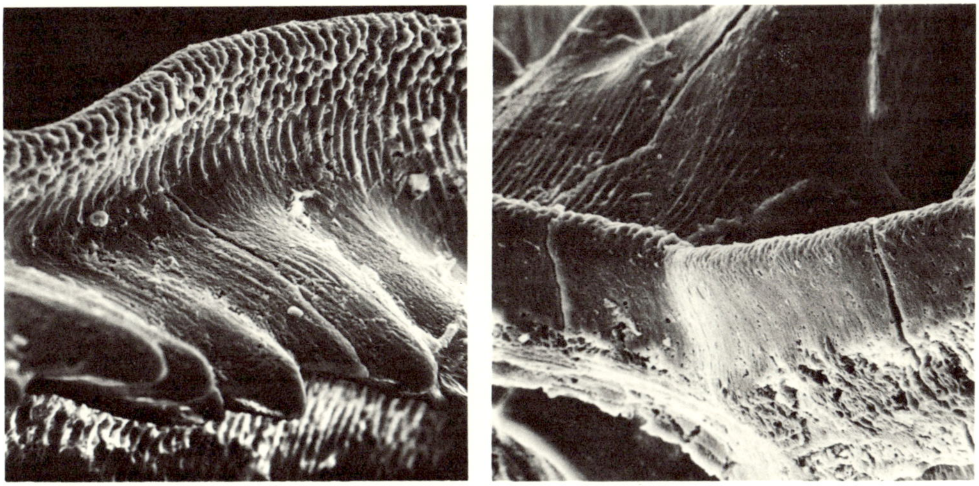

1 2

FIG. 35. Coarse striation and reticulation in *Amorphognathus* sp., M. Ord., Wales (Lindström & Ziegler, n).——*1.* Striations along denticles and reticulation on platform (oblique upper view), about ×380.——*2.* Termination of coarse pattern on lower side (oblique lower view), about ×410.

or platform ledges), including carina denticles and other coarse ornamentation (Fig. 35). Phylogenetically the reticulation is a secondary pattern. It is believed that its systematic investigation will lead to taxonomic use.

Six types of primary micro-ornament were first described by LINDSTRÖM and ZIEGLER (1971) from Panderodontacea, extended by LINDSTRÖM, MCTAVISH, and ZIEGLER (1972) for Prioniodontidae, and rediscussed by BARNES, SASS, and MONROE (1973) for Panderodontidae and by BARNES and SLACK (1975) for some Acanthodontidae.

Smooth surface.—Surfaces that are smooth even under high electron microscopical power occur in localized areas in many coniform elements (LINDSTRÖM & ZIEGLER, 1971; BARNES & SLACK, 1975) as well as in ramiform elements. They are usually along the anterior and posterior margins and on the apices of coniform elements, as well as between denticles and along both sides of pectiniform carina and adcarinal grooves in advanced elements.

Fine striation.—Fine striae (<0.7 μm) seem to be a common micro-ornament in many elements. As a pattern, fine striae are concentrated on certain areas, but are not continuous over the whole length of the element (Fig. 36). Fine striae are composed of crystallites that are aligned one after the other (Fig. 37). Most run parallel to the length axis and the denticles; however, some run oblique or transverse to this direction.

Coarse striation.—Coarse striae (>0.7 μm, ridges and edges) occur in coniform elements as a more continuous feature. In pectiniform and ramiform elements they usually run parallel along denticles (see above) and may converge toward the apices (Fig. 38). PIERCE and LANGENHEIM (1970) observed superficial similarity with RETZIUS' striae in human tooth enamel; however, the structure is different. Coarse striae, carinae, and ridges in coniform elements show traces of the crystallite facettes ($(10\bar{1}0)$) in cross section. Like fine striae, the coarse ones are formed by alignment of crystallites. Thus, crystallites and ornamentation coincide, which may have saved energy during growth. Relation between crystallites and coarse outer striation is shown in Figure 39.

Coarse striae may occur in coniform elements along the anterior margin (e.g.,

Belodina), along both sides of the longitudinal furrow (e.g., *Panderodus*), or on other areas. They may be parallel or converge in apical direction. Width of coarse striae ranges from 0.7 to 6.0 μm (BARNES, SASS, & MONROE, 1973). BARNES and SLACK (1975) stated that coarse striae not only represent a type of surface ornament but have an internal continuation that they called radial lamellae (see below).

Basal wrinkles.—Basal wrinkles occur in a 50- to 100-μm zone around the basal margin. Individual wrinkles are lengthwise, 1 to 2 μm wide, and form bundles. The wrinkles (as yet observed only in *Panderodus*) represent a basal enlargement of surface for secretion of hyaline matter. As the element grew, the zone of wrinkles moved to remain associated with the basal margin (Fig. 40).

Longitudinal furrow.—In some Panderodontacea (typically in *Panderodus*) a longitudinal furrow or groove extends the entire length of the element on the inner side (Fig. 41). This furrow penetrates almost to the basal cavity as a deep narrow slit (LINDSTRÖM & ZIEGLER, 1971). Its width is about 2 to 3 μm on the outer surface,

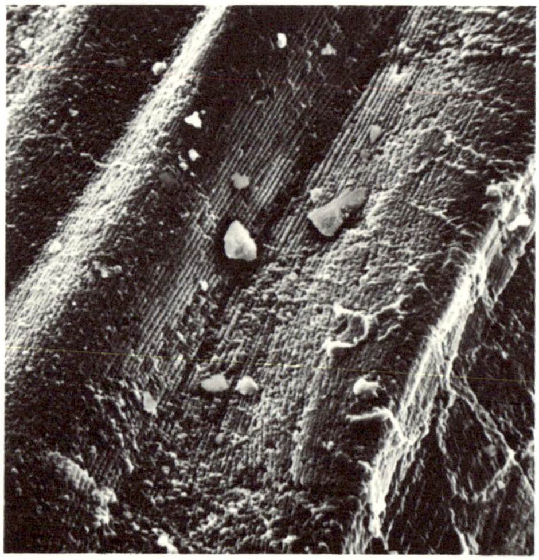

FIG. 36. Fine striae on anterolateral face of *Protopanderodus varicostatus* (SWEET & BERGSTRÖM); note that fine striae occupy such coarse features as costae and grooves; ×450 (Lindström & Ziegler, 1971).

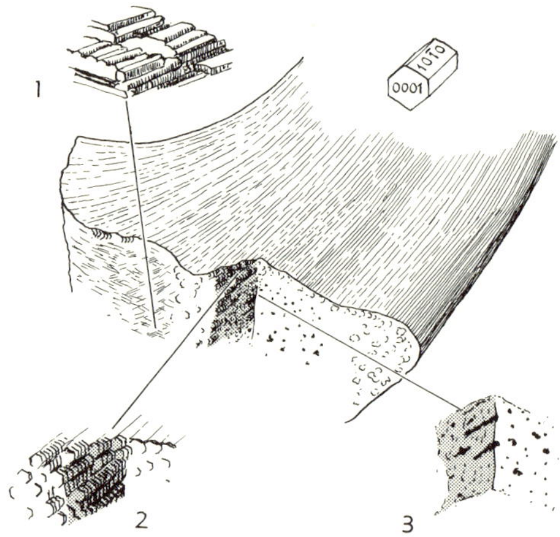

FIG. 37. Fracture surfaces and crystallite structures of *"Acanthodus" iowenus* (FURNISH) observed by SEM; enlargements approximately ×700 (after Lindström & Ziegler, 1971).——*1.* Elongate crystallites of hyaline matter, arranged as lamellae.——*2.* Crystallites of the surface aligned one after the other.——*3.* White matter at core of elements; black equals holes or cavities.

narrowing rapidly inward and then closing completely within a few microns of the basal cavity. Basally its deepest part swings in a tangential direction whereby the anterior margin of the furrow strongly overlaps. This overlapping part wedges out within the zone of basal wrinkles, and the furrow terminates here. Apically the furrow comes to an end shortly below the tip of the cusp. On the outer surface the furrow is bordered on both sides either by coarse striae that may converge toward it (Fig. 42,*3*), or by a narrow, smooth zone. These coarse striae were interpreted as expressions of radial lamellae by BARNES, SASS, and MONROE (1970). Internally the furrow is smooth, too. This was explained by LINDSTRÖM and ZIEGLER (1971) by the assumption that during growth this furrow was held open by tissue and therefore no secretion could take place. LINDSTRÖM and ZIEGLER suggested that the furrow may have functioned for the insertion of muscles.

Microdenticles and dental pits.—Some Prioniodontacea possess tiny projections along the posterior margin (Fig. 41,*1*).

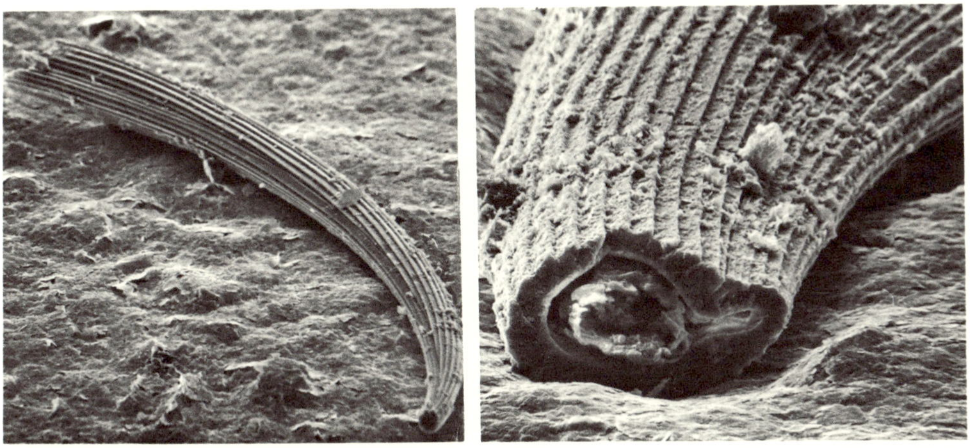

1 2

Fig. 38. Coarse striation in *Neopanderodus perlineatus* Ziegler & Lindström.——*1*. Nearly complete element, ×141 (Lindström & Ziegler, 1971).——*2*. Cross section near midlength, obverse side up; note longitudinal furrow near right (posterior) margin, ×357 (Ziegler & Lindström, 1971).

These, as well as minute elongate dental pits on the trailing edge of the first proximal denticles in ramiform elements, are believed to have functioned as muscle-attachment sites. Similar pits have been observed on nodes of pectiniform elements (e.g., *P* element of *Icriodus* and *Ancyrodella*; Fig. 41,*2*).

STRUCTURAL OBSERVATIONS

Lamellar Structure

The principal constructional elements are concentric lamellae (Pander, 1856) that may normally be 0.2 to 1.2 μm thick (Barnes, Sass, & Poplawski, 1973). At points of rapid growth, lamellae have been observed to be 5 μm thick (Barnes, Sass, & Monroe, 1970). The elements grew through outward secretion of lamellae, each new lamella forming a more or less complete envelope on the exterior of the element. Hence, the innermost lamellae represent early growth stages (Fig. 43). Early lamellae could be obliterated during growth through the formation of white matter (see below). At points of rapid growth, the lamellae may be separated by interlamellar spaces (Hass, 1962; Lindström, 1964).

The lamellae consist of apatite crystallites that have been observed to be arranged with their prism surfaces parallel to the direc-

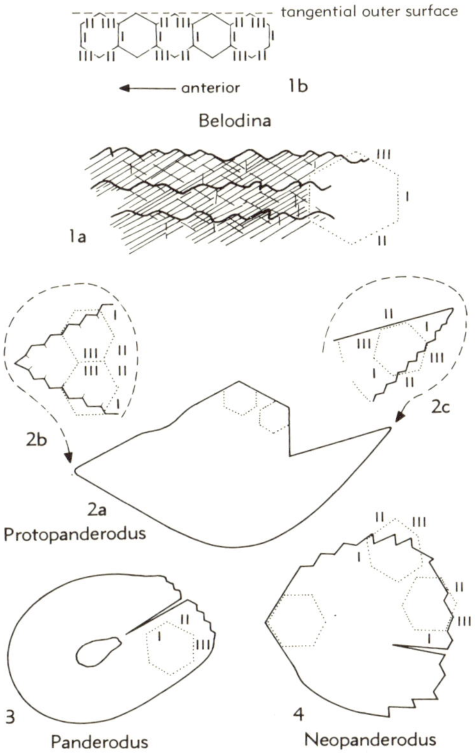

Fig. 39. Schematic cross sections of some panderodontids showing orientation of crystallites (after Lindström & Ziegler, 1971). Prismatic faces indicated as I, II, III; radial prismatic faces (=normal to outer surface) are indicated as I; the tangential outer surface is the assumed line that connects all outer edges of elements; generalized crystallite outlines are dotted to show orientation.

tion of growth (Lindström & Ziegler, 1971; Barnes & Slack, 1975). For the flanks of the cusp and denticles, this means that the prism axis of crystallites is oriented parallel to the outer surface, the ornamentation of which may be influenced by crystallinity (fine lengthwise striae, and a tendency for longitudinal facets in certain elements to meet at angles of about 120°) (Fig. 37, 39).

The lamellae envelop the outer surface of the element except for basal zones early described as "inverted basal cavity" and "escutcheon" (Fig. 44). Exceptions to this have been reported by Müller and Nogami (1971) and by Barnes and Slack (1975). The former authors regarded the attenuation of lamellae on the upper surface of certain pectiniform elements as due to resorption. Barnes and Slack observed that lamellae of acanthodontine coniform elements may run discordantly to the outer surface and disappear where they meet this surface. The thickness of any lamellae in any particular direction is correlative with the rate of growth in that direction. Also, it follows from the outer geometry of elements that growth must be zero or near zero in certain directions. Thus, the local

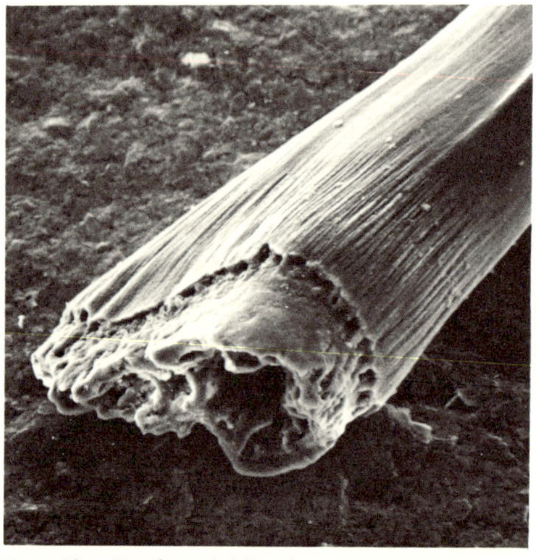

Fig. 40. Basal wrinkles in *Panderodus gracilis* (Branson & Mehl); note zone of basal wrinkles, sutural pits, and basal filling; distal parts of the latter are broken away; ×192 (Lindström & Ziegler, 1971).

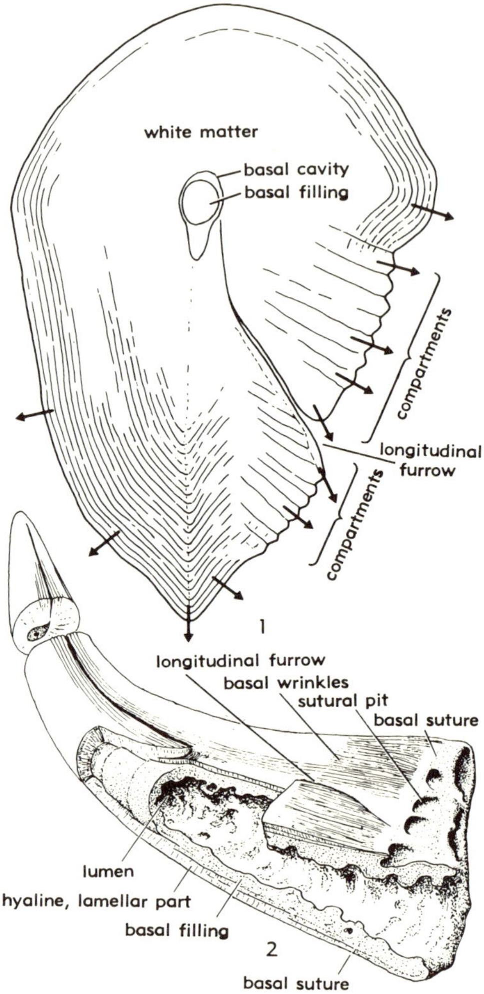

Fig. 41. Morphology of *Panderodus*.——*1*. Generalized transverse section based on Barnes, Sass, and Poplawski (1973) and Lindström & Ziegler (1971) to illustrate centrifugal growth of element; arrows indicate growth direction (Lindström & Ziegler, n).——*2*. Diagram of *P. unicostatus* (Branson & Mehl) showing principal construction and some surface features, about ×140 (Lindström & Ziegler, 1971).

attenuation of lamellae may be explained through other processes as well as by resorption. Barnes, Sass, and Poplawski (1973) reported on a feature they called "radial lamellae." It occurs in *Panderodus* as an expression of the coarse, lengthwise striae adjacent to the longitudinal furrow

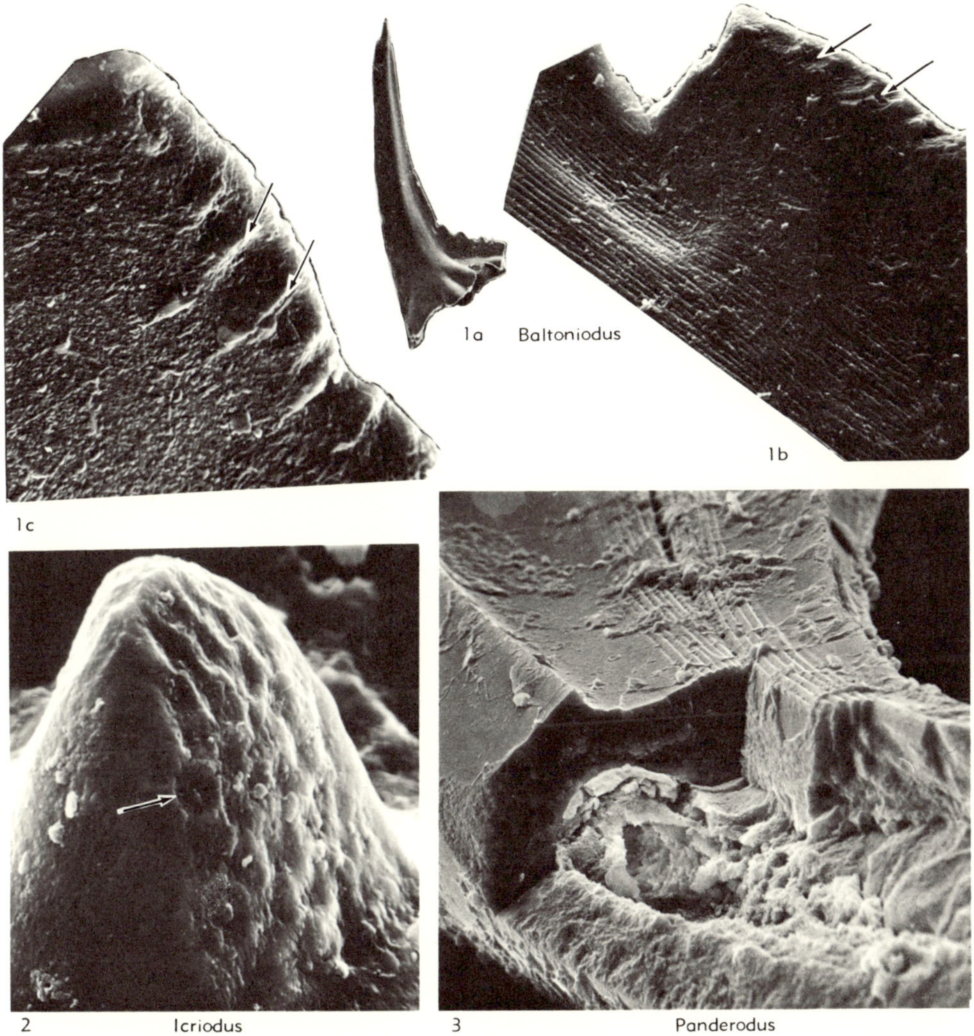

FIG. 42. Dental pits and longitudinal furrows.——*1.* Element of *Baltoniodus* n. sp. A (Lindström, McTavish, & Ziegler, 1972); *1a,* Anterolateral view showing denticles on posterior upper margin, ×31; *1b,* Magnification of posterior upper margin showing dental pits (arrows) on posterior edge of first proximal denticle, ×600; *1c,* Magnification of *1b* showing dental pits separated by minute ledges (arrows), ×1,270.——*2.* Lateral row node of *Icriodus* sp. showing dental pits (arrow), ×1,710 (Lindström, McTavish, & Ziegler, 1972).——*3.* Longitudinal furrow in *Panderodus unicostatus* (BRANSON & MEHL) exposed by artificial fracture in midlength of element; note coarse striae on either side and "compartments" in section on right side of furrow; compare with Figure 41; ×500 (Lindström & Ziegler, 1971).

on the cusp. The concentric lamellae are indistinct in this part and the most obvious structural feature is a set of radial and longitudinal planes of separation. These planes of separation originate as grooves between longitudinal ridges. During radial growth of the element, each groove retains its location, the radial plane of separation remaining as a record of its position as successive stages of growth (see Fig. 42, 43). The designation radial lamellae should be abandoned, because it suggests a relation-

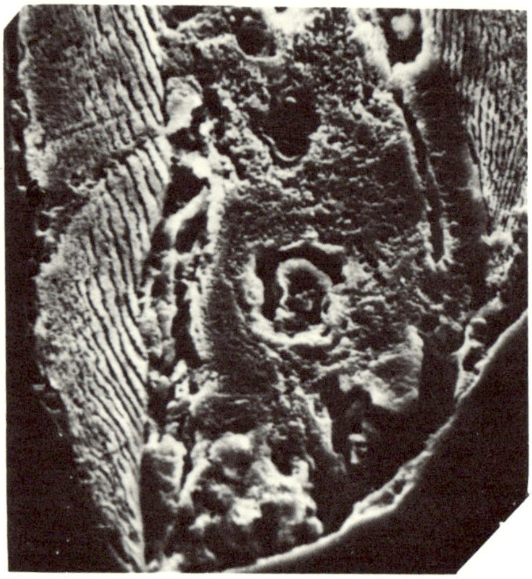

FIG. 43. Section through basal region of element of *Panderodus compressus* (BRANSON & MEHL) showing basal fillings in center and lamellae on both sides, ×575 (Barnes, Sass, & Poplawski, 1973).

ship to the growth lamellae. If a special term is required, "aligned compartments" may be preferable.

White Matter

This structure was known by early authors as cellules, cancellated structure or postmortal vesicles (PANDER, 1856; GROSS, 1954, 1957, 1960; HASS, 1962). Its significance for element growth was discussed by LINDSTRÖM (1964). These early investigations were based on optical methods.

Under the electron microscope, white matter typically shows numerous subspherical or irregular voids ranging in size from 0.1 to 0.5 μm, only rarely reaching 1 μm. The voids or holes are not obviously connected (Fig. 45). They are randomly distributed or may be arranged in rows that may run perpendicular to each other (PIETZNER & others, 1968). Some of these voids can be identified as thin canals in axial direction, for example, in *Acontiodus* (LINDSTRÖM & ZIEGLER, 1971). As yet, it remains unresolved whether these voids were empty or filled with gases or organic matter. Round bodies, 0.1 to 0.7 μm in diameter, possibly crystallites, and referred to as spheres, were described in chemically etched white matter by BARNES, SASS, and MONROE (1973) in the form-genera *Ambalodus* and *Cordylodus,* where they form a pebbly pattern. Holes in the white matter tend to be larger toward the axis of albid elements. Narrow gashes described by BARNES, SASS, and MONROE (1973) as structural features within the white matter are most probably small fractures grown together or parting surfaces of crystallites that were artificially enhanced by their etching method.

White matter appears at the tip of the basal cavity and thence continues into the apical part of the element cusp at the expense of hyaline matter. The boundary between white and hyaline matter appears sharp in optic views but is rarely sharp in electron microscopical views. A zone of incipient white matter usually forms a transition. Because white matter is regarded as recrystallized, the transition zone may be regarded as partly recrystallized. Also, in lamellar matter, holes somewhat smaller than those in white matter occur (LINDSTRÖM & ZIEGLER, 1971). They may have been filled with organic material and thus mediated the migration of substances required for the process of recrystallization.

Another characteristic feature of the white matter is that it consists of more finely crystalline matter (PIETZNER & others, 1968). During the recrystallization process the orientation of the 0001 crystal surfaces (basal pinacoid) and the prisms ($10\bar{1}0$) is retained. In hyaline matter the former are arranged transversely to the length axis of the conodont element. The basal pinacoid surfaces (observed in panderodontids by LINDSTRÖM & ZIEGLER, 1971) grow together so that the white matter is transected by numerous practically continuous transverse surfaces. Along these surfaces breakage occurs very easily (PIETZNER & others, 1968). Such surfaces may appear linear in some sectional views. They were regarded by BARNES and SLACK (1975) as being roughly parallel to the interlamellar spaces of hyaline matter and were interpreted as remnants of these.

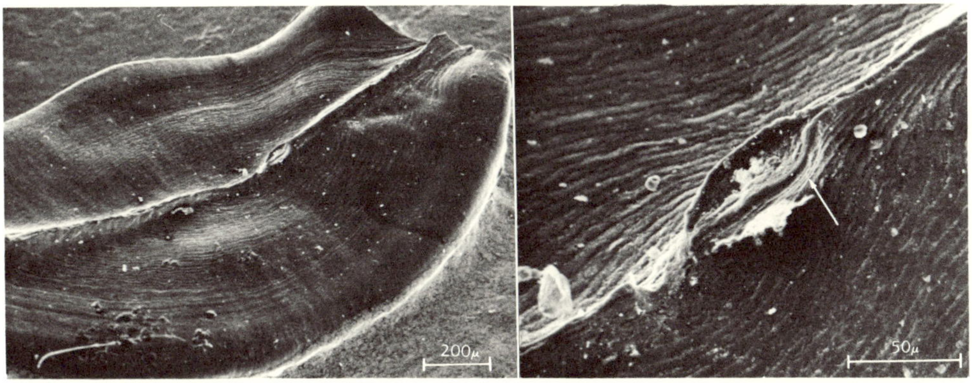

1 2

Fig. 44. Lamellae exposed on lower element surface of *Polylophodonta* sp.; arrow in 2 points to a number of very thin lamellae deposited immediately following formation of basal pit (Pietzner & others, 1968).

Origin, formation, and function of white matter within the conodont element has been discussed widely. Because it appears in all growth stages and is always covered by lamellae, Lindström and Ziegler (1971) believed that it was formed during the growth of the conodont. On the other hand, Barnes and Slack (1975) suggested that it is formed after the conodont element achieved its entire growth. Some interpretations of white matter are highly contradictory. Thus, Barnes, Sass, and Monroe (1973) believed that white matter increased the strength of elements, whereas Lindström (1964) and Lindström and Ziegler (1971) indicated that it actually had reduced the strength, which may have had a selective advantage for the conodont.

Fibrous Structure

Branson and Mehl (1933) described certain Ordovician conodont elements (in particular, species of the genus *Chirognathus*) as "fibrous," alluding to the frayed, woodlike structure exhibited by broken specimens (Fig. 46,*1, 2*). These elements are largely hyaline, that is, they have little or no white matter. Rhodes and Wingard (1957) hypothesized that fibrous elements, called Neurodontiformes by them, were distinct from other elements because they lacked lamellar structure; however, Hass (1962) and Lindström (1964) found that so-called fibrous elements have lamellae like other conodont elements. Hence, the observed fibers must belong to the fine structure of lamellae. This was confirmed by Ziegler and Lindström (1972) and Barnes, Sass, and Monroe (1973), who described lamellae composed of long, needlelike crystallites oriented parallel to the growth axis. On fractures across the growth axis, the fibrous, or hyaline elements show a pattern of closely packed prisms broken at different levels, rather like pavements formed by pillar basalt.

Growth Axis

Pander (1856) and Hass (1941, 1962), among others, have described the growth axis of element denticles as a line of opaque material (now referred to as white matter), or as a succession of inflexion points of lamellae, separated by interlamellar spaces. The growth axis is the trace of successive growth stages of the denticle tip. Usually it is the first locus of formation of white matter. Barnes, Sass, and Monroe (1973) and Barnes and Slack (1975) referred to a canal running along the growth axis from the tip of the basal cavity; however, this observation has not been repeated by other investigators. Assuming that the observed feature, called "growth canal," is not

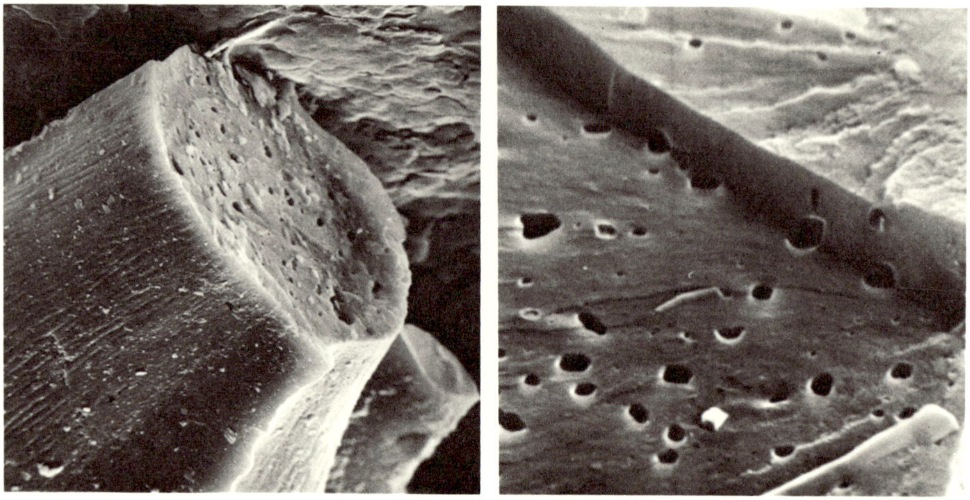

1 2

Fig. 45. White matter with numerous holes and absence of a growth canal; seen in sections of Ordovician conodont elements (Lindström & Ziegler, n).——*1. Phragmodus undatus* Branson and Mehl, ×1,440. ——*2. Cordylodus flexuosus* (Branson & Mehl), ×7,850.

ubiquitous, it can be interpreted according to three different models that agree with known circumstances.

1. In unetched elements the growth canal is represented by a great number of more or less interconnected small cavities. Such cavities normally occur in the white matter. In unetched elements they might be too small and too poorly interconnected to appear as a canal. On etching, the cavities widen and merge into a canal.

2. The growth axis is a zone of unusually high free crystallization energy in the elements, because crystallites have to diverge at the point of lamellar inflexion. It is particularly susceptible to chemical etching. Hence, the points of inflexion are removed, and the canal appears.

3. The growth axis may be developed as a line of interlamellar spaces separated by very thin lamellae (Hass, 1941, 1962; Lindström, 1964). The thin lamellae are readily dissolved, with a continuous canal as a result.

It is suggested that all of these models are applicable in individual instances. Also, it is probable that the degradation of lamellar material to form the white matter may have proceeded to a point where the growth canal formed during life in some conodonts.

Basal Filling

Basal fillings of elements were reported by Branson and Mehl (1933a), who described them as bonelike (Fig. 46,*3, 4*). Lindström (1955) reported on two kinds of fillings in diverse and well-preserved Lower Ordovician elements. One kind is blackish brown and amorphous or, in one case, obviously lamellate. This filling was not dissolved in dilute HCl but received a carbon coating during boiling for five hours in concentrated H_2SO_4. It was reduced to white ash when strongly heated. The other kind shows the same optical properties as the conodont elements and was dissolved by HCl. It is composed of very thin (about 1 μm) undulating lamellae that can give rise to globular structures. In the latter kind of filling, cellular structures with about 7 μm diameter were found near the apex of the basal cavity in one specimen. Both kinds of basal filling were found in the same species.

Gross (1957) and Lindström (1964)

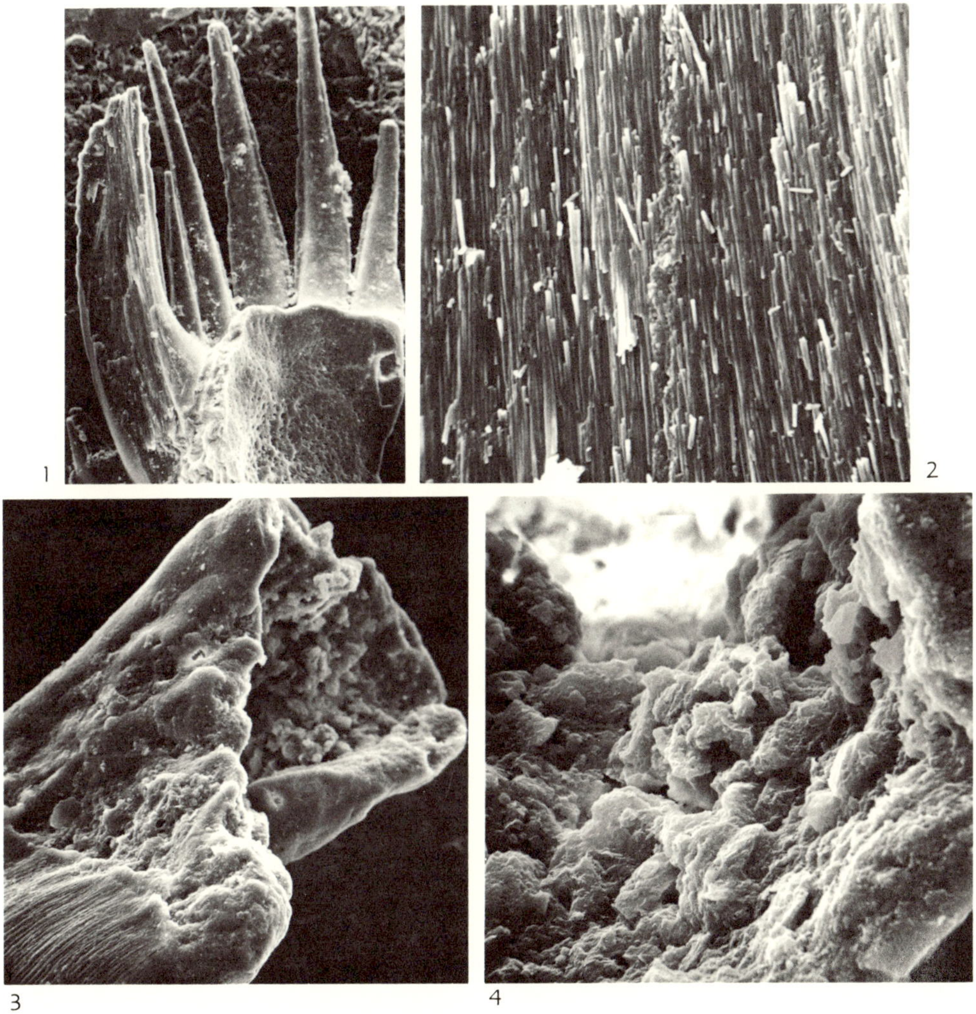

Fig. 46. Fibrous structure and basal filling in conodont elements (Lindström & Ziegler, n).——*1,2. Chirognathus* sp., M. Ord., showing artificial fracture along main denticles. *1.* View of entire specimen, ×141. *2.* Enlargement of fibrous structure of main denticle; crystallites forming fibers are about 0.2 μm in diameter and about 6 to 7 μm in length; trace of lamellae extending obliquely from lower left to upper right; about ×3,500.——*3,4.* Basal filling of *Panderodus* sp. *3.* About ×315. *4.* Enlargement showing foamy texture of basal filling, about ×1,220.

found growth lamellae in the basal fillings (or basal plates, *Basiskörper* of Gross) of pectiniform elements. Gross (1957) suggested that the lamellae of the basal filling were out of phase with those of the element and that this was caused by resorption of lamellae at the junction between the element and the filling. Lindström (1964) reported no evidence for such resorption, and showed that the basal filling of *Panderodus* elements can be divided into segments perpendicular to the growth axis of the conodont.

Pietzner and others (1968) showed the crystallites of the filling to be smaller and more isodiametrically shaped than those of the conodont element. Furthermore, the fillings concentrated more rare earth ele-

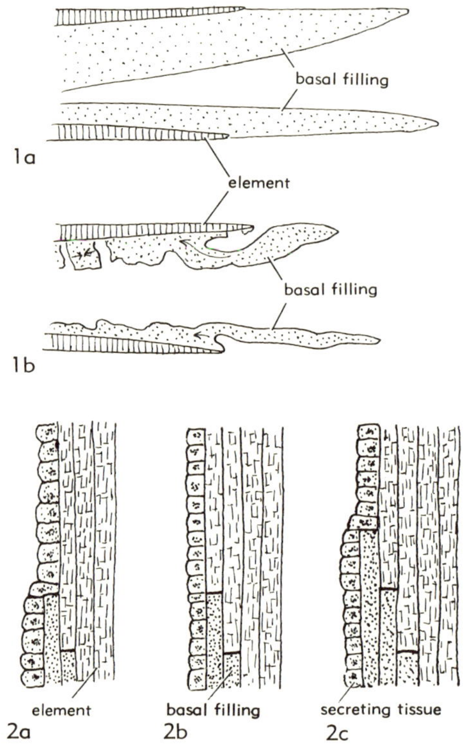

FIG. 47. Structure and formation of basal fillings (after Lindström & Ziegler, 1971).——*1.* Interpretation of structure of a basal filling before *(1a)* and after *(1b)* shrinkage.——*2.* Successive steps in formation of zone of recessive basal margin. It is suggested that the same tissue secreted both element and basal-filling lamellae. During growth, tissue secreting basal filling encroached on that secreting element, thereby creating an offlap of element lamellae and an inverted basal cavity. Diagrams *2a, c* illustrate the possibility that basal filling and element were not secreted contemporaneously. The offlap of element lamellae becomes visible when basal filling falls off or is destroyed during fossilization.

ments. These authors described holes penetrating the filling in some specimens. Most illustrated specimens show a much more distinct and regular lamellation in the elements than in the attached fillings.

With the aid of a great number of superb thin sections, MÜLLER and NOGAMI (1971) confirmed several observations of earlier studies. Among other things, they observed that the element is harder than the basal filling, owing to the smaller crystallites and the more abundant organic material of the filling. The lamellae of the basal filling can be wrinkled, perhaps owing to shrinkage. Lamellae of the basal filling can show additional irregular structures, including spherulites. The basal filling can contain voids of different shape and format (Fig. 43).

LINDSTRÖM and ZIEGLER (1971) found the crystallite fabric of panderodontid basal fillings to be isotropic (without preferred orientation) in contrast to that of the element; growth lamellae were found to be continuous with those of the element; and irregularities such as sutural pits at the junction (suture) between the element and the basal filling were interpreted as due to shrinkage. In one instance, the structure at a breakage of the basal filling suggested some flexibility of the filling material during life (Fig. 40, 41, 46, 47).

Several observations on the basal filling suggest that it consisted of more or less flexible, chemically resistant, but partly shrinkable organic material, the phosphatization of which may be at least partly postmortal. Some apparent lamellation, as well as the formation of spherulites, may be explained as *Liesegang*-banding formed during postmortal phosphatization; however, original lamellation appears to be well documented.

CHEMICAL AND MINERALOGICAL PROPERTIES

By WILLI ZIEGLER and MAURITS LINDSTRÖM

[Forschungsinstitut Senckenberg, Frankfurt and Philipps Universität, Marburg]

The hope that knowledge of the chemical composition of conodonts[1] could aid in recognizing their biological affinities was the stimulus for early research. ELLISON (1944), HASS and LINDBERG (1946), PHILLIPS in RHODES (1954), and RHODES

[1] Authors of this chapter prefer the term "conodont" for an individual component of the skeletal apparatus rather than "element," as used in other chapters of this volume. —Editor

and WINGARD (1957) published chemical data. PIETZNER and others (1968) and BRADSHAW, NOEL, and LARSON (1973) revised the early work and produced additional data in more precise detail by use of modern research techniques (electron microprobe, scanning electron microscope, IR-spectroscopy, different kinds of X-ray apparatuses, neutron activation).

INORGANIC CHEMISTRY AND MINERALOGY

Conodonts are composed mainly of calcium phosphate which belongs to the apatite group. Because of this, they are soluble in hydrochloric, sulphuric, and nitric acids, and insoluble in acetic, formic, and citric acids (ELLISON, 1944; BECKMANN, 1952). Extended exposure to monochloracetic and formic acids, however, may lead to solution or at least corrosion (ZIEGLER, LINDSTRÖM, & MCTAVISH, 1973). The successful separation of conodonts from most indurated sediment is thus made possible. According to ELLISON, apatite of conodonts has the hardness 3 to 5 on MOH's scale and a specific gravity of 2.84 to 3.10.

Percentage composition of Devonian conodonts without basal fillings, based on wet chemical analyses, was given by PIETZNER and others (1968) as: moisture, 0.16; insoluble residue, 1.59; annealing loss, 0.25; and total CO_3, 2.00. The same analyses showed percentage chemical compositions of: PO_4, 53.30; Ca, 37.28; Sr, 0.40; rare earths (as Ce), 0.42; Al, 0.09; Fe, 0.04; Ka, 0.03; Na, 0.62; H_2O, 2.85; CO_3, 1.84; and F, 2.60. Based on atomic weights, the ratio of PO_4 to Ca, Sr, and rare earths is 3:5, as it is known from minerals of the apatite group. Difference between total CO_3 and carbonate CO_3 is due to the presence of organic matter (see below). PIETZNER and others (1968) arrived at the following formula for conodont matter without basal fillings:

$$Ca_5Na_{0.14}(PO_4)_{3.01}(CO_3)_{0.16}F_{0.73}(H_2O)_{0.85}$$

They regarded this to be a carbonate apatite, francolite. The OH ions together with the

TABLE 3. *Comparison of Chemical Elements Detected in Conodont Elements and Basal Filling by X-ray Spectral Analysis (Pietzner & others, 1968) and Neutron Activation (Bradshaw, Noel, & Larson, 1973).*

Chemical element	X-ray spectral analysis: Basal filling	X-ray spectral analysis: Conodont element	Neutron activation; Undifferentiated material
Y	x	x	–
Sr	x	x	–
Ba	x	–	x
Rb	x	–	x
Th	x	–	–
Cu	x	x	x
U	?	–	–
Pb	x	x	–
Zn	x	x	x
Ni	x	x	–
Fe	x	x	x
Co	x	x	x
Mn	x	x	x
Ti	x	x	–
Cr	x	x	x
Sc	?	–	x
La	x	x	x
Ce	x	x	–
Pr	x	–	–
Nd	x	x	–
Sm	x	–	–
Eu	?	–	–
Gd	x	x	–
Tb	?	–	–
Dy	x	?	–
Ho	?	–	–
Er	x	–	–
Yb	?	–	–
Ge	–	–	x
Mg	–	–	x
V	–	–	x
Br	–	–	x
Sb	–	–	x
Au	–	–	x
Zr	–	–	x
In	–	–	x
Ag	–	–	x
Bi	–	–	x
Mo	–	–	x

CO_3 substitute for phosphate and do not occupy lattice positions as in the hydroxyapatite. MCCONNELL (in ELLISON, 1944) believed that dahllite, lewistonite, or dehrnite may also occur in conodonts, whereas HASS and LINDBERG (1946) considered apatite in conodonts to be a dahllite.

X-ray and electron radiation diffraction prove that the lamellar part, white matter,

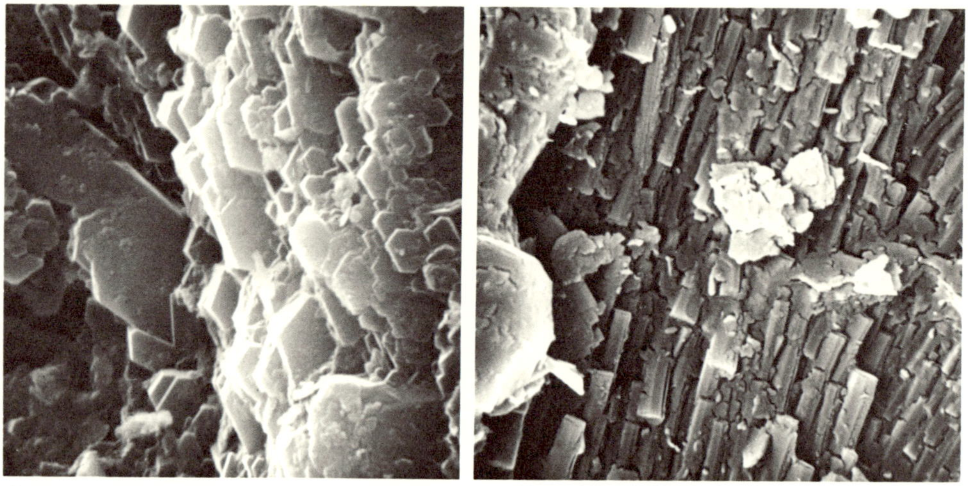

1 2

FIG. 48. Shape and arrangement of apatite crystallites in conodont elements.——*1.* Platy crystallites of *Panderodus simplex* (BRANSON & MEHL), ×4,960 (Lindström & Ziegler, 1971).——*2.* Elongate prisms of *Oepikodus evae* (LINDSTRÖM), about ×2,800 (Ziegler & Lindström, n).

and basal filling, the three major constituents of the whole conodont, are identical with respect to crystal chemical properties (PIETZNER & others, 1968). This is supported by the lattice dimensions of the carbonate apatite:

conodont:
$a_0=9.37$ Å, $c_0=6.91$ Å; $c_0/a_0=0.737$
basal filling:
$a_0=9.35$ Å, $c_0=6.90$ Å; $c_0/a_0=0.738$

Although BRANSON and MANKIN (1964) identified conodonts studied by them as francolite, the lattice dimensions given by them

($a_0=8.085$ Å, $c_0=6.888$ Å; $c_0/a_0=0.852$)

are different and not in agreement with data in the literature.

X-ray spectral analyses (elements above

TABLE 4. *Distribution of Strontium and Yttrium in Conodont Elements and Basal Filling (from Pietzner & others, 1968).*

Structure	Element		
	Sr(ppm)	Y(ppm)	Sr : Y
Conodont elements	4,400	300	14.7
Basal filling	2,900	5,300	0.53

atomic number 20) show significant differences in chemical content of basal filling and conodont. PIETZNER and others (1968) indicated generally a larger number and higher concentrations of chemical elements in the basal filling than in the conodont proper (with strontium as the only exception; see Table 3):

Strontium- and yttrium-concentration ratios of conodont and basal filling are especially significant (Table 4).

BRADSHAW, NOEL, and LARSON (1973), using neutron activation analysis, found several of the same chemical elements as PIETZNER and others (1968), but they found others in addition (see Table 3). The most striking difference is the absence of strontium and yttrium. PIETZNER and others (1968) described the yttrium content of the basal filling to be about 20 times greater than that in the conodont. Accordingly, basal fillings show a distinct yellow-greenish fluorescence under ultraviolet radiation (PIETZNER & others, 1968, pl. 18, fig. 2).

Microprobe investigations (X-ray intensity spectra) yield evidence that element distribution in basal plate and conodont is not homogeneous. Calcium and phos-

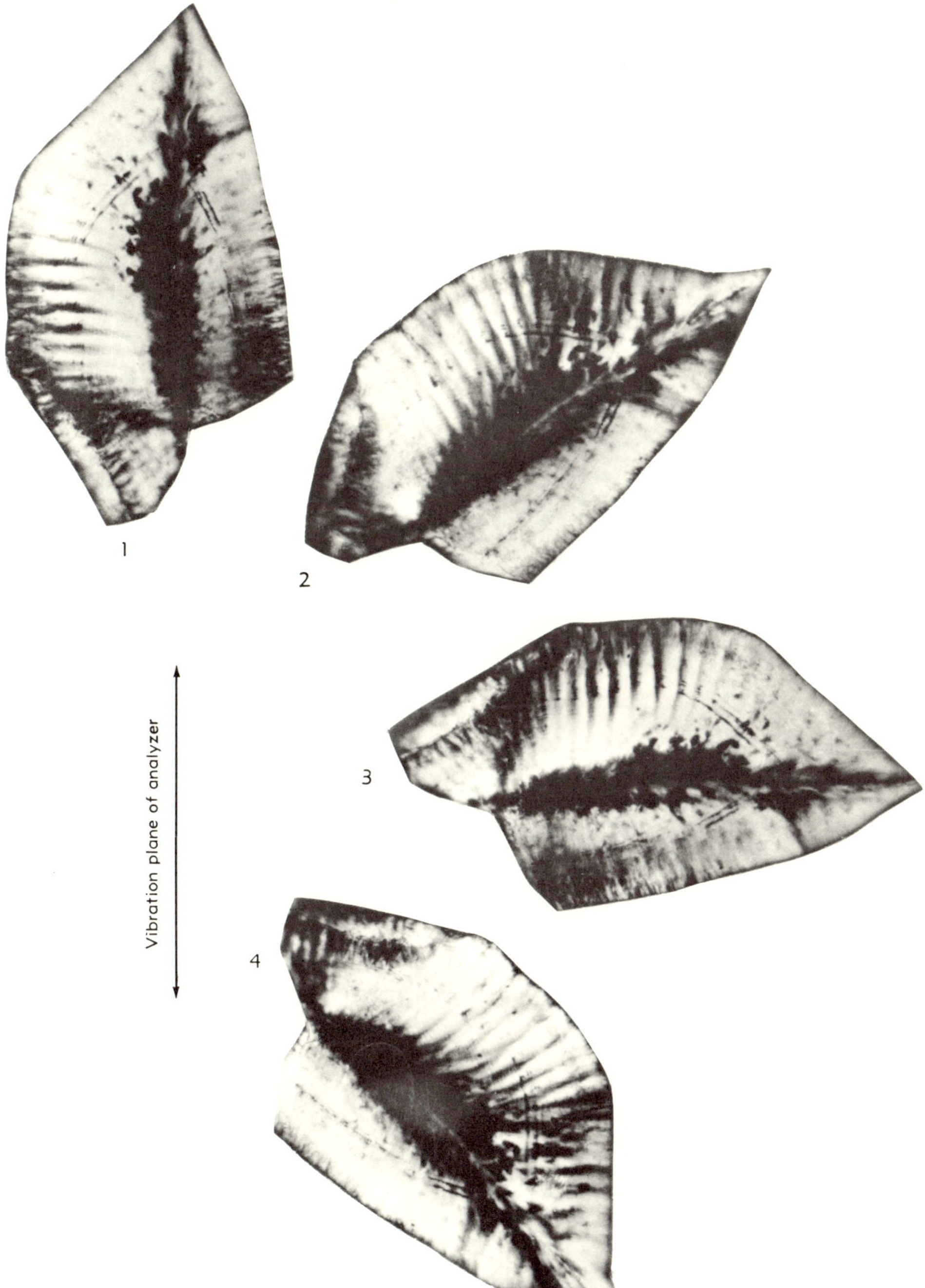

FIG. 49. Extinction of crystal units in a pectiniform element of *Siphonodella* sp. photographed at 45° intervals, ×55 (Hass, 1962).

phorus may form a slightly greater proportion of the conodont than of the basal filling (PIETZNER & others, 1968, pl. 18, fig. 4-6; pl. 10, fig. 1-3). Fluorine seems to decrease in concentration from the margin of the basal plate toward the conodont; however, the greatest yttrium concentrations are probably along the outer margin of the basal plate. This suggests postmortal shifting of the fluorine, yttrium, and rare-earth element concentrations. Assumably, fluorine and yttrium were built into the lattice of the apatitic matter. In part, adsorption is conceivable too, because of the smaller size of the crystallites and the variable, though mostly high, content of organic matter in the basal filling. PHILLIPS in RHODES (1954) suggested increase of fluorine through geologic time by adsorption.

Lamellar parts and white matter of the conodont are generally of the same composition, but the lamellar parts are chemically distinguished so far by a larger content of carbonate and organic matter. Elemental distinction between lamellar and fibrous conodonts is indicated by neutron activation analysis (BRADSHAW, NOEL, & LARSON, 1973).

X-ray and microprobe analyses (HASS & LINDBERG, 1946; PHILLIPS in RHODES, 1954; PIETZNER & others, 1968) showed the crystallinity of the conodont apatite, which is formed by innumerable small crystallites (Fig. 48). The orientation of their crystal lattices can be inferred from optic studies with the polarization microscope as well as from X-ray diffraction photographs. Their optic *c*-axes were found to be parallel to the direction of growth, that is, generally at right angles to the lamella. This means that for any orientation of the conodont under the polarization microscope with crossed nicols there are certain rays of crystallites in extinction position and these dark rays run in the growth direction across sectors of the element (HASS & LINDBERG, 1946; LINDSTRÖM, 1964) (Fig. 49).

ORGANIC CHEMISTRY

If the mineral matter of a conodont is dissolved by a weak reagent, the shape of the unit may be preserved as an ephemeral "ghost conodont" (LINDSTRÖM, 1964). This ghost probably consists of organic material that forms a thin film between the apatite crystallites. The internal recrystallization and dissolution of apatite required for the formation of white matter would have been mediated through this organic film surrounding the crystallites (LINDSTRÖM & ZIEGLER, 1971). According to CLARK and MILLER (1969), organic material may have predominated in Cambrian conodonts.

Little is known about the precise nature of the organic material. PIETZNER and others (1968) identified probable amino acids, and ARMSTRONG and TARLO (1966) identified leucine, phenylalanine, and lysine, as well as traces of proline, alanine, and glutamic acid; however, the conodonts containing the latter array of organic components came from a sediment with a long and complex sedimentologic history. Therefore, the organic composition of the conodonts can be regarded as essentially unknown.

COLOR AND ALTERATION: AN INDEX TO ORGANIC METAMORPHISM IN CONODONT ELEMENTS

By ANITA G. HARRIS
[U.S. Geological Survey, Washington, D.C.]

Unweathered conodont elements are pale yellow, light to dark brown, black, gray, opaque white, or crystal clear. Although conodont elements have been studied for more than 125 years, until recently their color variation largely has been ignored and left unexplained. The darkening of elements has been noted primarily as an unfortunate impediment to morphologic study. It has also been noted that elements from thin, undistributed cratonic sequences generally are pale yellow, those from thick geosynclinal sequences (that also happen to be complexly deformed) generally are brown or black, and that elements from carbonates interbedded with garnet-grade metamorphic rocks are crystal clear.

ELLISON (1944) was the first experimentally to induce color alteration in elements. He reported that elements heated in a closed tube gave off water and turned dark gray. Two decades later, LINDSTRÖM (1964) correctly attributed this color alteration to a carbon-fixing process, but it was not until the end of the sixties that CLARK and MILLER, then PEITZNER and others verified that elements do contain trace amounts of organic matter (chiefly in their basal part). Regional mapping of element color in samples from outcrops and drill holes has shown that elements become darker with increasing depth and duration of burial (EPSTEIN, EPSTEIN, & HARRIS, 1974, 1975, 1977). This field evidence also has been confirmed in laboratory experiments showing that color alteration of conodont elements is time and temperature dependent and virtually unaffected by pressure (EPSTEIN, EPSTEIN, & HARRIS, 1974; 1975, 1977). Thus, element color alteration is the result of carbon fixing within trace amounts of organic matter dispersed chiefly in the basal part of the apatite conodont element and depends on the depth and duration of burial and the geothermal gradient. The color of a conodont element can, therefore, be used as a metamorphic index to assess depth of burial, maximum paleotemperatures, and even oil and gas potential.

EXPERIMENTALLY INDUCED COLOR ALTERATION

EPSTEIN, EPSTEIN, and HARRIS (1977) reproduced all "field" colors of conodont elements in controlled high-temperature long-term runs in open air, with and without water. Their data show (Fig. 50):

1. The sequence of color change from pale yellow to black found in field collections is the same as that produced by heating alone.

2. Color alteration is progressive, cumulative, and irreversible.

3. Color alteration is time- and temperature-dependent. At 300°C, color alteration begins after 350 hours of heating (at color alteration index, CAI, of 1½); at 400°C, after 5 hours of heating; at 500°C, after only half an hour of heating.

During higher temperature runs, not shown on Figure 50, black elements became gray, then opaque white, and in the last stage before decomposition, crystal clear. This same sequence occurs in elements from increasing grades of metamorphic host rocks in the field (H. P. SCHÖNLAUB, written commun., 1974, 1975). During this process, the fixed carbon of the black element perhaps is driven out of the element (volatilized), thus clearing it. The change from opaque white to crystal clear may result from the release of water of crystallization as well as from recrystallization. Crystal-clear elements can be produced by heating in open air at 950°C for four hours.

Experiments at high temperature (550°C), confined pressure (1 kbar), with and without water, using methane (reducing) and argon (inert) as pressure media, indicate that: (1) confined pressure and anoxic conditions neither retard nor accelerate color alteration, and (2) water in combination with confined pressure considerably retards color alteration. Consequently, open-air experimental temperature data cannot be applied to wet sealed systems at relatively high pressure.

APPLICATION OF COLOR ALTERATION

The color alteration index (CAI) for conodont elements can be used as a geothermometer and as a tool for metamorphic and structural analysis as well as for assessing oil and gas potential.

Experiments by EPSTEIN, EPSTEIN, and HARRIS (1977) established temperature ranges for each CAI. These same workers have correlated CAI with other organic metamorphic indexes. Palynomorph trans-

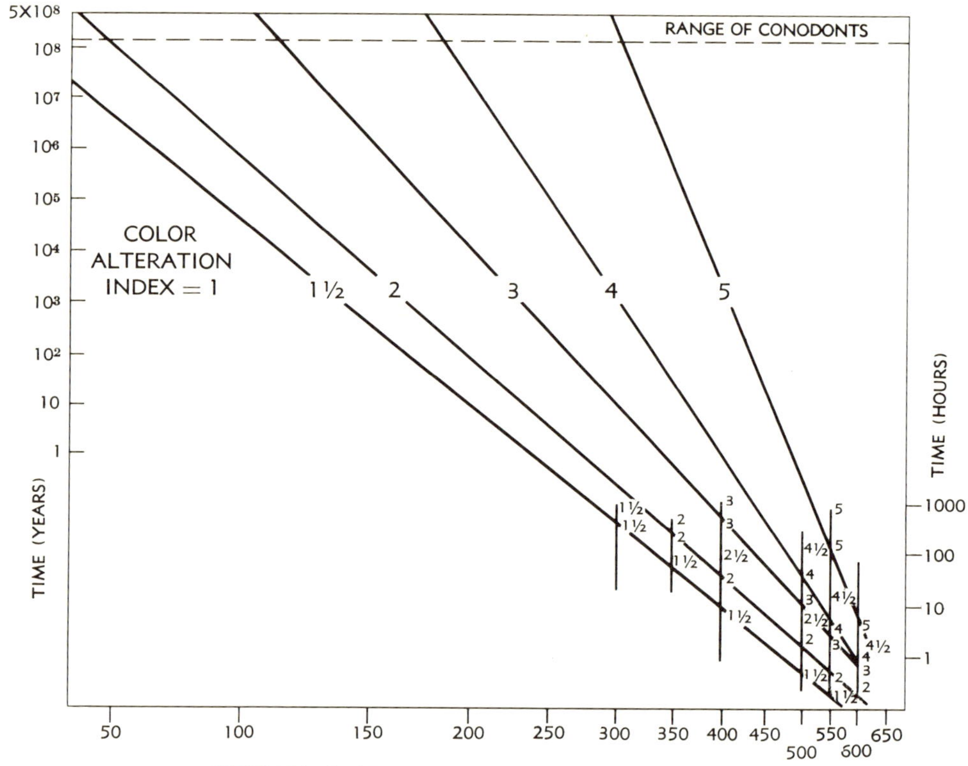

FIG. 50. Arrhenius plot of heat-induced, open-air, conodont element color alteration data (after Epstein, Epstein, & Harris, 1977). Vertical bars represent duration of heating runs at constant temperature. For the 300 to 400°C runs, elements were pulled every 24 hours; for higher temperature runs, specimens were pulled every half hour or hour for the first 24 hours and at 24-hour intervals thereafter. A comparison of the experimental data with field collections has resulted in the recognition of five color intervals. These are numbered and termed color alteration indexes (CAI); the diagonal lines bound CAI fields.

lucency and vitrinite reflectance measurements (two indexes commonly used by the petroleum industry for assessing organic metamorphism) were made on material from the same rock samples from which conodont element CAI had been determined. Thus, element CAI has been tied to chemically determined fixed carbon values via two other optical organic indexes (Frontispiece). Comparison of element CAI isograd maps with known oil- and gas-producing fields, comparison of element experimental time-temperature data with known hydrocarbon time-temperature generation curves, and correlation of CAI with other organic metamorphic indexes, all show that: (1) element color alteration begins beyond the threshold of hydrocarbon generation; (2) a CAI of 1.5 to 2 is at the deadline for oil and condensate production; and (3) a CAI of 4.5 is near the deadline for dry gas production.

In order to compile CAI isograd maps or to use conodont element CAI as a geothermometer, a field collection of elements must first be indexed. To ensure consistent indexing, elements should be matched to color standards. This can be done by using color chips (Munsell Color Company, 1971), an element color chart (EPSTEIN, EPSTEIN, & HARRIS, 1977, fig. 5), sets of standards assembled using induced color alteration specimens (time-temperature requirements for cooking standards can be determined

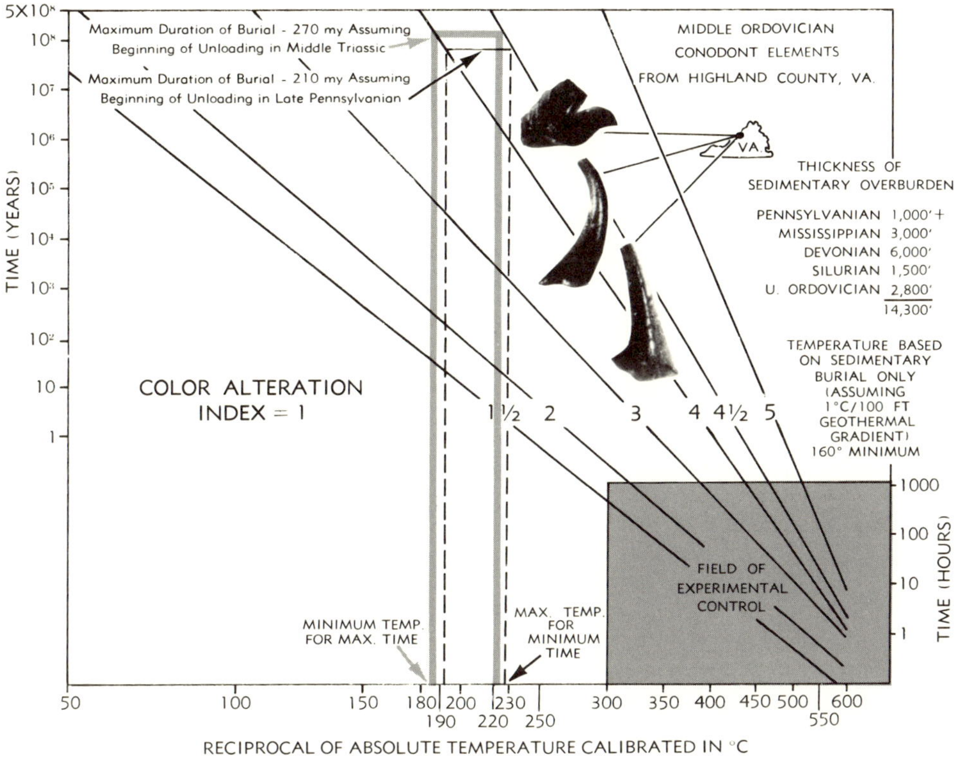

FIG. 51. Example of use of conodont element CAI as a geothermometer (after Epstein, Epstein, & Harris, 1977). Arrhenius plot of data from open-air heating runs (same plot as in Fig. 50) showing best color fit for Middle Ordovician specimens from Virginia; use of this plot for determination of minimum and maximum temperature ranges of conodont elements is explained in text.

from Fig. 50), or specimens from field collections.

The shape and size of specimens affect their color. Even though the experimental specimens shown on the frontispiece are all elements of *Phragmodus undatus* BRANSON & MEHL, there is still some range in color within each index because of the robustness or delicateness of the various specimens. If, however, the same element is compared for all indexes (for example, the thin-bar fragment in the lower right of each box on the frontispiece), the index sequence is obvious. Robust elements should not be compared with delicate elements. At CAI 5, however, all morphotypes are black. Additional procedures and problems for indexing specimens have been described by EPSTEIN, EPSTEIN, and HARRIS (1977).

An example of the use of conodont element CAI as a geothermometer is illustrated by Figure 51. Specimens from near Monterey, Virginia, in the Appalachian Mountains, were compared to standards and determined to have a CAI of 4.0 to 4.5. The specimens are of Middle Ordovician age; thus, the maximum possible time for burial and heating could be as long as 470 m.y. In this part of the Appalachian Mountains, however, Middle Triassic is the latest possible time for beginning of unloading by erosion. Therefore, 270 m.y. is the maximum possible time for burial and heating, thus providing the lowest possible temperature for elements of CAI 4.0 to 4.5 (heavy lines). Projecting the 4.0 to 4.5

field segment of the 270 m.y. line to the X axis yields a temperature range of 185 to 220°C for these specimens. If, however, unloading began in the Late Pennsylvanian, which is the earliest possible time for unloading in this region, the maximum time for burial and heating is 210 m.y. (dashed lines), thus a maximum temperature range of 190 to 230°C. This temperature range is very close to the first determination, showing that time becomes less important beyond durations of about 10 million years. The 185 to 230°C temperature range is also compatible with other geologic data. In the vicinity of Monterey, at least 14,300 feet of rock is known to overlie the Middle Ordovician. Assuming an average geothermal gradient of 1°C per 100 feet, known overburden alone can account for a temperature of 160°C. In this calculation, neither thickening of section by folding and faulting nor restoration of missing section has been considered. In spite of this, the temperature range determined from the Arrhenius plot is compatible with isopach data. Moreover, Paleozoic rocks in the Monterey area are cut by Cenozoic intrusions. The missing 25 to 70°C can be easily accounted for by a higher-than-average Cenozoic geothermal gradient.

In summary, the color alteration index of conodont elements is a valuable tool for assessing organic metamorphism.

1. It is a rapid and inexpensive method requiring only standard laboratory techniques and a binocular microscope. Other chemical and optical methods require expensive and complicated procedures and optical equipment.

2. Standards can be easily assembled and reproduced. A color chart can be used for index determination or a set of standards can be assembled from field collections or produced in the laboratory.

3. Conodont element CAI provides thermal cutoffs for oil, condensate, and dry gas production.

4. Conodont elements extend into thermal and age ranges that cannot be analyzed by other organic metamorphic indexes.

5. Conodont elements are most abundant and most easily concentrated from marine carbonate rocks in which palynomorphs are generally poorly preserved, and in which other organic materials (vitrinite and kerogen) are least abundant and often absent.

GLOSSARY OF MORPHOLOGICAL AND STRUCTURAL TERMS FOR CONODONT ELEMENTS AND APPARATUSES

By Walter C. Sweet

[Ohio State University]

Terms used to describe features of conodonts or their skeletal elements are included in the following glossary only if their meaning is different, or more restricted, than that given in an English dictionary, or if their precise meaning in the description of conodonts is not clear in the context of associated descriptive terms. Terms printed in boldface are preferred, or recommended; those printed in italics have been used, but are not preferred or recommended. German equivalents for many terms are included in parentheses.

aboral. Toward the lower, or under side of an element or process.

aboral attachment scar. Attachment surface situated on lower, or under side of element.

aboral cavity. Same as basal cavity.

aboral edge. Same as basal margin.

aboral extension. Same as basal extension.

aboral groove (Basalfurche). Same as basal groove or basal furrow.

aboral margin. Same as basal margin; has also been used to refer to lower, or under side of element or process.

aboral process. Same as linguiform process.

aboral projection. Same as anticusp; has also been used to refer to basal extension (=aboral extension).

aboral side. Same as lower, or under side.

aboral surface (Aboralfläche). Same as lower, or under side.

aboro-lateral groove. Same as basal furrow or groove.

accessory lobe. Nodose projection of anterior part of platform of some pectiniform elements, situated between posterior end of blade and that part of platform crossed by transverse ridges.

adcarinal groove *(Adcarinalgrube).* More or less elongate depression or trough on either side of fixed blade or carina of some planate and scaphate pectiniform elements.

adenticulate. Without denticles.

alate element. A bilaterally symmetrical ramiform element that lacks an anterior process, but has a posterior process and a lateral process on each side of the cusp.

albid element. An element containing white matter; contrasts with hyaline element.

angulate element. An arched pectiniform element with two primary processes, which are anterior and posterior.

anguliplanate element. An angulate element with lateral platform extensions and the attachment surface characteristic of planate pectiniform elements; see angulate, planate.

anguliscaphate element. An angulate element with lateral platform extensions and the attachment surface characteristic of scaphate pectiniform elements; see angulate, scaphate.

anterior *(Vorne).* Toward front end of element in conventional orientation. In coniform elements, convex side of cusp or side of element opposite that toward which cusp apex points. In ramiform and pectiniform elements with discernible cusp, convex side of cusp or side of element opposite that toward which denticles incline. In platformed pectiniform elements with blade, side toward which distal end of blade is directed. In some elements, convex side of basal cavity defines anterior side.

anterior arch. Archlike structure formed by junction of lateral processes at anterior end of alate ramiform element.

anterior bar. Same as anterior process.

anterior blade. Same as blade.

anterior deflection. Downturned distal end of anterior process.

anterior denticles. Denticles on anterior side of element.

anterior edge. Sharp edge on anterior side of element.

anterior face *(Vorderseite).* Front side of element or feature of an element in conventional orientation. For determination, see anterior.

anterior inner bar. Same as anterior inner-lateral process.

anterior inner-lateral bar. Same as anterior inner-lateral process.

anterior inner-lateral process. Anteriormost of two or more lateral processes on inner side of asymmetric element.

anterior limb. Same as anterior process.

anterior margin *(Vorderrand).* Trace of anterior side of an element or feature of an element in lateral view.

anterior outer bar. Same as anterior outer-lateral process.

anterior outer-lateral bar. Same as anterior outer-lateral process.

anterior outer-lateral process. Anteriormost of two or more lateral processes on outer side of asymmetric element.

anterior process. A process that is continuous at its proximal end with the anterior side of the cusp; upper surface commonly denticulate, but may be adenticulate.

anterior projection. Same as anterior process.

anterior side. Same as anterior face.

anterior trough margin *(Vorderer Trogrand).* Portion of platform in planate elements of some *Polygnathus* species in front of the geniculation point.

anterobasal corner. Area immediately adjacent to the junction of the anterior and basal margins of an element.

anticusp. Downward projection of basal margin beneath cusp; commonly only downwardly spatulate anterobasal corner and doubtfully distinct from anterior process.

apex. Tip of basal cavity, cusp, or denticle.

apical denticle. Same as cusp.

apical lamella. Small expansion or lip on anterior side of basal part of cusp of elements of *Apatognathus* (see Fay, 1952, p. 9).

apparatus (skeletal apparatus) *(Apparat).* The entire complement of elements composing the endoskeleton of a conodont.

appressed denticles. Same as fused denticles.

apron. Term sometimes used for the laterally flaring basal portion of certain types of elements.

arched *(gewölbt, gebogen).* Descriptive of elements in which the longitudinal axis is convex upward.

assemblage *(Conodonten-Gruppe).* Physical association, on a bedding plane, of several types of discrete conodont elements interpreted as skeletal parts of one animal.

attachment scar. Area of palmate pectiniform element to which basal plate was attached; commonly consists of small basal pit with groovelike extension(s), the whole surrounded by zone of recessive basal margin and forming flat or broadly concave or convex surface on under side of element.

attachment surface *(Haftnarbe).* Area of conodont element to which basal plate or basal funnel is, or was, attached; coextensive with that part of the under surface and sides of an element on which the edges of lamellae are visible.

axis *(Achse, Längsachse).* Midline of an element, a process, a cusp, or a denticle.

azygous node *(Zentralknoten, Zentralknötchen, Azygus-knoten).* Node, commonly larger than adjacent ones, situated directly above basal pit or growth center in certain palmate pectiniform elements; forms part of carina.

bar. Same as process; has most commonly been used in reference to processes of ramiform elements.

bar tooth. Same as denticle.

basal *(Basal).* Toward the base, basal margin, or under side of an element.

basal attachment scar. Attachment surface on lower, or under side of element; see aboral attachment scar, attachment surface.

basal canalules. Minute spherical or tubular voids in basal portion of certain coniform elements.

basal cavity *(Basalhöhle).* An upwardly concave, basically conical invagination of the under side of a conodont element, about which the element was built through successive accretion of lamellae; opens toward under side, and apex is directed toward upper side of element. Inner surface of basal cavity may be coextensive with attachment surface, but if invaginated portion of attachment surface is bordered by zone of recessive basal margin, invaginated portion is a basal pit and forms only part of attachment surface.

basal cone. Same as basal funnel; has also been used to refer to base of certain types of elements.

basal extension. Evagination of basal margin that extends downward below level of under side of posterior process; has also been termed aboral extension.

basal excavation. Same as basal cavity or basal pit.

basal filling *(Basisfüllung).* A general term for various types of basal material; see basal cone, basal funnel, basal plate.

basal funnel *(Basistrichter).* An infrequently preserved portion of certain conodont elements, consisting of a conical, phosphatic basal plate formed of lamellae, and concave on its under side; tip or apex fits into that of the basal cavity or pit; structure formed concurrently with the more generally preserved part of the element, which formed by accretion of lamellae about the basal cavity; commonly red-brown and of bony aspect.

basal furrow. Same as basal groove.

basal groove *(Basalfurche).* Groovelike extension of basal cavity or basal pit beneath a process or beneath some part of the carina of palmate pectiniform elements.

basal margin *(Unterrand).* Lower edge of last-formed lamella of an element; coincides with trace of under side of unit in lateral view if attachment surface lacks zone(s) of recessive basal margin, otherwise coincident with outer margin of attachment surface.

basal pit *(Basalgrube).* An upwardly concave, basically conical invagination of the under side of a conodont element, which is bordered or surrounded by a zone, or zones, of recessive basal margin and, together with these, forms an attachment surface. If entire attachment surface is upwardly concave and there are no contiguous zones of recessive basal margin, the term basal cavity is used.

basal plate *(Basalplatte, Basiskörper).* A tabular or broadly conical homologue of a basal funnel, but attached to the under side and formed concurrently with the growth of palmate pectiniform elements.

basal suture *(Basis-Sutur).* Boundary line between lower edge of the lamella of a conodont element and the upper edge of the simultaneously secreted lamella of its basal plate or funnel.

basal wrinkles *(Basisrunzeln).* Coarse longitudinal ribs that form a narrow band, or zone, adjacent to the basal margin of elements proper of *Panderodus, Neopanderodus, Belodina* and related genera.

base *(Basis, Basalteil).* Portion of conodont element that includes the basal cavity or pit; commonly used to refer to the part of an element adjacent to the under side, but has also been used to refer to the under side itself.

bimembrate (skeletal) apparatus. A conodont skeletal apparatus consisting of two morphologically distinct element types, regardless of the total number of elements that may have composed the apparatus.

bipennate element. A ramiform element with an anterior and a posterior process; posterior process commonly longer than anterior process, which may also be curved or deflected toward one side and may even be adenticulate.

blade *(Blatt, Klinge).* (a) A carminate or angulate pectiniform element. (b) In palmate or scaphate pectiniform elements, that part of the longitudinal axis anterior to the basal cavity or pit. That part of the blade bordered by platform extensions is commonly termed a fixed blade; any portion of the blade that protrudes anteriorly beyond platformed portions of the element is a free blade.

blade parapet. High narrow inner-platform structure, to which the blade of platformed *P* elements of *Cavusgnathus* is joined.

boss structures. Small raised features distributed over part of the upper surfaces of basal plates or basal funnels; most have an oblong outline and, at high magnification, a rough surface texture.

bowed. Descriptive of elements in which the longitudinal axis is convex toward one side and concave toward the other.

buttress. Same as linguiform process.

carina *(Carina, Kamm, Hauptknötchenreihe).* (a) A broad longitudinal ridge on one of the sides of a coniform element. (b) A row of denticles or nodes on upper side of palmate or scaphate pectiniform elements. (c) That part of the longitudinal axis of pectiniform elements flanked by platforms.

carminate element. A pectiniform element with two primary processes, which are anterior and posterior, and a longitudinal axis that is straight, or essentially so, in lateral view.

carminiplanate element. A carminate element with

lateral platform extensions and the attachment surface characteristic of planate pectiniform elements. See carminate, planate.

carminiscaphate element. A carminate element with lateral platform extensions and the attachment surface characteristic of scaphate pectiniform elements; see carminate, scaphate.

central node (Zentralknoten). Same as azygous node.

compound elements *(Zusammengesetzte Conodonten).* General term for ramiform and pectiniform elements; contrasts with simple elements.

coniform element. Basically conical elements without greatly produced or denticulated margins; in much of the literature, coniform elements are termed "simple cones."

conodont. Individual representative of the Conodonta (cf. brachiopod, mollusk, mastodont, etc.), but most commonly used to refer to a disjunct skeletal element (cf. brachiopod valve, mollusk shell, mastodon tusk, etc.) of the Conodonta. This dual usage was begun by PANDER (1856), who used the term for an entire individual of the Conodonta in every place in his monograph except in his formal definition of "Conodonten," which refers only to the isolated skeletal elements on which he based his concept of the group as a whole.

conodont apparatus. Same as apparatus.

conodont-bearing animal. Same as conodont.

conodont element. Same as element.

conodontifer. Same as conodont.

costa *(Costa, Rippe, Leiste).* A narrowly rounded or sharp-edged longitudinal ridge on one or another of the sides of a coniform element.

crimp *(Umschlag).* Marginal area on under side of planate pectiniform element, between edge of last-formed lamella and outer edge of platform.

cristula. Same as rostral ridge.

crown. A conodont element minus its basal funnel or plate; see element proper.

cup *(Gewölbe).* Term sometimes used for the greatly expanded basal cavity of scaphate pectiniform elements.

cusp *(Hauptzahn, Grosser Zahn).* The spinelike, fanglike, or conical denticle above the apex of the basal cavity or pit; in coniform elements, the entire element above the base. In ramiform and pectiniform elements, the denticle above the apex of the basal cavity or pit has commonly been termed "main cusp," but this term is inappropriate for it implies that there may also be minor denticles somehow situated above the basal cavity apex.

denticle *(Zähnchen, Zacken, Kleiner Zahn, Dentikel).* Individual spinelike or conical element of the serrate edge or surface of a process or platform; the denticle situated above the apex of the basal cavity or pit is termed a cusp.

denticulate. Bearing denticles.

dextral element. Component of a skeletal apparatus on the right-hand side of the plane of bilateral symmetry.

digyrate element. Bilaterally asymmetric ramiform element with short, adenticulate posterior process and longer, denticulate inner- and outer-lateral processes, the distal extremities of which commonly twist in opposite directions.

discrete denticles *(Getrennte Zähnchen).* Denticles separated from adjacent ones by a space.

discrete element. Same as element, but refers to one that has become dissociated from the skeletal apparatus of which it was once a part.

dolabrate element. A ramiform element consisting of a cusp and a posterior process; commonly pick-shaped in lateral view.

element (conodont element, skeletal element). An individual component of the skeletal apparatus of a conodont; consists of the commonly preserved "element proper" and the less commonly preserved basal funnel or plate; holoconodont is a synonym.

element proper. A conodont element minus its basal funnel or plate; the most commonly preserved portion of an element.

erect *(Aufrecht).* When used to describe the cusp or denticles of an element, this term implies that the axis of the cusp or of the denticles is normal to that of the upper edge of the base or of a posterior process.

escutcheon. Same as attachment surface. Has commonly been used to describe a scarlike attachment surface displaced toward or onto the side of a process or processes.

fang. Same as cusp.

fibrous *(fibrös).* When used in connection with conodont elements, this term customarily alludes to a structural condition that causes elements to break with longitudinally fibrous fracture; elements so described also commonly lack white matter except along thin longitudinal growth axis; elements that lack appreciable white matter but are not known to fracture with fibrous habitus are best described as hyaline.

fixed blade *(Festes Blatt).* Portion of longitudinal axis on upper side of platformed pectiniform elements between free blade and carina; fixed blade is flanked by platforms.

flange. Shelflike lateral projection of the side of a process; wider than a lateral ridge, but narrower than a platform.

free blade *(Freies Blatt, Klinge).* Portion of blade that protrudes anteriorly beyond platformed portions of a pectiniform element.

furrow *(Furche).* Used in description of conodont elements for any longitudinal groove, trench, or depression; may be a feature of either upper or under side, or may describe an elongate depression on one of the faces of a coniform element (see longitudinal furrow); also used to describe a groove on the upper surface of a basal plate or funnel.

fused cluster. A cluster of elements that are fused or cemented to one another.

fused denticles. Very closely set denticles, each partly or entirely confluent with adjacent or adjoining denticles; contrasts with discrete denticles.

geniculate coniform element. Coniform element in which posterior margin of cusp joins upper edge of base to enclose an acute angle.

geniculation point *(Plattform-Knie)*. Point of inflection, in lateral view, of the platform margin of planate pectiniform elements of some species of *Polygnathus*; margin slopes upward from its posterior end to geniculation point, then is sharply deflected downward; geniculation point marks anterior limit of ornamented part of platform; anterior of it platform consists of adcarinal grooves and their upturned margins.

germ denticles. Inappropriate term for overgrown, or suppressed denticles.

growth axis *(Wachstumsachse)*. Imaginary line joining tips of successive conical lamellae in a cusp or denticle; may be outlined in some elements by concentrations of white matter, or internally by a narrow tube.

growth center *(Wachstumszentrum)*. Point about which a conodont element developed; that is, apex of basal cavity.

growth lamella. Same as lamella.

growth lines *(Wachstumslinien)*. Traces of lamellae in section; also used to refer to striae on an attachment surface.

heel. Upwardly convex posterior segment of the upper margin of elements of *Belodina*.

height *(Höhe)*. Any measurement taken in a direction normal to the upper or lower margin or surface of an element.

holoconodont. Same as element.

hyaline element. An element that lacks white matter except, perhaps, along thin growth axes in cusp or denticles; contrasts with albid element; typical fibrous elements are also hyaline, but it is not clear that all hyaline elements are fibrous in structure.

inferior side. Same as lower, or under side.

inner face *(Innenseite)*. Side of cusp or denticle on inner side of an element.

inner-lateral process. Lateral process on the inner side of an element.

inner platform. Platform on inner side of pectiniform element.

inner side. Portion of asymmetrical element on concave side of anteroposterior midline.

interior limb. Same as inner-lateral process.

interlamellar spaces *(interlamellare Spalten)*. Funnel-shaped cavities within lamellae along growth axes.

interlamellar stripes (or striae) *(interlamellare Streifen)*. Faint dark bands within, and parallel to the boundaries of lamellae; seen only in thin sections of exceptionally well-preserved conodont elements; thought to represent levels of higher than ordinary concentration of organic matter within lamellae and thus to indicate that formation of a lamella did not take place in a single short episode.

inverted basal cavity (umgestülpte Basalhöhle). Same as zone of recessive basal margin.

keel *(Kiel)*. Ridgelike longitudinal structure on under side of planate pectiniform elements; main keel is situated beneath blade-carina; secondary keels extend from basal pit beneath lateral processes. Term has also been used to describe longitudinal ridges or flangelike projections along anterior or posterior margins of coniform elements.

keel angle. Angle opening anteriorly between main keel and secondary keel.

kink *(Knick)*. A sharp lateral bend in the carina of some platformed pectiniform elements.

lamella *(Lamelle)*. One of the many thin layers of minute phosphatic crystallites of which an element is composed.

lateral *(seitlich, Seiten-, lateral)*. Term used to describe the situation of any structure or feature developed on or projecting from one of the faces of an element flanking the anteroposterior midline.

lateral bar. Same as lateral process.

lateral face *(Lateralseite, Seite)*. One of the sides of an element flanking the anteroposterior midline.

lateral process. Process projecting from one of the faces of an element flanking the anteroposterior midline; see inner-lateral process, outer-lateral process.

laterally confluent denticles. See fused denticles.

length *(Länge)*. Any measurement taken in the anteroposterior direction.

limb. Same as process.

linguiform process *(Zungenförmiger Fortsatz)*. A process that is tongue shaped when viewed from its upper or under side.

lip *(Lippe)*. Small lateral expansion flanking basal cavity; commonly projects downward.

lobe *(Lappen)*. Process flanked by flanges or platforms; commonly formed by lateral expansion of a lateral process; may bifurcate.

longitudinal. Used to describe the orientation of a structure, feature, or section whose long dimension is parallel to that of the element or the segment of the element on which it is situated.

longitudinal furrow *(Längsfurche)*. A deep longitudinal groove formed by infolding of the surface on one side of elements proper of *Panderodus, Neopanderodus, Belodina,* and related genera of the Panderodontacea; also termed panderodont furrow.

loop. Elliptical segment of attachment surface enclosing basal pit of segminate or segminiplanate pectiniform elements. Closed posteriorly; continuous anteriorly with attachment surface beneath anterior process; and commonly expressed as raised area of recessive basal margin around basal pit.

lower side *(Unterseite, Unterfläche).* Side of an element toward which the basal cavity or basal pit opens; under side is used synonymously.

lumen *(Lumen).* Central invagination in a basal funnel.

M position. In locational notation for conodont skeletal apparatuses, *M* designates the position occupied by pick-shaped dolabrate elements or by similar bipennate, digyrate, or coniform elements.

main carina *(Kamm, Hauptzähnchenreihe).* Portion of axis on upper side of planate or scaphate pectiniform elements posterior to apex of basal cavity or growth center; consists of a row of fused or discrete denticles, commonly nodelike.

main cusp. Same as cusp.

main keel. See keel.

main trough *(Haupttrog, Trog).* Furrow or depression parallel to anteroposterior midline on upper side of planate or scaphate pectiniform elements; has also been termed "trough," or "sulcus."

midplane. An imaginary surface that includes the apexes of the cusp and basal cavity and the uppermost and lowermost points on the basal margin.

multimembrate (skeletal) apparatuses. A general term for conodont skeletal apparatuses consisting of more than one morphologically distinct type of element, regardless of the total number of elements of all types in the complete apparatus; see bimembrate, trimembrate, quadrimembrate, quinquimembrate, seximembrate, septimembrate.

multiramate element. A ramiform element with more than four basic processes.

natural assemblage. Same as assemblage.

navel. Same as basal pit.

node *(Knoten, Knötchen, Tuberkel).* A low knob- or bumplike denticle.

nongeniculate coniform element. Coniform element in which there is a smooth transition from posterior cusp margin to upper basal edge; in lateral view, posterior margin and upper basal edge form a straight or smoothly arcuate line.

notch. Deep indention of margin of a basal plate or funnel; also used as a general term for any such marginal indentation in elements proper.

obverse side. Same as inner side.

oral. Same as upper. (This term is commonly used in preference to "upper," but it is inherited from a time when conodont elements were believed to be teeth and thus to have projected into a buccal, or oral, cavity.)

outer face *(Aussenseite).* Side of cusp or denticle on outer side of element.

outer-lateral process. Lateral process on the outer side of an element.

outer platform. Platform on outer side of pectiniform element.

outer side *(Aussenseite).* Portion of asymmetrical element on convex side of anteroposterior midline.

overgrown denticles *(Kiemzähnchen, Unterdrückte Zahnchen).* Denticles that formed in early stages of growth, but later fused completely with adjacent denticles; outlines of such denticles can commonly be seen through the thin, subsequently secreted lamellae by which they are enveloped; commonly (but inappropriately) termed "germ denticles"; may also be termed suppressed denticles.

P position. In locational notation for conodont skeletal apparatuses, *P* designates the positions occupied by pectiniform elements; characteristically, two types of pectiniform elements are in fully developed apparatuses and the positions occupied by these are designated *Pa* and *Pb*.

Pa position. See *P* position.

parapet *(Parapet, Brüstung).* Wall-like longitudinal structure on platform of pectiniform element, or on flange of ramiform element.

pastinate element. A pectiniform element with three primary processes, which are anterior, posterior, and lateral.

pastiniplanate element. A pastinate element with lateral platform extensions and the attachment surface characteristic of planate pectiniform elements; see pastinate, planate.

pastiniscaphate element. A pastinate element with lateral platform extensions and the attachment surface characteristic of scaphate pectiniform elements; see pastinate, scaphate.

Pb position. See *P* position.

pectiniform elements. A category of basically blade- or comb-shaped conodont elements, with two, three, or four processes that are commonly much higher than they are wide; one or more of the processes may develop lateral flanges or platforms; a general designation for the shape categories termed "blades" and "platforms" in much of the conodont literature.

planate element. A pectiniform element with conspicuous lateral ledges, brims, or platforms flanking one or more of its processes; and with an attachment surface on its under side distinguished by a zone of recessive basal margin, which at least partially surrounds a basal pit that has groovelike extensions beneath at least the primary processes.

plate. Same as platform.

platform *(Plattform, Tafel).* Laterally produced shelflike structure flanking a process in pectiniform elements; platformed segment is posterior, and its upper surface is commonly (but not invariably) pitted, nodose, ridged, or exhibits some combination of pits, nodes, and ridges.

posterior *(Hinten).* Toward the rear end of an element in conventional orientation. In coniform elements, concave side of cusp or side of element toward which cusp apex points. In ramiform and pectiniform elements with discernible cusp, concave side of cusp or side of element toward which denticles incline. In platformed pectiniform elements with blade, side on which platforms are developed. In some elements, concave side of basal cavity defines posterior side.

posterior bar. Same as posterior process.

posterior blade. Same as posterior process.

posterior deflection. Downturned distal end of posterior process.

posterior denticles. Denticles of the posterior process.

posterior edge. Sharp edge on posterior side of element.

posterior face *(Hinterseite).* Back side of an element or feature of an element in conventional orientation. For determination, see "posterior."

posterior inner bar. Same as posterior inner-lateral process.

posterior inner-lateral process. Posteriormost of two or more lateral processes on inner side of asymmetric element.

posterior limb. Same as posterior process.

posterior margin *(Hinterrand).* Trace of posterior side of an element or feature of an element in lateral view.

posterior outer bar. Same as posterior outer-lateral process.

posterior outer-lateral bar. Same as posterior outer-lateral process.

posterior outer-lateral process. Posteriormost of two or more lateral processes on outer side of an asymmetric element.

posterior process. A process that is continuous at its proximal end with the posterior side of the cusp; upper surface commonly denticulate, but may be adenticulate.

posterior side. Same as posterior face.

posterobasal corner. Area immediately adjacent to the junction of the posterior and basal margins of an element.

primary process. A process that projects from the cusp; under side occupied by an extension of the basal cavity or basal pit.

process *(Ast, Fortsatz).* Anterior, posterior, or lateral projections of ramiform and pectiniform elements; commonly, but not invariably denticulate.

proclined *(Proklin, Vorwärtsgeneigt).* When used to describe the cusp or denticles of an element, this term implies that the axis of the cusp or denticles forms an angle of more than 90 degrees with the upper edge of the base or of a posterior process.

prong. A projection of the margin of a basal plate or funnel; also used as a general term for projections of the basal margin of elements proper.

pseudokeel. Raised flat area on under side of planate elements of some species of *Siphonodella*; contains deep pit near anterior end and is bordered in elements of mature individuals by narrow zones of recessive basal margin.

pulpa. Same as basal cavity.

pulp cavity (Pulpahöhle). Same as basal cavity.

quadrimembrate (skeletal) apparatus. A conodont skeletal apparatus consisting of four morphologically distinct element types, regardless of total number of elements in complete apparatus.

quadriramate element. A ramiform element with anterior, and posterior processes, and a lateral process on either side of the midplane.

quinquimembrate (skeletal) apparatus. A conodont skeletal apparatus consisting of five morphologically distinct element types, regardless of total number of elements in complete apparatus.

radial ridge *(Radialrippe, -leiste).* Ridge on upper surface of platformed pectiniform element trending diagonally from longitudinal axis of element toward margin of platform; contrasts with longitudinal and transverse ridges.

ramiform element. Conodont element in which at least one of the sides or edges of the base is drawn out laterally, anteriorly, or posteriorly from the cusp into a process that is serrate on its upper edge.

ramp *(Rampe).* Sloping surface on some platformed pectiniform elements, between low- and higher-level areas of the platform.

reclined *(Reklin, Rückwärtsgeneigt).* When used to describe the cusp or denticles of a conodont element, this term implies that the axis of the cusp or of the denticles is straight and forms an angle of less than 90 degrees with the upper edge of the base or of a posterior process.

recurved *(Zurückgebogen).* When used to describe the cusp or denticles of a conodont element, this term implies that the cusp or denticle axis is not straight, but arcuate, and that the apex of the cusp or denticles points posteriorly.

reverse side. Same as outer side.

rostral ridge *(Rostralleiste, -rippe).* Ridge of more or less fused nodelike denticles extending posteriorly from anterior portion of platform; rostral ridges outline the rostrum and adcarinal grooves; they form parallel, collar- or rufflelike structures on pectiniform elements of some species of *Siphonodella* and *Polygnathus*.

rostrum *(Rostrum).* Narrow neckline or beaklike structure in anterior part of platform in planate pectiniform elements of most *Siphonodella* and some *Polygnathus* species.

S positions. In locational notation for conodont skeletal apparatuses, *S* designates the positions occupied by ramiform elements that form a symmetry-transition series; *Sa* denotes the position of bilaterally symmetrical alate elements; *Sb* denotes the position of digyrate or tertiopedate elements; *Sc* denotes the position of bipennate or dolabrate elements; and *Sd* denotes the position of quadriramate elements.

Sa position. See *S* positions.

Sb position. See *S* positions.

Sc position. See *S* positions.

scaphate element. A laterally elaborate or platformed pectiniform element with an under side marked by a capacious, commonly cuplike basal cavity.

Sd position. See *S* positions.

secondary carina *(Sekundärkamm).* Row of nodes or denticles on upper surface of palmate or scaphate pectiniform elements marking position of a lateral process that is surrounded by platform extensions.

secondary keel *(Sekundärkiel).* See keel.

secondary process. A branch of a primary process.

segminate element. A pectiniform element with only one primary process, which is anterior in all known forms.

segminiplanate element. A segminate element with lateral platform extensions and the attachment surface characteristic of planate pectiniform elements; see segminate, planate.

segminiscaphate element. A segminate element with lateral platform extensions and the attachment surface characteristic of scaphate pectiniform elements; see segminate, scaphate.

septimembrate (skeletal) apparatus. A conodont skeletal apparatus consisting of seven morphologically distinct element types, regardless of total number of elements in a complete apparatus.

seximembrate (skeletal) apparatus. A conodont skeletal apparatus consisting of six morphologically distinct element types, regardless of the total number of elements in a complete apparatus.

sheath. Expanded portion of base enclosing basal cavity and its extension.

simple cone (element, conodont). Same as coniform element; contrasts in much of the literature with compound element (or conodont).

sinistral element. Component of a skeletal apparatus on the left-hand side of the plane of bilateral symmetry.

skeletal apparatus. Same as apparatus.

skeletal element. See element.

slant. Pitch of anterior face of blade of pectiniform elements.

stellate element. A pectiniform element with four primary processes, one of which is anterior and another posterior.

stelliplanate element. A stellate element with lateral platform extensions and the attachment surface characteristic of planate pectiniform elements; see stellate, planate.

stelliscaphate element. A stellate element with lateral platform extensions and the attachment surface characteristic of scaphate pectiniform elements; see stellate, scaphate.

submerged denticles. See overgrown denticles.

sulcus. See main trough.

suppressed denticles *(Kiemzähnchen, Unterdrückte Zähnchen).* See overgrown denticles.

symmetry-transition series. The array of three or four morphologically different but intergradational types of coniform or ramiform elements that occupy *S* positions in the skeletal apparatus of a conodont.

tertiopedate element. Asymmetric ramiform element with a posterior process and a lateral process on each side of the cusp; posterior process commonly long and denticulate.

tip. See apex.

transverse ridge *(Querrippe).* Ridge on upper surface of platformed pectiniform element that trends more or less normal to longitudinal axis of element.

trimembrate (skeletal) apparatus. A conodont skeletal apparatus consisting of three morphologically distinct element types, regardless of total number of elements in a complete apparatus.

trough *(Trog).* See main trough.

under *(unten,* aboral). See under side.

under side *(Unterseite, Unterfläche).* Side of an element toward which the basal cavity or basal pit opens; lower side is used synonymously.

unimembrate (skeletal) apparatus. A conodont skeletal apparatus consisting of only one morphologically distinct element type, regardless of total number of elements in a complete apparatus.

upper *(oben).* See upper edge (or margin), or upper side.

upper edge. The denticulated edge of the process of a ramiform or pectiniform element.

upper margin. See upper edge.

upper side *(Oberfläche, Oralfläche).* Side or surface of an element opposite that toward which the basal cavity or pit opens; almost always the side or surface bearing denticles.

white matter *(Weisse Substanz).* Term applied to portions of an element that are more or less opaque in incident light and thus appear white or gray in contrast to adjacent or surrounding more or less clear and translucent areas. Areas of white matter correspond to internal parts of elements characterized by voids, which may be interlamellar spaces or concentrations of small, densely packed, irregularly shaped cellules.

width *(Breite).* Measurement at right angles to height and length.

wrinkles *(Runzeln).* Sets of complementary, but minor, ridges and furrows, with axes parallel to the longitudinal axis of an element; commonly form a zone along the basal margin in certain panderodont elements (e.g., those of *Panderodus*).

zone of recessive basal margin. Faintly striated portion of an attachment surface flanking some part of, or surrounding, a basal pit, and formed as the basal margins of successively younger lamellae grew no farther downward than those of older lamellae, or actually retreated upward as the elements grew larger; has been termed "inverted basal cavity."

NATURAL ASSEMBLAGES OF ELEMENTS: INTERPRETATION AND TAXONOMY

By F. H. T. Rhodes and R. L. Austin

[Cornell University; University of Southampton]

Since conodont elements were first described (Pander, 1856) several thousand papers have been devoted to them. Because of the vagaries of fossilization and the techniques employed in the extraction of microfossils, the great majority of elements are known as single isolated specimens. They have proved to be a varied and stratigraphically useful group, and a binominal system of classification established upon these single elements includes approximately 4,000 form-species.

Systematic treatment of single-element conodont genera and species formed the basis of taxonomy in the first edition of *Treatise* Part W by Hass (1962). More recently there has developed a multielement concept of conodont taxonomy. In this concept two or more disjunct and usually morphologically different elements are grouped in a multielement species because they are presumed to represent the remains of an individual conodont (see Klapper & Philip, 1971). Several more or less distinct methods have been used in the reconstruction of these multielement apparatuses. The three most important are the statistical or empirical association of elements having common stratigraphic and geographic ranges, the consistent presence of forms on the same bedding planes, and the occurrence of elements fused together in insoluble residues.

Ziegler (1974) defined a conodont apparatus as a multielement association reconstructed on statistical or empirical grounds (for examples, see Sweet & Bergström, 1969; von Bitter, 1972; Baesemann, 1973). He defined an assemblage as a natural multielement association found on bedding planes (for examples, see Scott, 1942, 1973; Rhodes, 1953b, 1962). Elements found fused together have been called fused clusters and have been described from the Cambrian (Landing, 1977), Ordovician (Barnes, 1967), Silurian (Rexroad & Nicoll, 1964; Pollock, 1969), Carboniferous (Austin & Rhodes, 1969; Higgins, 1975), and Permian (Behnken, 1975). Druce, Rhodes, and Austin (1972) have defined an alternative terminology for conodont assemblages.

Several workers (e.g., Eichenberg, 1930; Schmidt, 1934, 1950; Scott, 1934, 1942, 1973; Dubois, 1943; Rhodes, 1952, 1953b, 1962; Schmidt & Müller, 1964; Lange, 1968) have described conodont assemblages and have shown that a single assemblage, which they interpreted as representing the remains of an individual animal, may contain discrete elements classified in as many as five single-element conodont genera. A number of different genera have been based upon the recognition of these assemblages as taxonomic units, and though the particular basis of nomenclature has not been consistent, there has developed a second taxonomic framework based upon assemblages, rather than single elements. The consequent taxonomic problems are complex and are discussed later. Most students of conodonts have accepted the interpretation of these bedding-plane remains as natural assemblages, but a few (e.g., Branson & Mehl, 1936; Branson, 1957; Fay, 1957) have suggested that they may be coprolitic in origin.

The purpose of this chapter is to describe these natural assemblages of elements, to define their characteristics, and to examine the problems of their interpretation and taxonomic treatment.

INTERPRETATION OF ASSEMBLAGES

More than 500 element assemblages have been recorded from black shale of Cambrian and Carboniferous age. They have also been reported in Upper Devonian and Lower Devonian limestones (LANGE, 1968; MASHKOVA, 1972). Their relative abundance in black shale probably is the result both of the quiet conditions under which the sediment frequently accumulated, and of the fact that fissility of the shale lends itself to the microscopic study of its bedding planes. Other rock types from which elements have been collected (such as limestone and sandstone) often accumulated under more disturbed conditions. They usually are subjected to such violent chemical and physical methods of disintegration that there is little hope of recovering anything except isolated elements. The rarity of assemblages from strata other than those of Carboniferous age probably is more apparent than real, although there is perhaps a higher proportion of black shale in the Carboniferous System than in other systems in which elements occur (Cambrian-Triassic). Need exists, however, for careful study of such rich conodont-bearing black shale units as those of the Upper Devonian of eastern and central North America and of the Mississippian localities of Oklahoma described by ELIAS (1956).

The first assemblages to be reported were those from the Heath Formation (Mississippian) of Montana (SCOTT, 1934, 1942) and the lower Namurian of Germany (SCHMIDT, 1934), which were described in simultaneous but independent publications. SCOTT and SCHMIDT differed in their interpretations of the zoological affinities of the conodonts, but were in complete agreement that the assemblages they described were natural associations. DUBOIS' (1943) study of element assemblages from the McLeansboro Formation (Pennsylvanian) of Illinois convinced him that they were natural rather than random. RHODES originally was unwilling to accept the interpretation of natural element assemblages, but became convinced of their validity as a result of a study of Pennsylvanian assemblages from Illinois and Kentucky (RHODES, 1952).

These workers who independently have studied element assemblages are thus in agreement in regarding them as natural associations. "Natural occurrence" is used here to describe an association that is the direct result of the original association of a variety of individual elements within the body of one animal. A number of distinct aspects of the occurrence of element assemblages support their interpretation as natural associations.

Recurrent association of elements of similar shape.—From the same and from different localities and horizons, assemblages contain conodont elements of the same variety of shapes. For example, Upper Carboniferous assemblages from the Coal Measures of Britain (RHODES & AUSTIN, MS) are closely similar to those described from the Pennsylvanian of Illinois and Kentucky (RHODES, 1952). In both occurrences, the *Idiognathodus* assemblage contains elements of five shapes. Each assemblage may not contain all variants, for some are incomplete; sometimes only a single pair of components of similar shape is found. The degree of resemblance is found in the uniformity of association rather than its completeness. In a detailed study of more than 200 assemblages, RHODES (1953b) found that only two of them did not reveal the admixture of shapes commonly associated in an assemblage. Overall numbers of elements in these various assemblages are also broadly consistent. The shapes of individual elements from British assemblages agree closely with shapes in comparable assemblages from the Pennsylvanian of North America (RHODES, 1952, p. 891-895). In the latter, individual elements are variable in form, but it is not yet possible to assess the significance of this fact. These results have been strikingly confirmed by the study of Collinson and others (1972).

Elements of similar shape may be present in more than one kind of assemblage. A bipennate element is present in numerous assemblages, but the detailed morphology of specimens of this type differs among multielement genera. This complicates evaluation, but in no way detracts from the importance of the regularity of association.

Major shape-category ratio of isolated elements.—If the major shape categories of elements occur in a fixed proportion within an assemblage, it is probable that isolated elements in the major shape categories would also be found in fixed proportions. A lack of published data on this subject is conspicuous.

Scott (1942) studied 3,000 isolated elements from the Heath Shale and noted that "most of the different kinds of individual conodonts can be recognized in the assemblages; furthermore, the kinds found as individuals are proportional in numbers to those represented in the assemblages," that is, the bipennate element is found approximately three times as often as one of the pectiniform elements. Dubois (1943) studied 479 isolated elements from the fissile black shales below the LaSalle Limestone (Middle Pennsylvanian) of Illinois. Of these he identified 108 examples of one type of pectiniform element, 67 examples of another pectiniform element, and 304 examples of a bipennate element, or a ratio of roughly 1.6 : 1 : 4.5. This contrasts with the ratio 1 : 2 : 4 that Scott established by analysis of conodont assemblages. Dubois explained the apparent anomaly by the "different ability of the teeth to withstand fragmentation."

The results of Dubois' analysis are difficult to evaluate, but the deviation from the predicted ratio could be explained by the differential hazards of preservation. One of the pectiniform elements is more massive and undoubtedly more resistant to abrasion and probably less liable to transportation than the more delicate components. Indeed, the vagaries of fossilization are such that it may be doubted whether any consistent ratios should be expected. The ratio obtained from a limestone, for example, may be quite different from that obtained from a black shale. Our own preliminary studies of the ratios between isolated components are not conclusive. There is a need for an extensive study of the ratios of single-element genera. Collinson and others (1972) also gave counts of individual elements associated with natural assemblages, but the results are not conclusive.

Paired occurrence of components and their alignment.—Generally, assemblages can be recognized by the paired arrangement of their components. These components are not only of the same size and general form but may be paired in such a way that one is the mirror image of the other. Sometimes minor morphological differences are observable between two such paired components. These differences are no greater than those found, for example, between comparable complementary teeth in skulls. These paired elements are frequently aligned with others in such a way as to form an elongated series. It would be difficult to account for such alignment, and virtually impossible to account for the paired relationship, except by the acceptance of these associations as natural assemblages (see also Collinson & others, 1972).

Structural similarity of assemblages.—Sixteen genera are now known as natural assemblages, ranging in age from Cambrian to Permian. Of these, 9 multimembrate assemblages are closely similar in their general make-up; the remaining 7 (including a unimembrate assemblage) represent 3 structural types.

Geographic distribution of assemblages.—Natural assemblages are known from the United States, Germany, Russia, Ireland, Scotland, England, and Wales. This widespread geographical distribution is another

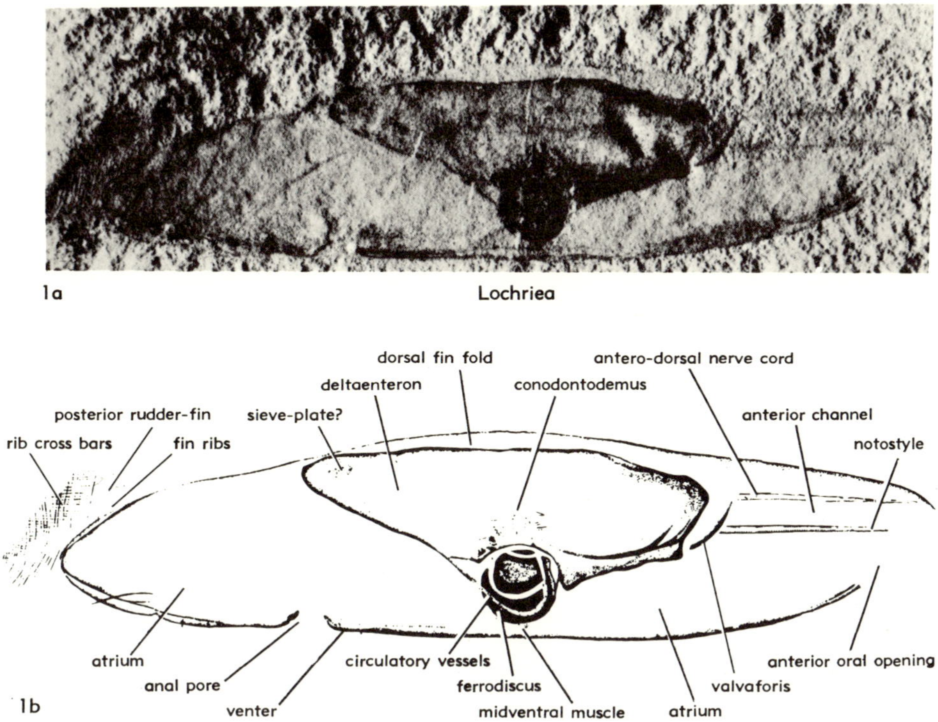

FIG. 52. A conodont-bearing organism, *Lochriea wellsi* MELTON and SCOTT, Carb. (Bear Gulch Ls.), USA (Mont.) (after Melton & Scott, 1973); *1a*, Specimen 6027, University of Montana collection, ×1.7; *1b*, Reconstruction showing terms applied to various anatomical features.

factor that supports their interpretation as natural associations. A single occurrence of an assemblage might be accepted as fortuitous, but the occurrence of several hundred assemblages, composed of similar components, many of them paired, in broadly similar numbers, in similar alignment, described by a dozen workers, from different parts of the geologic column in different continents, makes it difficult to maintain such a conclusion.

Coprolitic associations.—Coprolitic associations of conodont elements have been described from strata that also yield natural assemblages. Such associations may generally be distinguished from natural assemblages by three features: (1) a large number of isolated specimens: up to 150 compared with a maximum of 32 described from natural assemblages; (2) specimens lacking alignment or obvious pairing and sometimes representing more than one natural genus; and (3) a matrix showing a slight discoloration.

Zoological position and assemblages.—MELTON and SCOTT (1973) have described specimens from Carboniferous strata in central Montana that they interpreted as complete conodonts (Fig. 52). These specimens contain element assemblages. The element-bearing organisms are about seven centimeters long, elongated, and bilaterally symmetrical, with an anterior oral opening and structures interpreted as a dorsal nerve cord and a notostyle. A gutlike structure, the deltaenteron, and what was interpreted to be a circulatory system are also present. The element assemblage is believed to have functioned as a food-filtering system, and a posteroventral anal pore is also identified. These animals appear to have been free-swimming and soft bodied, and there is evidence of a reticulate membrane covering the body, which has a finlike posterodorsal projection.

Melton and Scott (1973) identified two species and genera of the organisms they described. In one of these, the outer membrane contains large numbers of microsclerites, which resemble sponge spicules in general form, and whose composition is similar to that of conodont elements.

The deltaenteron is the area in which the conodont elements are found in the fossil organism. Near the posterior end of the deltaenteron is a 2-mm-broad dark area which Melton and Scott (1973) suggested may have functioned as an exhalent sieve plate. A coiled structure, the ferrodiscus, contains a high proportion of residual iron, with relatively high phosphorous content near the anterior end.

As determined by microprobe analysis, the nerve cord is carbonaceous in composition, whereas the notostyle is phosphatic. Melton and Scott (1973) believed that the notostyle was probably flexible, though strong enough to give a degree of rigidity. They argued that the presence of a dorsal nerve cord, the notostyle, and internal phosphatic secretion all indicate that conodonts have close affinities with vertebrates, and may represent an ancestral group from which these organisms evolved.

Melton and Scott (1973) reconstructed the elements in the central midsection of the deltaenteron as assemblages attached along their basal margins to ligaments aligned dorsally to the ferrodiscus. They suggested that the element field (conodontodemus), may have functioned either in producing water currents with the deltaenteron, or in sieving out large particles from the deltaenteron. There is no widespread agreement on the significance of this report.

Conway Morris (1976) described a single specimen, with questionable conodonts, of a new genus and species *Odontogriphus omalus* from the Burgess Shale of Canada. He claimed that despite poor preservation, its affinities appear to be with the lophophorates.

TAXONOMY AND NOMENCLATURE OF ASSEMBLAGES

It has already been noted that an extensive "taxonomy" has been established upon isolated single-element conodont specimens. This nomenclature, which includes over 4,000 single-element form-species, has been established by workers who have rigidly observed the International Code of Zoological Nomenclature. The suggestion of Croneis (1938) for an independent "military classification" has not been generally followed.

The acceptance of natural assemblages of elements, containing up to five component single-element genera, led some workers (Eichenberg, 1930; Schmidt, 1934, 1950; Scott, 1942, 1973; Rhodes, 1952; Melton & Scott, 1973) to propose a new classification, based on the recognition of assemblages as the remains of individual organisms, and consequently as natural taxonomic units. Therefore, different classifications came into existence. The nomenclature of element assemblages has been established in several more or less distinct ways.

Method 1.—Assemblages have been assigned the earliest applicable name of an included component (e.g., Eichenberg, 1930; Schmidt, 1934; Sinclair, 1953).

Method 2.—Assemblages have been given new binomina, and the component specimens have been designated by descriptive technical terms. Scott (1942) followed essentially this practice, identifying single-element genera (but not single-element species) represented in natural assemblages, and describing the components by common nouns coined from the single-element generic names. Thus, specimens of *Hindeodella* were termed hindeodells, and so on. Scott emphasized, however, the desirability that the earlier "form-classification" should be retained (1934), even though he found it inconvenient to employ it for assemblages.

Method 3.—Assemblages have been given new binomina and the component elements have been designated by their previously established single-element generic and specific names, if any (e.g., RHODES, 1952).

Method 4.—The use of "parataxa" to accommodate form classification categories (e.g., MOORE & SYLVESTER-BRADLEY, 1957) was revived in a modified version in the first edition of this *Treatise.* MOORE and SYLVESTER-BRADLEY (1957) argued that the existing framework of single-element names should be retained as a scheme of parataxa, supplementary to a parallel multielement classification and nomenclature. Both, they argued, should be subject to the requirements and protection of the International Code of Zoological Nomenclature. They suggested that parataxa should be available for fragments or ontogenetic stages of organisms (e.g., spicules, aptychii, scolecodonts). MOORE (1962) offered one possibility of retaining two complementary schemes of nomenclature by insisting that the positive recognition of a particular single-element species in a multielement genus was not warranted because morphologically indistinguishable forms were thought to be present in more than one multielement species. Both methods of nomenclature have found little favor with taxonomists (see RHODES, 1957; FAY, 1957).[1]

Method 5.—Apparatuses have been named after the earliest applicable name of a single element present in the assemblage. The other elements of the association are referred to by the use of symbols (e.g., MOORE & SYLVESTER-BRADLEY, 1957; SWEET, 1970; JEPPSSON, 1971; VON BITTER, 1972; BAESEMANN, 1973).

In the systematic section of this volume, the earliest applicable name of a single element that has been determined to belong to an assemblage furnishes the name for that biologic unit.

In summary, a study of assemblages permits the following general observations:

1. Most known assemblages contain numerous specimens with a variety of shapes.

2. A few assemblages are composed exclusively of one shape of element.

3. Elements with similar shapes may be present in more than one assemblage. This implies a functional similarity both within and between these assemblages. We refer to such elements as being homologous.

4. Where the same shape of element is not present in two distinct assemblages it is sometimes found to be represented by a similar element of the same structural type. This is another example of homology.

5. It is often difficult to determine with certainty the numbers of kinds of elements and the total number of each type present in element assemblages. The total number of specimens probably ranges from a minimum of 9 to a maximum of 32.

6. Data are inadequate with respect to the extent of element variation within and between assemblages. RHODES (1952) has given details of extensive variation in components of *Idiognathodus* and has suggested that such variation may represent the extent of infraspecific variation within a single assemblage. Further collecting and study are needed to assess the true taxonomic significance of this variation.

7. *Idiognathodus, Cavusgnathus, Gondolella* and *Prooneotodus* have definite but different alignments and arrangements of elements within assemblages.

8. Overall size of assemblages is small. The largest are about 9 mm in length and 2 to 3 mm in width.

9. No assemblage yet discovered contains elements attached to any basal filling substance. In view of the undisturbed condition of the assemblages, it seems unlikely that such material was present in the enclosing body.

10. Many conodont assemblages have been preserved in bituminous material, some in black shale, and a few in limestone.

[1] For additional discussion of the parataxa problem in taxonomy, see *Bull. Zool. Nomencl.*, v. 38, pt. 1, 1981, p. 37-48.

MAJOR CATEGORIES OF NATURAL ASSEMBLAGES

Rhodes (1962) referred similar but distinct Carboniferous assemblages to a group that he designated as Class A genera. Other genera with a different structural plan, he considered to be distinct. Jeppsson (1971, 1972) also recognized two types of apparatuses. Sweet and Bergström (1969) suggested division of known Ordovician multielement apparatuses into three major categories. Lindström (1973) recognized four main types of apparatus. Klapper and Philip (1971) also described four types of conodont apparatus, each of which has a fundamental structure that is reflected by a constant association of elements. The elements were given symbol names. A similar system for description of the ramiform and pectiniform elements was devised by Jeppsson (1971) and Sweet and Bergström (1969).

The classification and terminology of skeletal apparatuses outlined by Sweet in this volume is based on a rigid application of element types and numbers. It is unambiguous in application for isolated elements. With assemblages, however, there is often no certainty as to the exact morphology of some of the contained ramiform elements and this, together with the often incomplete nature of the material, renders counts of the numbers of kinds of component elements difficult.

Similarity in number of kinds of elements present in an assemblage may not be all-important since Avcin and Norby (1973) have drawn attention to certain assemblages of *Idiognathodus* that appear consistently to lack an alate element. In assemblages, it is possible to recognize but not yet possible to assess the significance of style of denticulation. This has led to a separate notational scheme for apparatuses of each major group (e.g., Klapper & Philip, 1971), which gives greater information concerning the morphology of elements, particularly the ramiform elements. The Klapper and Philip approach, described by Sweet (p. W18), can be utilized for assemblage descriptions.

At least five categories of natural assemblages have been recognized: (1) unimembrate assemblages with coniform elements, (2) multimembrate assemblages with coniform elements, (3) multimembrate assemblages with ramiform and pectiniform but no coniform elements, (4) multimembrate assemblages with pectiniform and coniform but no ramiform elements, and (5) multimembrate assemblages with ramiform elements. Discussions of known examples of these natural assemblages follow.

UNIMEMBRATE ASSEMBLAGE WITH CONIFORM ELEMENTS

Genus *Prooneotodus* Müller & Nogami, 1971

This assemblage (Fig. 53,5), described by Miller and Rushton (1973, p. 338-339) as *Proconodontus tenuis,* is bilaterally symmetrical and consists of oppositely curved sets of coniform elements arranged like parentheses. Each side of the assemblage consists of 4 to 6 (perhaps 8) elements in which the tips are close together and the bases are spread slightly apart. Tips of elements on one side of the assemblage are opposite those on the other side. One grouping shows this model; two others, less well preserved, are similar. Two additional groupings seem to represent one side of an assemblage. Other specimens consist of single elements. All are from the Upper Cambrian of England. Recently, Müller and Andres (1976) reported similar material from Sweden and Landing (1977) described fused clusters of this type from the Upper Cambrian of New York.

MULTIMEMBRATE ASSEMBLAGE WITH CONIFORM ELEMENTS

Genus *Belodella* Ethington, 1959

Lange (1968) illustrated an assemblage from the Upper Devonian of Germany con-

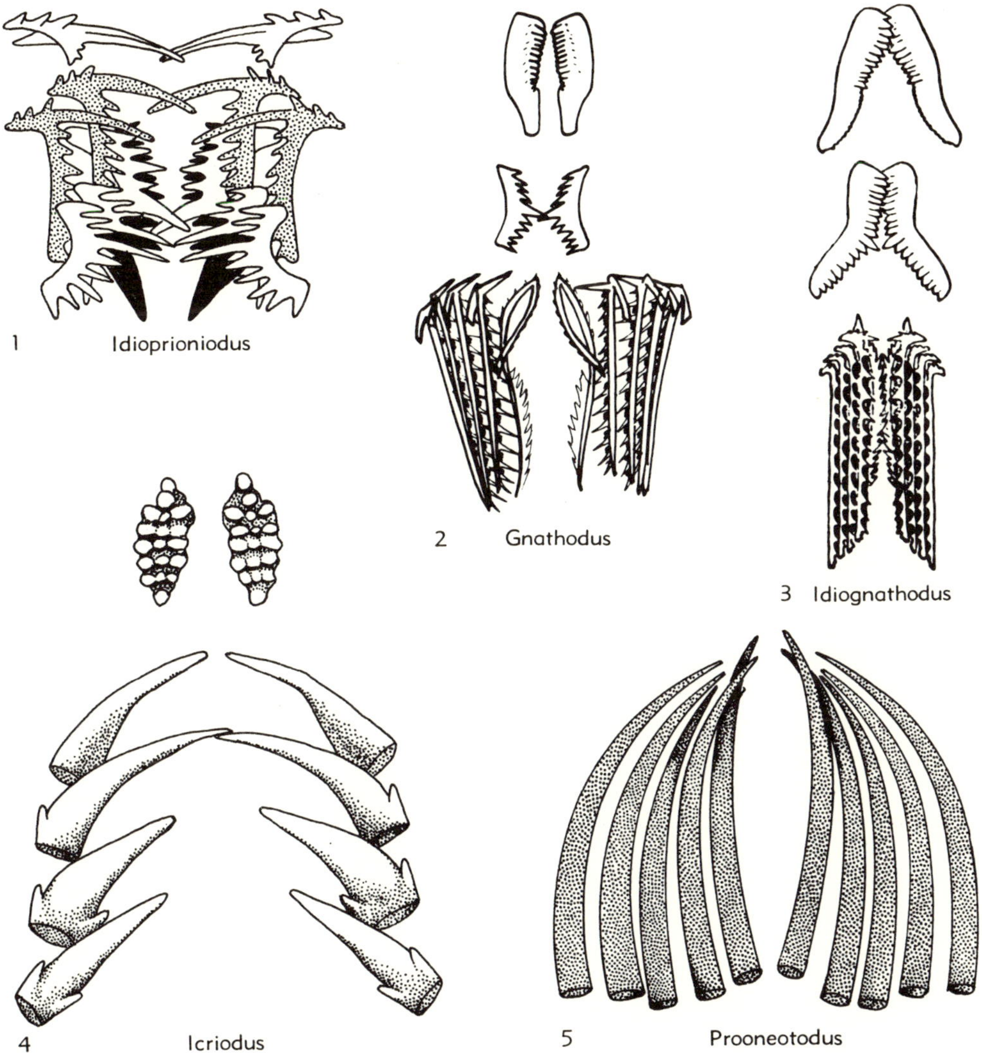

FIG. 53. Reconstructions of conodont assemblage-genera illustrating major types recognized in this chapter (Rhodes & Austin, n).——*1. Idioprioniodus* GUNNELL, a multimembrate assemblage with ramiform elements, based on reconstruction by RHODES (1952); some elements are shaded to distinguish them; approx. ×14.——*2. Gnathodus* PANDER, a multimembrate assemblage with ramiform and pectiniform but no coniform elements, based on the reconstruction by RHODES (1954); ×20.——*3. Idiognathodus* GUNNELL, a multimembrate assemblage with ramiform and pectiniform but no coniform elements; approx. ×14.——*4. Icriodus* BRANSON and MEHL *sensu* LANGE, a multimembrate assemblage with pectiniform and coniform but no ramiform elements; approx. ×18. Because we are not fully persuaded that this is a natural association, the two elements are shown together but are drawn at a slightly different scale with the coniform elements being relatively exaggerated and separated from the pectiniform elements.——*5. Prooneotodus,* a unimembrate assemblage with only coniform elements; based on reconstruction by MILLER and RUSHTON (1973); approx. ×20.

sisting of nine coniform elements. SERPAGLI (1967) was the first to give a detailed multielement reconstruction for *Belodella* (see also SWEET & BERGSTRÖM, 1969). An alternative reconstruction has been proposed by COOPER (1974a), who considered the

Belodella apparatus to contain two dominant element types, lenticular and triangular, with a possible transition series between them.

MULTIMEMBRATE ASSEMBLAGES WITH RAMIFORM AND PECTINIFORM BUT NO CONIFORM ELEMENTS

Genus *Cavusgnathus* Harris & Hollingsworth, 1933

Assemblages assigned to *Cavusgnathus* have been described as quadrimembrate types and contain pairs of the following distinct elements: *Pa* scaphate, *Pb* angulate, and two ramiform elements, one bipennate and the other dolabrate. Norby (pers. commun.) reports undescribed assemblages that are quinquimembrate with an alate element (see also von Bitter, 1972). Occurrence: Bear Gulch Formation (Namurian) of Montana.

Genus *Gnathodus* Pander, 1856

Assemblages with *Pa* and *Pb* pectiniform elements, a dolabrate element, and bipennate ramiform elements of the symmetry transition are referred to *Gnathodus* (Fig. 53,*2*). Norby (1974) commented on assemblages that contain *G. bilineatus.* Barskov, Alekseev, and Goreva (1977) and Lane and Ziegler (1979) have provided additional observations on the type of *Gnathodus.* Occurrence: Namurian of Germany.

Genus *Gondolella* Stauffer & Plummer, 1932

Trimembrate assemblages with four pairs of one type of ramiform element, two pairs of another type, and a pair of pectiniform elements are included in this genus. Von Bitter (1976a) has provided additional details. Occurrence: Pennsylvanian of North America.

Genus *Idiognathodus* Gunnell, 1931

Assemblages referred to *Idiognathodus* (Fig. 53,*3*) contain paired discrete elements and may be either seximembrate, septimembrate or octimembrate depending on the interpretation of the symmetry-transition series. Morphology of the scaphate *Pa* element is variable. The *Pb* element is angulate and form of the dolabrate element varies. A symmetry-transition series of bipennate elements is also present. The apparatus of *Idiognathodus* reconstructed by Baesemann (1973) is either seximembrate or septimembrate depending on the interpretation of the symmetry-transition series. Unpublished information (Norby, Avcin, & Austin) concerning ramiform elements associated with *Pa* elements of *Idiognathodus* in Ireland and Illinois suggests that the symmetry-transition series also contains a ramiform element that is characterized by a long posterior process and two lateral processes developed in the same plane, but at right angles to the posterior process. Baesemann (1973) considered *Idiognathodus* and *Streptognathodus* to be synonyms. Von Bitter (1972) reconstructed separate quinquimembrate apparatuses for *Streptognathodus* and *Idiognathodus.* As a convenience for description of assemblage material, we follow Baesemann (1973); however, we recognize that *Idiognathodus* and *Streptognathodus* might be distinct because of their different *Pa* pectiniform elements and possible restriction to different environments (see Merrill, 1975a; Heckel & Baesemann, 1975). Assemblages of *Idiognathodus* are known from the Pennsylvanian of North America and from the Visean to Namurian of Europe.

Genus *Ozarkodina* Branson & Mehl, 1933

Abduasimova and Koren found an *Ozarkodina* assemblage on a limestone surface in Central Asia that has been reported by Mashkova (1972). The assemblage contains two *Pa* and two *Pb* elements together with two dolabrate elements and one each of two distinct bipennate elements. Lange (1968) has described an assemblage that probably represents *Ozarkodina.* Occurrence: Devonian of Germany and Central Asia.

Genus *Palmatolepis* Ulrich & Bassler, 1926

LANGE (1968) recognized five kinds of elements in Upper Devonian assemblages from Germany. Pectiniform and four other elements are present. PUCHKOV (pers. commun.) reported similar assemblages in the Devonian of the Ural Mountains. The statistical reconstruction of the *Palmatolepis* apparatus by BOOGAARD & KUHRY (1979) is closely similar to the assemblages of LANGE (1968). Occurrence: Devonian of Germany and Russia.

Genus *Polygnathus* HINDE, 1879

Pectiniform and ramiform elements of several types in the original material of HINDE have indicated to some students that the assemblage may be coprolitic (e.g., SCHMIDT & MÜLLER, 1964). HUDDLE (1972) suggested that the specimens may represent food balls rather than coprolites. A portion of the collection of conodonts described by EICHENBERG (1930) is here assigned to *Polygnathus*. Occurrence: Devonian, New York; Lower Carboniferous, Germany.

Genus *Scaliognathus* BRANSON & MEHL, 1941

Some of the elements from the Culm of the Harz Mountains, Germany, illustrated by EICHENBERG (1930) are assigned to *Scaliognathus*. Occurrence: Lower Carboniferous, Germany.

This assemblage category also includes assemblages described as *Lochriea* by SCOTT (1942).

MULTIMEMBRATE ASSEMBLAGE WITH PECTINIFORM AND CONIFORM BUT NO RAMIFORM ELEMENTS

Genus *Icriodus* BRANSON & MEHL, 1938

Bimembrate assemblages from the Upper Devonian of Germany, which correspond to the Type 4 apparatus of KLAPPER and PHILIP (1971), are included in *Icriodus* (Fig. 53,*4*). It is possible that this association is a combination of two unimembrate multielement associations. Apparatus Type 5 of LANGE (1968), containing one pair of pectiniform elements and about 30 coniform elements, is also assigned to *Icriodus*. The *Icriodus* apparatus has been reconstructed by KLAPPER and PHILIP (1971).

MULTIMEMBRATE ASSEMBLAGES WITH RAMIFORM ELEMENTS

Numerous assemblages have been described that we assign to this category. LANGE (1968) described an assemblage characterized by the presence of two distinct elements, one bipennate and the other alate. SCHMIDT and MÜLLER (1964) described an assemblage containing five distinct ramiform elements. These assemblages are similar in plan to those described by RHODES (1952). The latter assemblages are characterized by the presence of four distinct ramiform elements. Assemblages described by BURNLEY (1938) and undescribed assemblages reported by NORBY (pers. commun.) belong to two additional distinct genera (cf. *Metalonchodina* and *Hibbardella*) with the same overall character as those previously mentioned. It is often difficult to identify the ramiform elements present in all of the above-mentioned assemblages and uncertainty remains concerning the correct identification of individual elements. It is therefore difficult to assign this group to a genus or genera within the multielement concept. The Type 3 apparatus reconstructed by KLAPPER and PHILIP (1971), the quadrimembrate *Neoprioniodus* apparatus reconstructed by VON BITTER (1972), and the seximembrate apparatus of *Idioprioniodus lexingtonensis* reconstructed by BAESEMANN (1973) are similar to known Carboniferous assemblages.

Genus *Idioprioniodus* GUNNELL, 1933

HASS (1962) regarded *Idioprioniodus* GUNNELL (Fig. 53,*1*) to be a junior subjective synonym of *Ligonodina* BASSLER, 1925; however, we follow current practice and apply the name *Idioprioniodus* to assemblages of Carboniferous age that are composed exclusively of ramiform elements.

Assemblages described by Burnley (1938), Jones (1941), Rhodes (1952), and Schmidt and Müller (1964) are probably all incomplete. Norby (pers. commun.) has described assemblages including examples of a septimembrate apparatus, which he refers to *Idioprioniodus*. Merrill and Merrill (1974) have reconstructed a septimembrate apparatus for *Idioprioniodus* on statistical reasoning. The reconstruction of the apparatus by Baesemann (1973) suggests that *Idioprioniodus* is seximembrate.

This assemblage category also includes an assemblage of ramiform elements from the Devonian of Germany described by Lange (1968), which we tentatively assign to the genus *Ligonodina* Bassler, 1925.

ZOOLOGICAL AFFINITIES OF CONODONTS

By Klaus J. Müller

[Friedrich Wilhelms Universität, Bonn]

The origin of conodonts is considered by many paleontologists to be one of the most fundamental unanswered questions in systematic paleontology. Since their first description in 1856, conodonts have been discussed in numerous publications. Although detailed evidence on the nature of the animal has accumulated, opinions are still extremely divided. Taxonomic assignment of conodonts has ranged from algae to high vertebrates (Fig. 54) and postulated affinities are summarized in Table 5.

Table 5. *Summary of the Postulated Affinities of Conodonts, 1856-1975.*

Source	Affinity
As plants	
1964 Fahlbusch, p. 189	algae
1969 Nease, p. 10	vascular plants, Conodontophyta
As Coelenterata (Procoelomata)	
1973 Bischoff, p. 158	Conularia
As Aschelminthes	
1944 Denham, p. 216	copulatory structures of Nematoda
1973 Missarzhevsky, p. 54	Priapulida
1974 Hofker, p. 29	dental apparatus of Aschelminthes (Rotatoria, Gastrotricha, Kinorhyncha)
As Gnathostomulida	
1969 Durden, p. 855	class of Gnathostomulida
1969 Ochietti & Cailleux, p. 2664	cf. Gnathostomulida
1969 Rodgers, p. 855–856	group of Gnathostomulida
As Mollusca	
1861 Owen, p. 118	"naked mollusks"?
1867 Owen in Murchison, p. 356, 545	Nudibranchia?
1875 Morse in Newberry, p. 42	progenators of Nudibranchia (cf. *Doris, Aeolis*)
1875 Stimpson in Newberry, p. 42	lingual teeth of Mollusca
1844 James, p. 146	mollusks
1898 Woodward, p. 2	Nudibranchia or Cephalopoda?
1936 Loomis, p. 663	Gastropoda
1937 Pilsbry, p. 101	molluscan teeth
As Annelida	
1861 Owen, p. 118	Annelida?
1867 Owen in Murchison, p. 356, 545	Annelida?
1878 Ulrich, p. 87, also cited in Ulrich & Bassler, 1926, p. 1	Annelida
1886 Zittel and Rohon, p. 135	Annelida
1898 Woodward, p. 2	Annelida?
1934 Scott, p. 455	annelid jaws
1943 Dubois, p. 158	Annelida
1952 Rhodes, p. 890	tend to support association with Annelida

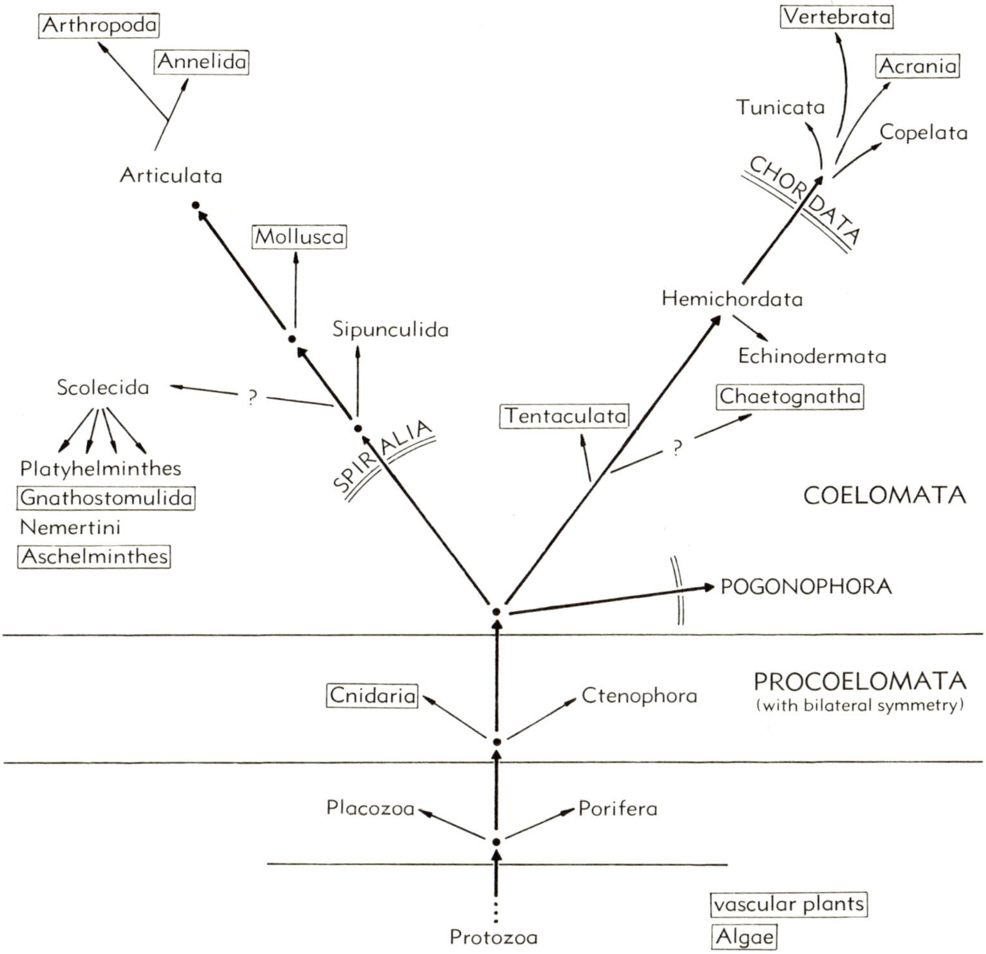

FIG. 54. Diagram of possible phylogenetic connections of larger taxonomic units in the animal kingdom (after Siewing, 1976). Frames indicate units to which conodonts have been assigned or with which a relationship has been postulated. They demonstrate the wide disparities in assignment of conodonts among various authors.

TABLE 5. *(Continued.)*

Source	Affinity
As ARTHROPODA	
1861 HARLEY, p. 549	cf. *Ceratiocaris* (Malacostraca, Phyllocarida)
1867 BARRANDE and others in MURCHISON, p. 365	Crustacea and other lower animals, including trilobites
As TENTACULATA	
1973 LINDSTRÖM, p. 100	related to Brachiopoda
As CHAETOGNATHA	
1973 RIETSCHEL, p. 417	related to Chaetognatha
As CHORDATA	
1856 PANDER, p. 9	fishes
1875 NEWBERRY, p. 43	teeth of Cyclostomi
1875 AGASSIZ in NEWBERRY, p. 42	teeth of Selachii
1879 HINDE, p. 356	low type of fishes?
1882 ROLLE, p. 408	related to *Amphioxus* [=Branchiastoma, Actania]
1885 CLARKE, p. 40	myxinoid fish

TABLE 5. (*Continued.*)

Source	Affinity
1921 BRYANT, p. 12	primitive type of fish
1923 MACFARLANE, p. 36–39	Cyclostomata
1926 ULRICH and BASSLER, p. 5	Pisces
1928 HOLMES, p. 2	Pisces
1929 KIRK, p. 495	Ostracodermi
1932 STAUFFER and PLUMMER, p. 22	primitive fishes
1934 SCHMIDT, p. 81	Placodermi
1936 BRANSON and MEHL, p. 233	fish affinities
1939 DEMANET, p. 215	*Coelacanthus lepturus* AGASSIZ (Crossopterygii)
1949 BECKMANN, p. 162	fishes
1950 SCHMIDT, p. 15	Aphetohyoidea (Placodermi)
1954 GROSS, p. 83	Agnatha
1954 RHODES, p. 450	wormlike creatures or primitive vertebrates
1957 RHODES and WINGARD, p. 454 (for Neurodontiformes)	primitive vertebrates related to *Archeognathus*
1964 SCHMIDT in SCHMIDT and MÜLLER, p. 128	fishes
1968 HALSTEAD, p. 11	planktonic protovertebrates
1969 SCOTT, p. 423	Agnatha
1973 MELTON and SCOTT, p. 52	subphylum Conodontochordata

SYSTEMATIC CRITERIA AND PRESUMED SIGNIFICANCE

The following criteria either have been utilized for a systematic assignment by various authors or may be useful for this purpose:

Mineralogical composition.—Carbonate apatite is present in hard parts of such systematic groups as brachiopods, conularids, hyolithelminthids, annelids, arthropods, and vertebrates. It is particularly widespread in early Paleozoic fossils. Several of the groups that contain apatite in their hard parts are not closely related, and composition of hard parts is of little taxonomic value. Nevertheless, because of the molluscs' presumed inability to form phosphatic hard parts, the possibility of a close systematic relationship between the molluscs and conodonts has been rejected (YOCHELSON, 1975).

A chemical evolution for conodont elements has been postulated, from predominantly carbonaceous material toward phosphatic matter (CLARK & MILLER, 1969), which would support the idea of a closer relationship with entirely carbonaceous fossils. As all later conodont elements are uniformly composed of phosphatic matter, the connection, if any, must have existed early in the history of the conodonts.

Gross outer morphology of individual elements.—Various theories regarding affinity have been based entirely on similarities between the morphology of conodont elements and organs of various animal groups; however, other criteria have been neglected. Outer shape obviously developed independently in various groups, and for recognition of relationships this aspect has been overemphasized. For example, the grasping apparatus of some spiders contain elements strikingly similar to those of the conodont *Belodina* (Fig. 55).

Morphology of apparatus.—Several theories have been based on the composition and shape of assemblages. SCHMIDT (1934, 1950) attempted to homologize the various elements of an Upper Carboniferous apparatus with the hyal teeth, mandibles, and branchial basket with gill-rakers of the Placodermi; however, as shown by subsequent studies, this apparatus is specialized and cannot be considered typical for the conodonts as a group.

Regardless of the fact that there is no general agreement concerning the function of the apparatus, it is likely that there is a relationship between the morphology and the function of the apparatus. Nevertheless, it is important to remember that the recognition of a function for conodont elements (e.g., as a grasping apparatus) may

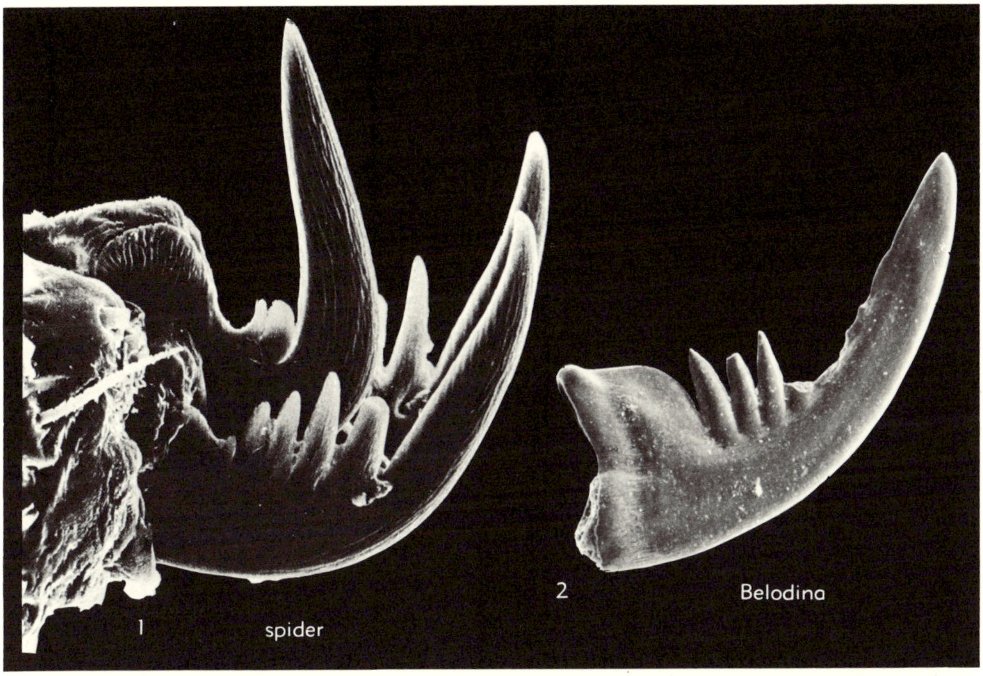

FIG. 55. Detail of a grasping apparatus of a recent spider with elements *(1)* convergent on that of the Upper Ordovician conodont *Belodina (2)*; *1,* ×134; *2,* ×67 (*1,* courtesy of Ernst Kullmann, Köln; *2,* Müller, n).

not lead to the revelation of their zoologic affinities, as the structure may show only functional convergence. An interesting functional interpretation of the conodont apparatus as a food-gathering mechanism has been summarized by LINDSTRÖM (1974).

Histology.—The internal structure of primitive conodont elements is rather complex. Its characteristic features could be important in systematic comparisons, but the internal structure evolution is now only partly understood.

Nevertheless, histology can be useful in negative determinations. For example, fundamental differences in histology indicate that postulated systematic affinity with the scolecodonts is unlikely (Fig. 56).

Soft parts.—Various workers have claimed to have discovered preserved soft parts of the conodont, but none of these can be accepted as reliable. FAHLBUSCH's (1964) theory regarding conodonts as "Zellenverbände, die sporangienartige Funktionen hatten" has been rejected totally. MELTON and SCOTT's (1973) animal may well have been a conodont predator rather than the soft parts of the conodont. The interpretations of function and physiology of the Montana specimens is yet to be confirmed (see RHODES & AUSTIN, this volume). It is hoped that the animal may eventually be found somewhere embedded entirely, possibly in sediments connected with submarine volcanism or in a small lagoon that evaporated rapidly. The Middle Cambrian Burgess Shale of British Columbia may be an ideal lithotype, and recently CONWAY MORRIS (1976) has described a single specimen that he interpreted to be a lophophorate animal with conodont element–appearing structures as internal supports for possible feeding tentacles. The specimen, *Odontogriphus omalus* CONWAY MORRIS, is poorly preserved and only impressions and molds of the possible coniform elements remain. Compared with lophophores of living bra-

FIG. 56. Internal structure of a scolecodont, differing fundamentally from that of conodont elements because of its channel system and lack of layered deposition; *Paulinites* sp., Ord. or Sil. (glacial boulder), Pol.; SEM photograph, ×3,720 (courtesy of Friedrich Strauch, Köln).

chiopods, the elements seem too small in relation to the size of body to permit an interpretation as supports for tentacles, a theory proposed by LINDSTRÖM (1974). A grasping organ is a possible alternative suggestion.

Amino-acid spectrum and protein structure.—This line of investigation could produce important clues; the difficiencies of preservation due to the geological history and age are serious problems to overcome. Examinations of excellently preserved elements, using as a control associated phosphatic fossils of known relationships, such as selachian teeth or horny brachiopods, may give additional results.

Size range.—Considerable deviation in size range between conodont elements and elements of similar shape from other organisms may be difficult to explain if a similar function is presumed. This point weakens HOFKER's (1974) theory, for example. In spite of their small size, possible relationships to Gnathostomulida, as discussed by OCHIETTI & CAILLEAU (1969) should be clarified by detailed investigations of ultrastructures of this group.

Stratigraphic distribution.—The stratigraphic range of conodonts from uppermost Precambrian or lowermost Cambrian to Upper Triassic yields little evidence for a taxonomic assignment. General occurrence with fish remains has been regarded by several authors (e.g., PANDER, 1856; KIRK, 1929; BRANSON & MEHL, 1936) as reason for placing conodonts in the Vertebrata; however, this is a matter of sedimentary sorting and deposition rather than of natural relationship.

Examples given demonstrate the common iterative origin of characters in various unrelated branches of evolution. Comparisons of the various groups with conodonts, therefore, have to be based on a combination of all available characteristics. It seems that at the present time no living or extinct group of organisms can be matched with the conodonts, which, therefore, have to be considered as a separate group of high taxonomic rank.

BIOLOGICAL CONSIDERATIONS AND EXTINCTION

By David L. Clark

[University of Wisconsin]

ELEMENT VARIATION

Conodont elements grew and enlarged by addition of layers of carbonate apatite. Complete ontogenies have been described in a few species and, in general, changes during ontogeny are not so drastic as to be confusing in identification. Certain Triassic Xaniognathidae are an exception, however, and the ontogenies of *Epigondolella* and *Paragondolella* illustrate change from more or less ramiform elements to definite pectiniform ones at maturity.

Variation is common among elements and may be manifest by range in number of denticles, length of platform, and kind and arrangement of ornamentation. The pectiniform *Pa* element of *Ancyrodella rotundiloba* (Bryant) from the lower Upper Devonian shows such variation. As illustrated in Figure 57, its earliest stages of ontogeny are rather uniform and the platform, dominant on most stages, is the last major structure to develop (Fig. 57,*1-5*). At a stage *(7)* considered to represent some important threshold, if not sexual maturity, three distinct varieties of elements are produced. In one *(8-13)*, the major nodes of the platform are aligned with the carina as in a cross. In a second *(14-19)*, a normal random pattern of nodes forms on the same shaped platform but irregular distribution is the rule. The third pattern *(20-23)* is a distribution of nodes that approaches bilateral symmetry during growth. The extremes of the three groups might be considered distinctive elements, except for the fact that they always occur together in the same samples and therefore probably represent a range of ornamentation within a single element.

Specific variation in multielement conodonts involves differences in number and morphology of elements, factors not well understood at present. One promising area of investigation for future work will be documentation of ontogenetic change in assemblages and quantification of specific variation among congeneric elements.

HOMEOMORPHY

Homeomorphy in conodonts has been recognized since at least 1940, when Branson and Mehl noticed there is no major morphologic difference between the *P* elements of *Taphrognathus* and *Streptognathodus* but there is possible chronologic difference. This case of homeomorphy later was documented firmly by Rexroad (1958a). Through the years, instances of homeomorphy in conodonts have become well known. It is probably important to differentiate between **structural homeomorphy** in unrelated taxa and **evolutionary homeomorphy** in which similar structures evolved at different times in phylogenetic sequences. This has been discussed by several authors (e.g., Jeppsson, 1971; Clark, 1972a). A third variety of conodont homeomorphy is **positional homeomorphy,** which involves similar elements evolving to fill similar positions in a conodont's multielement apparatus.

Specific cases of structural homeomorphy are known throughout the Paleozoic and Triassic. Following the Early Permian evolutionary crisis (Clark, 1972a), homeomorphy is easily interpreted for the *Pa* elements of such genera as *Neogondolella* and *Gondolella, Neostreptognathodus* and *Streptognathodus, Neospathodus* and *Ozarkodina, Xaniognathus* and *Ozarkodina s. l.,* and *Paragondolella* and *Neogondolella.* Carboniferous homeomorphy includes at least some of the taxa referred to by Rhodes, Austin, and Druce (1969) and Austin (1973a), but many of Austin's ex-

FIG. 57. Ontogeny and variation illustrated by pectiniform *Pa* elements of *Ancyrodella rotundiloba* (BRYANT), U. Dev. (Squaw Bay Ls.), USA (Mich.), all from same layer, all ×26 (after Müller & Clark, 1967).——*1-7.* Sequence from earliest stage having only slight projection of platform to stage with well-developed platform.——*8-23.* Adult developmental stages in three variations: cross pattern *(8-13)*, normal pattern corresponding to descriptions of most students *(14-19)*, and symmetrical pattern *(20-23)*. Specimen *24* appears intermediate between symmetrical and normal groups.

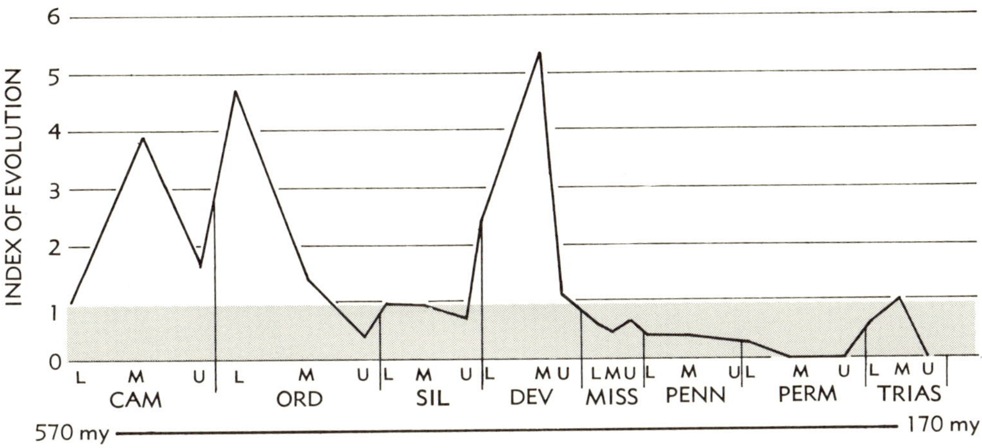

Fig. 58. Index of evolution among conodont elements. Index is based on ratio of new form-genera appearing to form-genera becoming extinct. Value greater than 1 indicates diversification and expansion in conodont element evolution; value of 1 indicates that as many conodonts were evolving as becoming extinct; value of 0.1 to 0.9 (shaded area) is at extinction threshold (more form-genera were becoming extinct than were evolving); value of 0 indicates crisis or extinction (after Clark, 1972a).

amples appear to be evolutionary homeomorphs rather than structural homeomorphs of unrelated taxa.

Similar species of *Icriodus* in Devonian rocks may have had distinctive ancestors, and similar early Paleozoic coniform elements (e.g., *Drepanodus*) may have evolved from different ancestors (Clark, 1972a).

Positional homeomorphy may be used in describing multielement conodont apparatuses in which slightly different morphologic types occupy more or less identical positions in a complete apparatus. Thus, Lane (1968) described a lineage of Mississippian and Pennsylvanian conodonts in which each stage in the evolution is distinguished by a distinct symmetry type. Similarly, Jeppsson (1969; 1971) has described several different apparatus types from the early and middle Paleozoic consisting of homologous elements that may have had similar functions. This positional homeomorphy has been alluded to by a number of imaginative students in defining multielement species (Bergström & Sweet, 1966; Sweet & Bergström, 1969). This concept is valuable especially in differentiating homeomorphs such as those of the "hibbardellid" form-genera *(Hibbardella, Roundya, Diplododella, Ellisonia, Trichonodella)* that had similar positions and functions in both unrelated and related multielement species and ranged through most of the Paleozoic and Triassic.

CRISES AND EXTINCTION

Crises in conodont evolution have been documented (Clark, 1972a). Diversity of elements has been used as a measure of evolutionary rates for conodonts (Fig. 58). This evolutionary index shows that a very limited diversity of conodonts was present during the Late Ordovician to Silurian and post-Devonian intervals. In fact, if this is

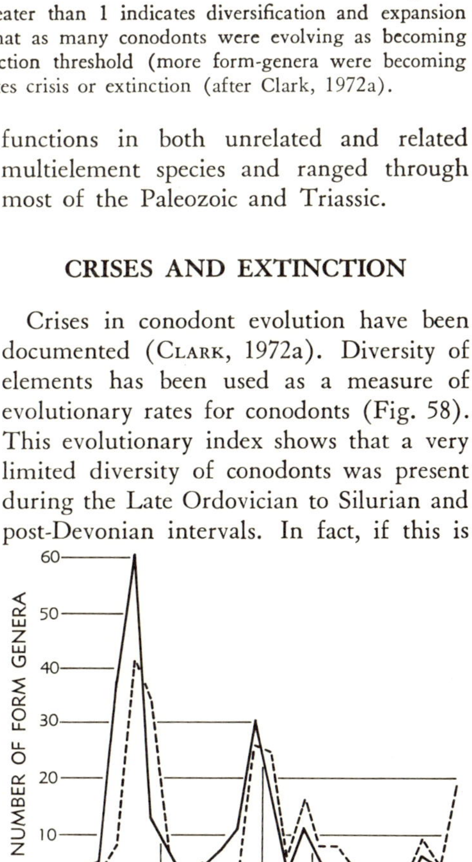

Fig. 59. Total new form-genera (solid line) appearing compared with total number (dashed line) becoming extinct (after Clark, 1972a).

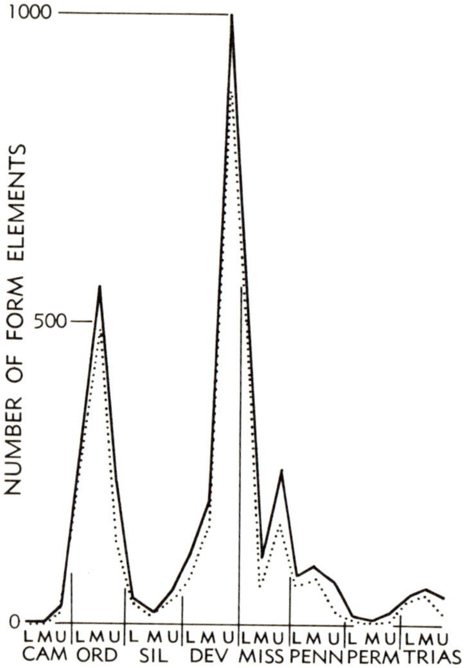

FIG. 60. Total form-species or elements (solid line) and number (dotted line) of new elements appearing (after Clark, 1972a).

an accurate estimation of evolution, conodonts were at the threshold of extinction continually after the Late Devonian. That is, more kinds were becoming extinct than were evolving. Except for the Late Triassic extinction, the most profound crisis experienced by conodonts may be that which occurred during the Early Permian, when most of the Paleozoic taxa were extinct and only a few forms survived to support a very minor Triassic resurgence. This is expressed in elemental diversification on the form-taxa level (Figs. 59 and 60). The relationship of evolution of new morphotypes to extinction of older forms is significant.

Late Triassic taxa show a marked decrease in number from a high in the Early Triassic. Approximately 30 form-species (Fig. 60), now organized into 7 or 8 natural or biologic genera, comprise the Middle and Late Triassic faunas. As these genera became extinct, they were not replaced and only 2 or 3 remained in the latest Triassic. There is data to suggest a slight decrease in the similarity index of European and North American species during the decline of conodonts (Fig. 61). Whether this is related to the Triassic opening of the North Atlantic Ocean or to the subtleties of extinction is unknown (CLARK, 1977). Rhaetian species are not known in North America but several species survived in the Tethyan region. These Tethyan conodonts were extinct prior to the Jurassic. Thus, the youngest conodont fauna is one from the Rhaetian of Austria that is characterized by some of the same species as those in the upper Norian *Epigondolella bidentata* Zone; i.e., *Cypridodella conflexa, "Hindeodella" suevica, Cypridodella delicatula,* and *Neospathodus lanceolatus.*

Post-Triassic reports of conodonts are of reworked or misidentified material and stratigraphy, or both (DIEBEL, 1956; NOHDA & SETOGUCHI, 1967). This has been reviewed by MOSHER (1967) and MÜLLER and MOSHER (1971).

The reasons for extinction of conodonts are enigmatic. Because the rise and fall of conodonts during the Paleozoic and Triassic follows a known pattern, the suggestion that extinction is an expected result of evolution and not of deficiencies shared by all species of a group has merit. Unknown factors of nutrient availability and preda-

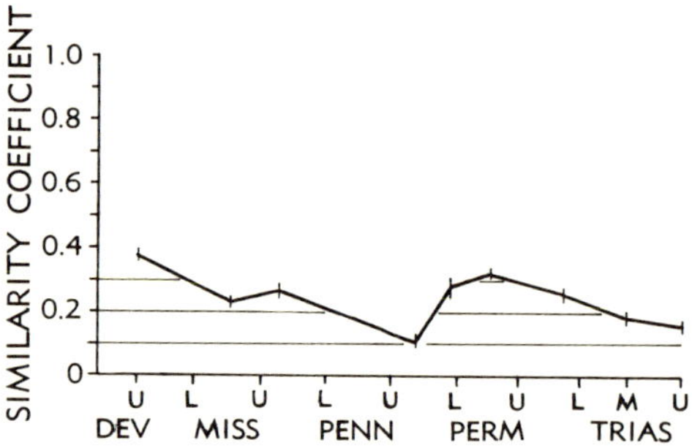

FIG. 61. Element diversity during the upper Paleozoic and Triassic shown by plots of similarity coefficients $\left(\frac{2w}{a+b}\right)$ for North America and Europe [w = number of conodont form-species common to both continents, a+b = total number of form-species on both continents] (after Clark, 1977).

tion may well have been involved in extinction of individual species. However, the idea that conodont extinction resulted from the chance accumulation of completely unrelated species-level extinctions must also be considered.

PALEOECOLOGY

By David L. Clark

[University of Wisconsin]

Determining the ecologic factors that influenced a group of organisms that has been extinct for 180 million years and whose biologic affinities are uncertain is a problem that still challenges students of conodonts. The fact that conodonts were widespread for 400 million years and are superb tools of biostratigraphy during this Cambrian to Triassic interval has compounded the problem. Because identical sequences of conodonts have been identified in different sediment types and in widely separated areas, more attention has been given to their stratigraphy than to their sedimentary relationships. Now that a basic biostratigraphy has been established, paleoecologic data are accumulating. These data are neither so voluminous nor so detailed as that available for many other groups of organisms. Less than a dozen serious studies on conodont paleoecology have been published since conodonts first were reported in 1856, and important work has been done only during the past few years. Ideas about conodont paleoecology are changed from the time when conodonts were considered to be organisms ubiquitous in the marine environment. Slightly more sophisticated modern hypotheses are reviewed in the following paragraphs.

MODE OF LIFE

The widespread occurrence of conodonts in a variety of coeval marine rocks has led most students to the conclusion that conodonts were planktic or nektic animals. Except for this widespread occurrence, little direct evidence supports a pelagic life style; however, study of depositional environments has strengthened this view in recent years (Seddon & Sweet, 1971; Chamberlain & Clark, 1973; Druce, 1973; Clark, 1974; Heckel & Baesemann, 1975; Clark & Rosser, 1976). The observation that most conodonts were bilaterally symmetrical, a feature advantageous for active pelagic organisms, is important; but bilateral symmetry is also common in benthic organisms.

Evidence for other than a pelagic mode of existence is known for at least certain conodont faunas. For example, some from the Ordovician are apparently restricted to distinct facies (Titus, 1974), and Barnes and Fåhraeus (1975) have suggested that at least some of these conodonts were nektobenthic rather than pelagic. Fåhraeus (1975) further concluded that the relationship between continental shelf extent and conodont diversity, as suggested during times of major crustal adjustment, is indicative of a nektobenthic mode of life. Because this kind of argument could be made for any organism living in the water column affected by shelf reduction, pelagic or benthic, it does not significantly alter previous ideas. Moreover, the fact that conodonts were relatively unaffected during the time of greatest shelf reduction (Permian to Triassic) when benthic groups were affected so strongly, might be interpreted to support a pelagic mode of life, at least during that time.

Conodonts probably ranged from pelagic to benthic environments. Most probably lived from just off the sea floor to much higher in the water column. Present evidence suggests that all were active, free-moving organisms.

DEPTH AND WATER ENERGY

Conodonts generally were fairly shallow-water, nearshore dwellers. This fact is demonstrated by their abundance in strata that can be safely interpreted as having been deposited under relatively shallow water, as well as by their virtual absence from the few deposits of deep-water Paleozoic and Triassic basins available for study. Specific depth and water-energy data are available for conodonts of different ages.

Ordovician.—Conodonts are generally absent from such extremely shallow-water, high-energy, algae-rich Lower Ordovician rocks as the dolomitic portion of the Prairie du Chien Group (HARDIN, 1972). This unit probably was never more than a few feet under water and, at times, was definitely emergent. The shale partings of this same unit, representing a slightly deeper and quieter environment, yield some conodont elements (FURNISH, 1938). Other stromatolitic facies, perhaps deposited in slightly deeper water, have yielded good conodont faunas.

According to HARDIN (1972), *Oneotodus* and *Scolopodus* probably represent deposition in shallow-water Lower Ordovician sediment, and BARNES and FÅHRAEUS (1975) reported that most conodonts with hyaline neurodont elements in the Middle Ordovician were nektobenthic littoral. Vertical stratification is well defined by a *Plectodina-Belodina* group that thrived in shallower water than a contemporary *Phragmodus* group (SEDDON & SWEET, 1971; BARNES & FÅHRAEUS, 1975). There is some agreement that *Drepanodus, Panderodus,* and a few other Ordovician taxa were pelagic (HARDIN, 1972; BARNES & FÅHRAEUS, 1975).

Devonian.—Early Devonian *Ozarkodina* (=*Spathognathodus*) faunas have been interpreted as belonging to sublittoral, lagoon to reef, and crinoid-meadow habitats. BARNETT (1971) also determined that, in general, Early Devonian conodont elements are less numerous seaward in deeper water. This nearshore dependence is emphasized in his distribution diagrams, which show a transition from no conodonts in supralittoral and upper littoral environments, to great abundance in sublittoral environments, to rare or absent at greater depths. FERRIGNO (1971) recognized five microfacies from lagoonal to deeper sublittoral Devonian rocks. Distribution of elements suggested vertical segregation of a planate deeper water *Plectospathodus-Polygnathus* group and a shallower water *Ligonodina-Lonchodina* group.

Icriodus and *Polygnathus* of the Late Devonian are known to be shallow-water (even near-reef) types. *Palmatolepis* and accompanying *Ancyrodella* and *Ancyrognathus* represent slightly deeper water but contemporary faunas (MÜLLER & CLARK, 1967; SEDDON, 1970; SEDDON & SWEET, 1971). SZULCZEWSKI (1971, p. 78) reported that elements are rare to absent in Upper Devonian stromatoporoid-coral facies, a very shallow-water association.

Carboniferous.—Vertical stratification is suggested in Lower Carboniferous rocks by a deeper water *Siphonodella-Pseudopolygnathus* fauna (including *Dinodus, Doliognathus, Dollymae, Scaliognathus, Staurognathus*) and a contemporary shallower water *Spathognathodus-Polygnathus-Clydagnathus* group.

Upper Carboniferous faunas consisting of *Cavusgnathus* apparently were dominant in shallow marginal marine waters and *Streptognathodus* and *Idiognathodus* predominate in more offshore, deeper waters (VON BITTER, 1972; MERRILL, 1973; CHAMBERLAIN & CLARK, 1973). These same faunas are commonly associated with trace-fossil assemblages which add more details on such factors as depth and nutrient supply. That is, the *Adetognathus* group occurs with very shallow-water *Cruziana* trace fossils in the Pennsylvanian of Utah (CHAMBERLAIN & CLARK, 1973). Younger Pennsylvanian *Zoophycos* strata are characterized by *Idiognathodus* and conodont elements are absent in abyssal (~2,000 m) *Nereites* strata. Contemporaneity of the conodont faunas is demonstrated by the occurrence of *Idiogna-*

thodus elements in debris flows and other exotic material that moved downslope into the abyssal *Nereites* facies of the Oquirrh Basin (Chamberlain & Clark, 1973).

Heckel and Baesemann (1975) interpreted Upper Pennsylvanian megacyclothems in eastern Kansas to support definite depth zonation in a pelagic model suggested by Seddon and Sweet (1971). They related black shale, indicating the deepest water during a transgressive sequence, with the greatest diversity of conodont elements, whereas shallow-water parts of the cyclothem show lowest element diversity. From shallowest to deepest, their depth-stratified cyclothem included: *Adetognathus, Ozarkodina, Aethotaxis, Idiognathodus, Idioprioniodus,* and *Gondolella.*

Permian.—Lowest Permian strata, deposited in water 4 to 10 meters deep, with corals, brachiopods, algae, echinoderms, and bryozoans, are commonly rich in idiognathodids, gnathodids, and adetognathids. *Neogondolella* and *Sweetognathus* preferred deeper water (>50 m) that was quiet and had limited nutrients (Clark, 1974). Upper Permian strata commonly have an inverse relationship in the presence of deeper water *Neogondollella* and shallow-water *Neostreptognathodus* (Behnken, 1975). *Hindeodus* and *Ellisonia* evidently were surface dwellers and occur in both shallow and deeper water deposits.

Babcock (1976) reported that Upper Permian elements in west Texas had a maximum abundance at intermediate depths in the basin and were rare in the shelf-edge facies and in the deepest part of the basin. This may confirm the nearshore dependency suggested for Pennsylvanian-Permian elements by Chamberlain and Clark (1973). Behnken (1975) considered *Hindeodus* to be an uppermost photic-zone dweller during parts of the Late Permian in Wyoming, perhaps living in abnormal salinities.

Triassic.—The rather persistent occurrence of Triassic conodont elements with pelagic ammonoids suggests that many of the stratigraphically important taxa may have had similar depth tolerance (Mosher & Clark, 1965; Mosher, 1968; Sweet, 1970). Most Triassic ammonoids were moderately shallow-water dwellers. Other Triassic conodont elements have been found in very shallow-water carbonates that intertongue with red beds (Clark & Rosser, 1976), giving additional evidence of the great ecologic range of conodonts.

SALINITY

Conodonts were confined to the marine environment. Most were probably stenohaline. This observation is based on their association with such stenohaline organisms as cephalopods as well as their general absence in rock that could be interpreted as other than normal marine. For example, in intertonguing marine carbonates and nonmarine red-bed sequences of Triassic age in western North America, elements are present only in the marine units and are rare to absent in all except bioclastic (normal marine) rock. Euryhaline adaptations also were possible, and Barnes and Fåhraeus (1975) suggested that pelagic Ordovician conodonts of the Midcontinent province are of this type.

Abnormally high, perhaps hypersaline conditions have been suggested for Mississippian species (e.g., *Cavusgnathus*) in the evaporitic sequence of the Windsor Group in Canada (von Bitter, 1976b), and Behnken (1975) suggested higher salinity tolerances for species of *Anchignathodus* (=*Hindeodus*) in the Minnekahta Limestone (Permian) of Wyoming.

On the basis of equatorial position and proximity to evaporite deposition, Barnes and Fåhraeus (1975), suggested that Ordovician faunas of the Midcontinent province may have tolerated higher than normal salinity. They further suggested that Middle Ordovician neurodont conodonts of the Midcontinent province may have tolerated hypersaline conditions.

TEMPERATURE

Little has been written concerning temperature preference of conodonts other than as it is related to the general shallow-water nature of their distribution. Plots of Paleozoic and Triassic element abundance against paleomagnetically determined latitudes suggest that conodonts were most abundant in low latitudes. BEHNKEN (1975) proposed that Permian speciation was fastest close to the equator and that time lag in distribution of species away from the equator could be detected in western North America.

Added to this general stenothermal theme is the fact that conodonts are most poorly known in the southern continents that are thought to have been most distant from the Paleozoic equator (South America, Africa). Also, NICOLL (1975) concluded that the absence of conodonts in otherwise fossiliferous normal-marine Permian rocks in western Australia is due to low water temperature in a high latitude. If this conclusion is valid, the temperature tolerance of Permian conodonts was below the level of brachiopods, bryozoans, crinoids, ostracodes, and foraminifera, which are invertebrates known to have a eurythermal range.

OXYGEN

The occurrence of conodont elements in black shale is explained by pelagic habits rather than by low-oxygen tolerance. Deep, and at times anaerobic basins, such as the Permian Las Delicias basin of Coahulia, Mexico, and the Permian Oquirrh basin of Utah, apparently had no conodonts. This is attributed, in part, to distance from shore for parts of the basins studied. It also suggests that the occurrence of elements in black shale may be evidence of the shallow, protected nature of the basin in which the sediment accumulated and that black Paleozoic and Triassic shale with no elements may indicate extremely deep and far offshore basinal deposition.

The Permian basin of west Texas was anaerobic for at least the later part of the Permian. Conodont elements are present at least 25 kilometers into the basin from the Permian shore line. Their occurrence in anaerobic sediment probably is evidence of a pelagic existence.

NUTRIENTS

Conodont feeding patterns have not been established. A correlation between conodont elements and their abundance with trace-fossil assemblages in Upper Paleozoic rocks probably is interesting but somewhat ambiguous for defining conodont nutrient demands (e.g., CHAMBERLAIN & CLARK, 1973; CLARK, 1974). The *Nereites* trace-fossil assemblage is indicative of a very low nutrient supply on the deepsea floor. The absence of conodont elements in sediment of this environment supports the idea that conodonts did not venture too far offshore. Whether or not this was because of a less dependable nutrient supply is not known. Some elements are moderately abundant in intermediate *Zoophycos* strata that are indicative of a limited to moderate nutrient supply. Most conodonts preferred shallow, warm water, in which nutrients generally were abundant (*Cruziana* facies). To what extent the general abundance of nutrients in the shallow-water environment was a limiting factor for conodonts is unknown.

LINDSTRÖM (1973) concluded that conodonts used their hard elements as an aid in food gathering, and probably fed on microplankton strained from sea water.

ORGANIC ASSOCIATIONS

Earlier references to conodont paleoecology included observations on conodont associations (e.g., MÜLLER, 1956c, 1962b). SEDDON and SWEET (1971) analyzed these reports and concluded that because most conodonts "lived above, rather than in or on the bottom" with other marine organisms, their associations with other organisms are by no means invariable. Their pelagic mode of life probably explains the common reports of associations with cephalopod and fish remains, in particular. In vertically segregated marine water of shallow and greater depths, they should be associated with most groups of marine organisms, at least someplace. The record confirms association with organisms as diverse as algae and arthropods, and all of the intermediate phyla as well.

MORPHOLOGY AND PALEOECOLOGY

The possibility that certain mega- and micromorphological structures of conodont elements may reflect ecologic factors has not been thoroughly investigated.

Barnes, Sass, and Monroe (1970) concluded that Ordovician elements that are robust, have a shallow basal cavity, are laminated throughout, and have little or no white matter are characteristic of nearshore, hypersaline, carbonate environments. Müller and Nogami (1971) have related details of white matter, regeneration of denticles, and other features to ecologic and physiologic stress. Also, microborings of elements described by these authors, and interpreted as a form of predation, may be a key in ecologic work.

Clark (1974) suggested that it may be possible empirically to establish a relationship between element architecture and ecology. Thus, some ramiform elements may represent shallow-water genera whereas broad, flat, pectiniform types associated with other ramiform types may represent organisms that lived in deeper water. Moreover, such important genera as *Amorphognathus, Palmatolepis, Siphonodella, Neogondolella, Platyvillosus,* and *Epigondolella* may be deeper water forms, whereas *Idiognathodus, Gnathodus,* and *Adetognathus* may be shallower water forms. Austin (1975) suggested that in the Carboniferous of England, specimens with a large basal cavity, such as *Patrognathus,* are probably shallow-water taxa and that specimens with small basal cavities, such as *Siphonodella,* are deeper water types.

The relationship of morphology to paleoecology clearly deserves much more study.

APPROACHES TO CONODONT PALEOECOLOGY

The stratigraphic record provides a number of paleoecologic models that can be studied in an attempt to relate conodont distribution to previously interpreted facies. For example, the Permian basin of west Texas has a well-understood lateral sequence of backreef, reef, forereef, and basin facies. Water depth of about 600 meters for the basin is known with some confidence. Conodont distribution for one interval within this model has been determined (Babcock, 1976), and although the conclusions may be specifically valid for this example only, the paleoecologic framework may have wider application. Conodont elements are absent in the lagoonal back-reef sequence of the uppermost Permian unit, the Lamar Limestone. Association analysis and ordination techniques allow recognition of two biofacies: a reefward biofacies within 6 kilometers of the basin edge and a basinward biofacies 6 kilometers or more into the basin. The reefward or *Ellisonia* biofacies contains *Ellisonia triassica* and *Lonchodina festiva.* The basinward or *Neogondolella* biofacies contains *N. serrata* and *Ellisonia gradata.* Babcock concluded that distance from basin edge rather than absolute depth probably was the key environmental gradient. The analysis of this data indicated that, first, conodonts were rare to absent in the reef and back-reef areas. (This may have been the result of too high a temperature and salinity.) Second, conodonts are widespread in the Permian basin, at least 22 kilometers from the shore, which supports the idea of a pelagic life style. (The black laminated Lamar mudstones are considered anaerobic.) Third, conodont diversity decreases basinward.

A second model study involves the Thaynes Formation of Early Triassic age in the western United States (Clark & Rosser, 1976). This unit thins from the deeper part of the Triassic miogeosyncline and interfingers with nonmarine red beds in eastern Utah and Wyoming. The *Para-*

chirognathus-Furnishius fauna can be traced across the Triassic basin to the red-bed intertongues. A multiple discriminant (canonical) analysis of all factors of petrology, associated fauna and flora, sedimentation rate, paleogeography, and stratigraphy suggested that these faunas were depth-stratified. Unpublished information (T. CARR, pers. commun.) based on different quantitative studies has confirmed the presence of different biofacies for *Parachirognathus* and *Furnishius*. In Lower Triassic rocks of western North America, *Parachirognathus* is largely restricted to the shallowest inner shelf facies and *Furnishius* is most abundant in slightly deeper water outer shelf facies. These genera were contemporaneous with deeper (basinal facies) water *Neogondolella* species.

A wealth of data is becoming available concerning conodont biofacies. The most promising of these paleoecologic investigations are based on application of mathematical analysis to conodont community study. It is apparent already that different biostratigraphies will be developed to characterize contemporaneous conodont biofacies.

BIOSTRATIGRAPHY AND EVOLUTION

By WALTER C. SWEET and STIG M. BERGSTRÖM

[Ohio State University]

Conodonts were abundant in seas of the past and flourished from latest Precambrian or earliest Cambrian into the latest Triassic, when they apparently became extinct. The group evolved rapidly and evidently passed unscathed through most of the crises that beset the organic world in, for example, the Late Devonian (Frasnian) and latest Permian. Upon death, the conodonts contributed literally millions of tiny skeletal elements to sediment accumulating on the sea floor. Because these elements are exceptionally resistant, and because they survive harsh physical or chemical treatment of the rocks that enclose them, they are easily and inexpensively isolated from those rocks in large numbers. Further, in many facies, conodont elements are the only identifiable or stratigraphically diagnostic fossils to be found in any abundance. Not surprisingly, then, the value of these elements as guide fossils is unexcelled in Paleozoic and Triassic rocks and their biostratigraphic and paleoenvironmental significance have been the principal factors involved in the extensive research summarized in this volume.

Conodonts are rare in most Cambrian rocks, except those of the Upper Cambrian, and a formal scheme of biostratigraphic units has been proposed only for the uppermost part of the system (MILLER, 1975). The vertical distribution of distinctive conodont species in post-Cambrian and pre-Jurassic strata, on the other hand, enables recognition of a sequence of more than 100 biostratigraphic units, which have been variously described as subzones, zones, or faunal units. Distinctive features of most of these units are elucidated in a readily available volume, *Symposium on Conodont Biostratigraphy* (SWEET & BERGSTRÖM, eds., 1971); hence, only major aspects of the stratigraphic distribution of conodonts are summarized here, along with biostratigraphically significant information published since 1971 and up to 1975, when this contribution was completed.

CAMBRIAN

Early and Middle Cambrian conodont faunas, known from scattered localities in North America, Europe, Asia, and Australia, are characterized by species of the Paraconodontida, most or all of which apparently formed skeletal apparatuses of weakly phosphatized coniform elements. In the middle of the Late Cambrian, paraconodonts were joined by the earliest species of the Conodontophorida, which are distinguished by more stoutly phosphatized elements, thinner lamellae, and other differences in internal structure. Late Cambrian conodontophorid genera are also characterized by coniform skeletal elements, united in most forms into unimembrate skeletal apparatuses. A few, however, formed multimembrate apparatuses that, in this respect at least, herald the more complexly developed ones of Ordovician and later species.

Although much remains to be learned about the stratigraphic ranges and worldwide distribution of Cambrian conodonts, they clearly have great biostratigraphic potential. For example, MÜLLER (1973) was able to recognize, but did not name, seven assemblage zones in strata astride the Cambrian-Ordovician boundary in Iran. The lower three of these zones are probably uppermost Cambrian. MÜLLER was not able with certainty, however, to do more than suggest the equivalents of these zones in previously described sections in Queensland and western North America. More recently, MILLER (1975), in a report published thus far only in abstract form, has assigned rocks in western North America adjacent to the Cambrian-Ordovician boundary to two zones, the lower of which (*Proconodontus* Zone) is entirely within the upper Upper Cambrian *Saukia* trilobite zone. It seems likely that the *Proconodontus* Zone is the general equivalent of MÜLLER's (1973) Assemblage Zone 3, which is also characterized by *Proconodontus*.

ORDOVICIAN

In the few continuous sequences studied thus far, the earliest conodonts with multimembrate skeletal apparatuses composed of ramiform elements (*Cordylodus*) appear in the very highest Cambrian strata, and genera with an increasingly varied array of uni- or multimembrate apparatuses of coniform elements dominate Lower Ordovician faunas. Several of these genera apparently gave rise at various times in the Early and Middle Ordovician to forms with multimembrate apparatuses composed of arrays of ramiform and pectiniform elements. Representatives of the latter are stratigraphically useful in the Lower Ordovician, but dominate collections from Middle and Upper Ordovician rocks. The genera to which they are assigned clearly include stocks from which most of the stratigraphically useful species of Silurian and younger conodont faunas were derived.

Beginning in the Early Ordovician, and continuing until latest Ordovician time, development of conodont faunas was conspicuously different in North America and Europe, such that two major faunal provinces have been recognized (SWEET & BERGSTRÖM, 1974). The North American Midcontinent province includes all of interior North America, parts of the eastern Great Basin, western belts of the Appalachian Mountains, at least part of the Canadian Arctic Archipelago, and some or all of the Siberian platform. In at least some parts of the Ordovician, this province may also have included parts of New Zealand, Australia, and South Korea, as well. The North Atlantic province, on the other hand, includes practically all of what is now northwestern Europe, parts of South America, and, during at least a large part of the Ordovician, a segment of the eastern Appalachians that stretches south from Newfoundland to Georgia and Alabama. Because of provincial differences in the nature and distribution of Ordovician conodont faunas, separate zonal schemes have been developed for Ordovician rocks in each province.

In the Lower Ordovician of the Balto-Scandic district of the North Atlantic province, SERGEEVA (1966), LINDSTRÖM (1971), and VIIRA (1975) have discriminated a succession of nine or ten biostratigraphic units based on conodonts. Many of these are widely traceable and the biostratigraphic scheme as a whole may be generally applicable throughout the North Atlantic province. Differences in the ways in which the Balto-Scandic sequence is divided into zones and subzones by SERGEEVA, LINDSTRÖM, and VIIRA appear to be largely the results of individual taxonomic and stratigraphic philosophies. A much more detailed zonal succession, with Tremadocian and lower Arenigian rocks divided into 20 zones, has been introduced by VAN WAMEL (1974) on the basis of his study of sections in southeastern Sweden. It seems unlikely, however, that many of VAN WAMEL's zones will be useful for regional correlation, for he was able to recognize only six of them in the three main sections that he studied and nine were identified in only a single section.

Representatives of conodontophorid genera with uni- or multimembrate apparatuses of coniform elements *(Drepanoistodus, Oistodus, Paltodus, Paroistodus, Protopanderodus, Scandodus, Scolopodus, Stolodus)* dominate conodont collections from the Lower Ordovician of the North Atlantic province. However, *Cordylodus,* the first conodontophorid genus with a multimembrate apparatus of ramiform elements, characterizes Tremadocian rocks; and *Prioniodus, Periodon,* and *Microzarkodina,* which constitute an important group of genera with multimembrate apparatuses of ramiform and pectiniform elements, appear low in the upper part of the Latorpian Stage and, with *Eoplacognathus,* a likely later Early Ordovician derivative of *Prioniodus,* are especially useful in biostratigraphic subdivision of higher Lower Ordovician strata in the North Atlantic province.

Middle and Upper Ordovician rocks of the North Atlantic province were divided into a sequence of 5 zones and 10 subzones by BERGSTRÖM (1971, 1973) on the basis of the ranges of species of *Prioniodus, Pygodus,* and balognathids such as *Amorphognathus, Eoplacognathus,* and *Polyplacognathus.* Genera with multimembrate apparatuses of coniform elements (e.g., *Dapsilodus, Drepanoistodus, Panderodus, Protopanderodus*) are also represented in Middle and Upper Ordovician rocks, but most of these are less conspicuous as components of many Middle and Late Ordovician faunas than was the case in the Early Ordovician. An advantage of BERGSTRÖM's zonal scheme is that many of the zones he recognized have been tied in rather closely to the standard sequence of graptolite zones.

ETHINGTON and CLARK (1971) recognized five conodont faunas in the Lower Ordovician of the North American Midcontinent province, and DRUCE and JONES (1971) delineated six zones in the Lower Ordovician Ninmaroo Formation of Queensland on the basis of the vertical distribution of elements that are also of North American Midcontinent type. The six zones of DRUCE and JONES probably represent about the same stratigraphic interval as do the lower three faunal units of ETHINGTON and CLARK. Many of the same elements have also been reported by MOSKALENKO (1967, 1973) and ABAIMOVA (1972, 1975) from Lower Ordovician sequences on the Siberian platform, and MÜLLER (1964) and LEE (1970) have figured a number of specimens of Midcontinent type from the Lower Ordovician of South Korea. Thus, it appears from currently rather scattered evidence that elements of Midcontinent type are widespread and that Lower Ordovician rocks in this vast area may ultimately be divisible into at least eight biostratigraphic units of zonal rank.

Virtually all of the elements collected to date from Midcontinent Lower Ordovician rocks represent conodontophorid genera, the great majority of which, as in the North Atlantic province, formed uni- or multimembrate skeletal apparatuses of coniform elements. Many of the species have been assigned by various authors to genera (e.g., *Cordylodus, Drepanoistodus, Oistodus, Paltodus, Prioniodus, Scolopodus*) that are known best from Lower Ordovician strata of the North Atlantic province; however, the generic assignment of many of these species is open to question and a substantial number of others represent genera (e.g., *Acanthodus, Chosonodina, Loxodus, Ulrichodina*) that appear not to have been represented in contemporaneous faunas from the North Atlantic province.

According to SWEET, ETHINGTON and BARNES (1971), Middle and Upper Ordovician conodonts of the North American Midcontinent province are represented in a sequence of 12 distinctive faunas, which are characterized for the most part by species with multimembrate skeletal apparatuses consisting of ramiform and pectiniform elements. The lowest of these is doubtfully distinct from the highest of the Lower Ordovician faunas recognized by ETHINGTON and CLARK (1971) from Midcontinent strata, and is almost certainly of Early Ordovician age. Faunas 2 through 4 contain an abundance of dominantly hyaline coniform elements, the generic assignment of which is still uncertain, but they are especially characterized by several species of multimembrate *Histiodella* and *Multioistodus,* whose skeletal apparatuses include distinctive arrays of ramiform and pectiniform elements. The origin of these genera is not known, but it is likely that they developed from Lower Ordovician forms with multimembrate apparatuses composed of coniform elements. Faunas 5 through 12 are distinguished by a succession of species of *Phragmodus, Plectodina,* and *Aphelognathus* (Cyrtoniodontidae), and by several species of *Oulodus* (Hibbardellacea), all of which formed seximembrate skeletal apparatuses of ramiform and pectiniform elements. Such genera as *Drepanoistodus, Panderodus,* and *Belodina* with multimembrate skeletal apparatuses of coniform elements (or their posteriorly denticulated analogues) are commonly represented in the Midcontinent Middle and Upper Ordovician, and selected species have considerable stratigraphic utility. Many of these species have also been recognized in Australia (PACKHAM, 1967) and Siberia (MOSKALENKO, 1973), which suggests that the Midcontinent province was as widely distributed in the Middle and Late Ordovician as it had been in the Early Ordovician.

SILURIAN

WALLISER (1964) was the first to propose that Silurian marine strata might be divisible into zones on the basis of the distribution of conodont elements, and reported that 11 conodont units (one "Bereich" and 10 zones) could be discriminated in the famous Cellon section of the Carnic Alps in Austria. In 1972, WALLISER summarized additional data to indicate that Cellon-based zones were widely recognizable in Europe, and REXROAD and NICOLL (1971) noted that many of these biostratigraphic units could also be recognized in North America. Modifications since 1971 deal primarily with refinements in the ranges of several species used to define zonal bases, and with Llandoverian zones, which are not well represented (or are absent) in the standard section at Cellon.

In 1971, SCHÖNLAUB demonstrated that Ordovician and Silurian rocks at Cellon are separated by an unconformity and that at least part of the section included by WALLISER (1964) in "Bereich I" is Ordovician, not Silurian. In 1972, ALDRIDGE reported on Welsh Llandoverian conodonts and discriminated two conodont zones in rocks older than the oldest Silurian preserved at Cellon. Thus, in 1975, ALDRIDGE was able to report that European Silurian rocks could be divided into 12 conodont zones, and that, from the level of the late Llandoverian *Pterospathodus celloni* Zone upward, these were widely recognizable. It should be noted, however, that there is still no information on earliest Llandoverian conodonts, and there is substantial question as to how uppermost Ordovician and lowest Silurian rocks are to be differentiated through the use of conodonts.

Conodonts with multimembrate apparatuses of coniform elements (e.g., *Panderodus, Walliserodus*) are common in many Silurian strata and are closely related to older forms, which are especially abundant in

Middle and Upper Ordovician strata of the North American Midcontinent province. Stratigraphically, however, the most useful Silurian conodonts are characterized by multimembrate apparatuses of ramiform and pectiniform elements. These include the prioniodontaceans *Distomodus, Apsidognathus, Icriodella, Aulacognathus, Pelekysgnathus,* and *Pterospathodus*; the polygnathaceans *Ozarkodina, Kockelella,* and *Ancoradella*; and the hibbardellid *Oulodus.* Although phylogenies of most of these genera remain to be worked out, it is likely that the rather cosmopolitan Silurian conodont faunas were largely derived from Ordovician Midcontinent predecessors rather than from Ordovician North Atlantic ancestors. Whatever their origins, however, Silurian conodont faunas occur in essentially the same succession in North America, Europe, and Australia, and the Silurian zonal scheme first worked out in the Carnic Alps appears to have wide application.

DEVONIAN

Described collections of Lower and Middle Devonian conodonts seem mostly to have been derived from rocks that accumulated in relatively shallow-water environments. Although the bulk of these collections are dominated by elements of icriodontid species, mostly referable to *Icriodus,* many of them also include specimens of *Ozarkodina, Pandorinellina,* and *Polygnathus,* which represent important stocks of the Polygnathidae.

Largely on the basis of the vertical distribution of distinctive species and subspecies of *Icriodus,* Ziegler (1971) was able to recognize a succession of seven informal faunal units in the European Lower Devonian; and, following Wittekindt (1966), he described for the European Middle Devonian (Eifelian and Givetian) a sequence of five conodont zones based on the ranges of various species of *Icriodus, Pandorinellina,* and *Polygnathus.* Klapper and others (1971) discriminated nine informal faunal units in the Lower Devonian of western North America based on associations of *Icriodus, Ozarkodina, Pedavis,* and *Polygnathus.* For the Middle Devonian of North America they discussed the distribution of seven or eight additional faunas, also characterized by species of *Icriodus* and *Polygnathus.* Although there is considerable agreement between the sequences of Lower and Middle Devonian conodont faunas described by Ziegler (1971) and Klapper and others (1971), there are also appreciable differences and these differences still inhibit establishment of a formal sequence of conodont-based zones in the Lower and Middle Devonian.

In aggregate, conodont faunas of Late Devonian age exhibit an appreciably greater diversity at both the generic and specific levels than do those of Early and Middle Devonian age. This diversity is attributable primarily to two distinct intervals of accelerated differentiation in the polygnathid stock, one that began in the latest Middle Devonian and a second that began in the middle Famennian. The first of these intervals was characterized by the appearance and rapid diversification of *Mesotaxis, Palmatolepis* (Fig. 62), and *Schmidtognathus,* whereas the second was distinguished by the development of *Bispathodus, Scaphignathus,* and their closely (and complexly) related, but morphologically diverse derivatives. Genera that were established during the first of these epochs of accelerated diversification evolved rapidly and are represented in Upper Devonian strata by sequences of short-lived, widely distributed species, whose distinctive characters and short vertical ranges make them exceptionally useful in detailed biostratigraphic division of Upper Devonian rocks. None of the genera of this group continues into the Carboniferous, and none of them appears to have post-Devonian descendants. On the other hand, various species of *Bispathodus* and *Scaphignathus,* the initial stocks of the middle Famennian polygnathid radiation, are regarded as the ancestors of groups of species assigned to *Clydagnathus, Patro-*

gnathus, Protognathodus, Pseudopolygnathus, and *Siphonodella,* all of which had their principal development and reached their widest distribution and greatest diversity in the Early Carboniferous.

On the basis of many carefully detailed studies of the distribution of species now assigned to *Ancyrognathus, Bispathodus, Mesotaxis, Palmatolepis, Polygnathus, Protognathodus,* and *Scaphignathus,* ZIEGLER (1962, 1971) divided Upper Devonian rocks of the Rhenish Schiefergebirge and adjacent areas into a sequence of about 30 faunally defined units, assembled into 11 conodont zones. Correlatives or direct equivalents of a majority of these have been recognized at various places in North America by KLAPPER and others (1971) and earlier authors whose work they summarize. Further, because Late Devonian conodont faunas were apparently cosmopolitan, the zonal scheme elaborated and periodically up-dated and refined by ZIEGLER and his coworkers has found wide applicability and has enabled detailed correlation of Upper Devonian rocks on an essentially worldwide basis.

CARBONIFEROUS

In North America, Carboniferous rocks are divided into a succession of 37 named zones, which were described or summarized by COLLINSON, REXROAD, and THOMPSON (1971), LANE and others (1971), MERRILL (1972), and LANE and STRAKA (1974). By comparison, 13 zones, confined to the Dinantian, are formally recognized in a summary of European opinion by RHODES and AUSTIN (1971) and AUSTIN (1973b). MEISCHNER (1970) has provided a useful summary of the distribution of post-Dinantian Carboniferous faunas in Germany; and, subsequent to the RHODES-AUSTIN summary, MATTHEWS, SADLER, and SELWOOD (1972), BUTLER (1973), and MATTHEWS and NAYLOR (1973) have given additional important data on Lower Carboniferous conodonts from southwestern England and southwestern Ireland.

Kinderhookian or Tournaisian strata are characterized chiefly by conodont elements referable to *Bispathodus, Clydagnathus, Patrognathus, Protognathodus, Polygnathus, Pseudopolygnathus,* and *Siphonodella.* With the exception of *Polygnathus,* all of these genera had their origin in the middle Famennian phase of Late Devonian polygnathid diversification, and all but *Polygnathus, Pseudopolygnathus,* and *Siphonodella* disappeared before the end of the Kinderhookian-Tournaisian interval. It is likely, however, that this group of short-lived, stratigraphically useful genera included the ancestors of *Gnathodus* and *Cavusgnathus,* which appeared in the later Kinderhookian or Tournaisian and represent the root-stock of the major element, if not all of the idiognathodontids and cavusgnathids.

A sequence of 7 concurrent range-zones and 4 subzones have been recognized in late Visean (P_2) to early Westphalian (G_2) strata in the Central province of Great Britain (HIGGINS, 1975).

In Carboniferous strata above the base of the Visean (or Osagean), polygnathacean elements are dominant in described collections. *Gnathodus* and *Cavusgnathus* are of major biostratigraphic significance in division of Mississippian rocks in North America above the level of their middle Kinderhookian or Tournaisian appearance, and these genera are of similar biostratigraphic importance in correlative parts of the European Dinantian. In Pennsylvanian rocks, and in those few European Silesian strata from which elements have been described, conodont biostratigraphy is based on additional polygnathaceans such as *Adetognathus, Idiognathoides, Neognathodus,* and members of the *Idiognathodus-Streptognathodus* plexus. *Neognathodus* and *Idiognathoides* are especially useful in biostratigraphic zonation of the pre-Missourian part of the Pennsylvanian system, primarily because it has been possible to work out reasonably detailed phylogenetic sequences for these genera (see MERRILL, 1972; LANE & STRAKA, 1974). MERRILL (in LANE and

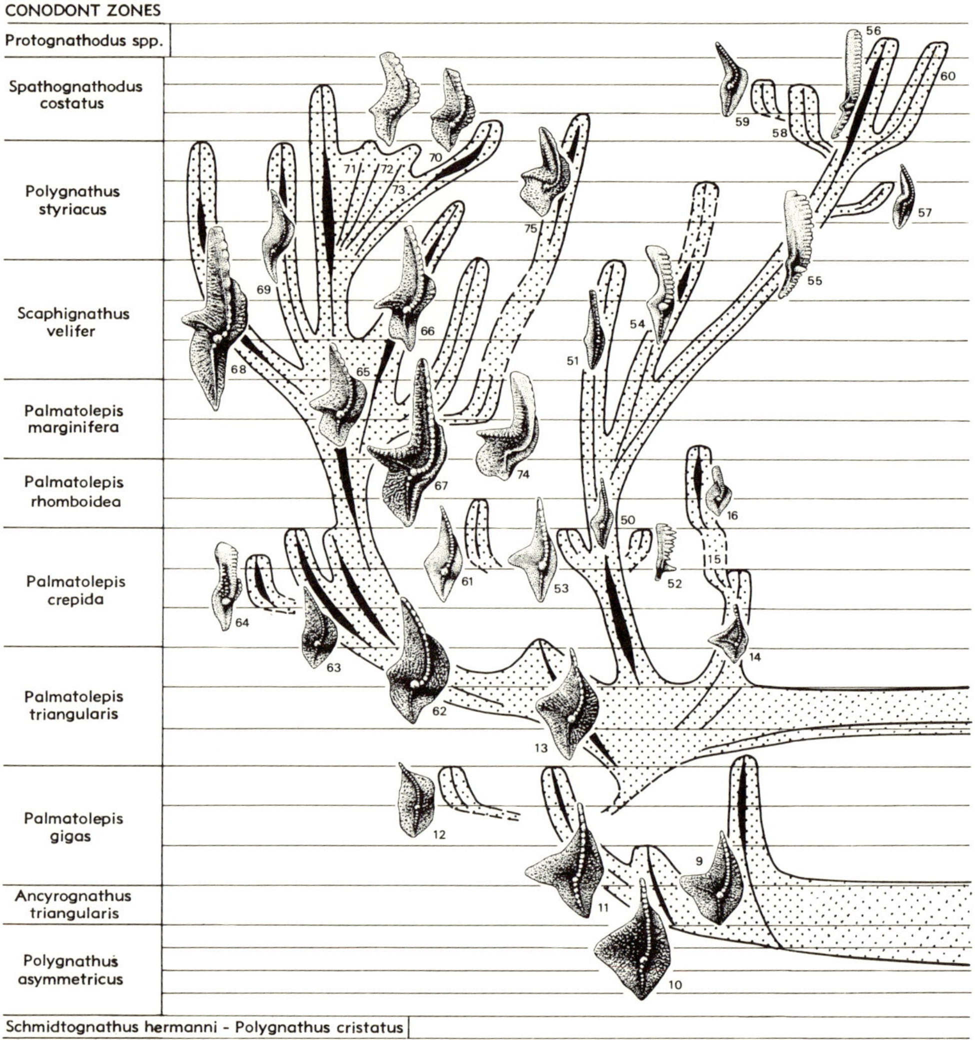

FIG. 62. Evolution of the *Pa* element in *Palmatolepis* ULRICH & BASSLER, 1926. The element originated from such wide-plated forms of *Polygnathus* as *P. asymmetricus ovalis (1)* and *P. asymmetricus asymmetricus (2)*. Associated species, *Palmatolepis disparalvea (a)* and *P. disparilis (b)*, with atypical large basal cavities, are descendants of *Polygnathus cristatus* and thus may not belong in *Palmatolepis*. Older species of *Palmatolepis (3, P. transitans; 4, P. punctata; 5, P. foliacea; 6, P. unicornis; 7, P. coronata; 8, P. proversa; 9, P. subrecta; 10, P. hassi; 11, P. gigas; 12, P. linguiformis)* show broad variability. *P. triangularis (13)* is the only direct descendant of the large-plated older group of *Palmatolepis*, originating in the transition between *P. subrecta* and *P. gigas*, and is ancestral to all younger taxa of the genus. Subsequent radiation gave rise to the following groups: *P. quadrantinodosalobata-P. subperlobata* branch, *14-20 (14, P. delicatula delicatula* and *P. delicatula clarki; 15,* not illustrated; *16, P. rhomboidea; 17, 21,* two morphotypes of *P. subperlobata; 18, P. quadrantinodosalobata; 19, 20,* not illustrated); *P. glabra* branch, *22-33 (22, P. tenuipunctata; 23,* not illustrated; *24, P. circularis; 25, P. glabra* n. subsp.; *26, P. glabra prima; 27, 28,* not illustrated; *29, P. klapperi; 30, P. glabra pectinata; 31, P. glabra lepta; 32, P. glabra acuta; 33, P. glabra distorta*); *P. regularis* branch, *34, 35* (two morphotypes); *P. quadrantinodosa inflexa-P. marginifera* branch, *36-47 (36-38, P. quadrantinodosa inflexa,* three early morphotypes of which *36* is the typical form: *39, P. marginifera marginifera; 40, 42-44,* four unnamed subspecies of

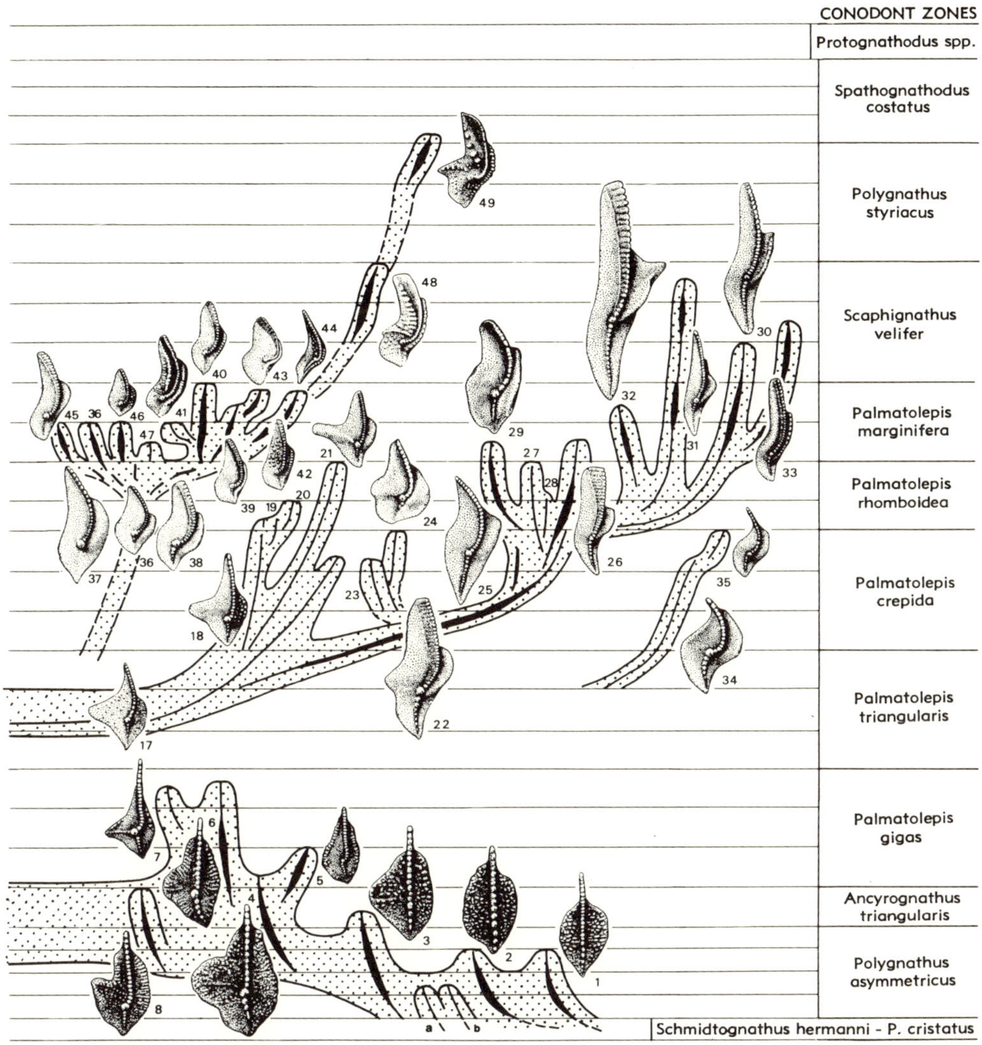

FIG. 62. *(Explanation continued from facing page.)*

P. marginifera; 41, P. marginifera duplicata; 45, P. quadrantinodosa inflexoidea; 46, P. quadrantinodosa quadrantinodosa; 47, not illustrated); *48, P. rugosa trachytera,* and *49, P. rugosa rugosa,* which are tentatively connected to the *P. marginifera* branch because of parapet development resembling that in *43* and *44* but may be related to *P. rugosa* cf. *P. r. ampla* and *P. rugosa ampla* in *74* and *75; P. minuta-P. gracilis* branch, *50-61 (50, 51,* two morphotypes of *P. minuta minuta; 52, P. minuta subgracilis; 53, P. minuta loba; 54, P. minuta schleizia; 55, 56, P. gracilis gracilis; 57, P. gracilis manca; 58,* not illustrated; *59, P. gracilis gonioclymeniae; 60,* not illustrated; *61, P.* n. sp.); *P. perlobata-P. crepida* branch, *62-64 (62, P. perlobata perlobata; 63, P. crepida; 64, P. termini)*; and *P. perlobata-P. schindewolfi* branch, *62, 65-75 (62, P. perlobata perlobata; 65, P. perlobata schindewolfi; 66, P. perlobata sigmoidea; 67, P. perlobata grossi; 68, P. perlobata maxima; 69, P. perlobata helmsi; 70, P. perlobata postera; 71-73,* several morphotypes of *P. perlobata schindewolfi* represented by one figure; *74, P. rugosa* cf. *P. r. ampla; 75, P. rugosa ampla*). Dotted areas indicate fields of transition between taxa or morphotypes; dashed margins mean relationship is uncertain. Solid lines within fields represent main phylogenetic trends and line width indicates relative abundance. Because of lack of known connecting forms, the relations of unconnected evolutionary branches are not clear (J. Helms & W. Ziegler, n).

others, 1971) has suggested a provisional zonation of fossiliferous post-Missourian rocks in the Appalachian basin, based on vertical changes in the character of the *Streptognathodus* complex; however, it is not clear that this provisional zonal scheme is applicable to Pennsylvanian strata elsewhere, and it is almost certainly incomplete because a substantial thickness of the uppermost Pennsylvanian in the Appalachian basin is nonmarine. von Bitter (1972) and Perlmutter (1975) have provided valuable descriptions of Virgilian elements, which are younger than any in the Appalachian basin, but neither writer has proposed formal biostratigraphic units for this youngest part of the Carboniferous in North America.

PERMIAN

Conodont elements of the Permian are known less well than are those of any other system. Using information largely from Nevada and Texas, however, Clark and Behnken (1971) and Behnken (1975) have divided Wolfcampian through Guadalupian strata into a sequence of 8 assemblage-zones; and Kozur (1975a), using the literature and largely undescribed collections from various localities in the Soviet Union, has defined 10 Permian zones. Although the two biostratigraphic schemes thus far proposed are similar in certain respects, they differ greatly in others. Neither can be regarded as much more than a provisional statement at present, for both are derived from observations on the distribution of elements in only a few, widely separated sections.

Either at the end of the Carboniferous (Kozur, 1975a) or within the earliest Permian (Clark & Behnken, 1971; Behnken, 1975), the major stock of the Polygnathacea *(Idiognathodus, Streptognathodus)* disappeared and was replaced by another *(Neostreptognathodus, Sweetognathus)*, probably derived from *Diplognathodus,* which had a long time range in the Carboniferous and continued into the Late Permian. In addition, *Hindeodus* continued on from the Carboniferous and vigorous development within the Gondolellacea, of the sort that gave rise to Late Carboniferous *Gondolella,* produced a sequence of distinctive Permian species of *Neogondolella.* Biostratigraphically useful elements of known Permian faunas seem thus to have been derived from two distinct ancestral stocks, the *Diplognathodus* lineage of the Polygnathacea and the Gondolellacea, a major group of conodonts that was rare until the Permian. The *Diplognathodus* lineage survived into the earliest Triassic, when it became extinct. The Gondolellacea, on the other hand, survived until the latest Triassic and seem to have produced all of the stratigraphically useful forms of that period.

TRIASSIC

Conodont elements have proved to be of considerable biostratigraphic utility in Triassic rocks and Sweet and others (1971) divided the system into a succession of 22 conodont zones. Kozur and Mostler (1972) and Kozur (1972, 1975b) have defined or redefined some 25 zones and an unnamed faunal unit in the same stratigraphic interval. In major features, these two zonal schemes are quite similar. They differ primarily in that the *Anchignathodus typicalis* Zone, which straddles the Permian-Triassic boundary in the scheme of Sweet and others (1971), is restricted through redefinition by Kozur and Mostler to the highest Permian. It must be noted, however, that Kozur and Mostler also regarded the *Otoceras concavum* and *O. boreale* zones to be highest Permian, whereas virtually all other stratigraphers have assigned those ammonoid zones to the lowest Triassic.

Polygnathacean conodonts such as *Hindeodus* and possible derivatives of the *Diplognathodus* lineage (e.g., *Isarcicella*) range upward from the Permian into rocks of earliest Triassic (Griesbachian) age, where they are useful in recognizing two or possibly three zones. Like *Otoceras,* however, these conodonts were survivors of stocks that had their greatest diversity in the late Paleozoic, and they were extinct by the end of the Griesbachian.

Post-Griesbachian Triassic rocks are distinguished by conodont elements that appear to represent two relatively conservative gondolellacean stocks, each with a long Permian (or pre-Permian) history. Neither the taxonomy nor the phylogeny of these stocks is well understood. However, it seems likely that each had a basically similar, seximembrate apparatus, and that each gave rise periodically, and perhaps iteratively, to species with unimembrate apparatuses composed solely of anguliplanate, segminate, or segminiplanate pectiniform elements.

One stock, represented in Triassic rocks by a succession of species of *Xaniognathus* and *Cypridodella,* probably included the ancestors of species assigned to *Neospathodus* and *Neogondolella,* which are useful in defining a sequence of Lower Triassic (Scythian) zones, and of species of *Epigondolella,* which together with additional species of *Neogondolella,* are of major importance in conodont zonation of the Middle and Upper Triassic.

The second gondolellacean stock, represented basically by seximembrate *Ellisonia,* probably includes the ancestors of *Furnishius* and *Hadrodontina,* which are stratigraphically useful in shallow-water Smithian rocks, and of long-ranging *Gladigondolella,* which may have biostratigraphic significance in the Smithian (as *"Gondolella" milleri*) and is used by Kozur and Mostler (1972) to define a *Gladigondolella tethydis* Zone in the lower Upper Triassic.

There is evidence to suggest that the two gondolellacean stocks, which include virtually all post-Griesbachian conodonts, had different distributions in Triassic seas. The *Ellisonia* stock, which included species with skeletal apparatuses composed mostly of robust, coarsely denticulate elements, is known best from rocks that were deposited in relatively shallow-water or nearshore environments, whereas the *Xaniognathus-Cypridodella* stock, which is characterized by species with small, fragile, closely denticulate elements, is best known from rocks that may have accumulated in somewhat deeper water, or more offshore environments. Derivatives of these two stocks, characterized in each case by reduced skeletal apparatuses, may well represent repeated adaptations to more specialized habitats within the major realms occupied by the parental stocks. Whatever the reasons for observed differences in distribution, it becomes increasingly obvious that parallel biostratigraphic schemes will probably have to be developed for major Triassic facies. This remains to be done.

The youngest conodont elements known are Rhaetic specimens from Austria referred by Mosher (1968) to species of *Cypridodella* and *Neospathodus.* The nature of elements upon which Kozur and Mostler (1972) and Kozur (1975b) based their Rhaetian "post-*hernsteini* faunal unit" is unknown, and Jurassic and Cretaceous conodont elements reported by Nohda and Setoguchi (1967) and Diebel (1956), respectively, are generally regarded as reworked Triassic specimens (Müller & Mosher, 1971).

CLASSIFICATION

By David L. Clark

[University of Wisconsin]

Phylum CONODONTA

Probably no other fossils have been assigned to so many different biologic groups as have the conodonts (see biologic affinity section, this volume). The rather systematic comparison of conodonts with various invertebrates, vertebrates, and plants has convinced most specialists that their morphologic differences support the assignment of conodonts to a separate phylum, Conodonta.

The recently discovered Pennsylvanian body impressions interpreted to be of "whole conodonts" by Melton and Scott (1973), as well as the Middle Cambrian Burgess Shale specimen of Conway Morris (1976), also support recognition of a distinct phylum in that neither the Melton and Scott nor the Conway Morris reports include data that suggest their animals could fit easily into any existing phylum. Although there are unanswered questions related to interpretations of both the Pennsylvanian and Cambrian specimens as conodonts, the eventual interpretations should not affect the assignment of conodonts to a separate phylum. Phylum status for conodonts can be justified as have been the various "worm" and algal phyla. Among extinct invertebrates, the Archaeocyatha now constitute a separate phylum even though a relationship with the Porifera is acknowledged. So close a relationship as this does not exist between the conodonts and any other group, extinct or living.

DEVELOPMENT OF CLASSIFICATION

Detailed classifications of conodonts generally have been based on a horizontal or structural assignment of form-taxa to various quite artificial groups. Thus, Ulrich and Bassler (1926) proposed that some 34 conodont form-genera could be grouped into 4 families that were organized on gross morphologic similarities of discrete elements. Almost 20 years later, Branson and Mehl (1944) recognized 8 families to which they assigned 73 form-genera. This organization of taxa also was based on form similarities. The classification in the first edition of *Treatise W* became more complex, and Hass (1962) recognized 7 families and 13 subfamilies for the 141 form-genera known to him. Hass actually proposed two classifications for conodonts, a utilitarian grouping with families and subfamilies based on similarities among discrete form-taxa, and a "biologic" classification for 6 assemblages of elements earlier referred to different genera. These 6 "genera" were not organized at a higher taxonomic level by Hass (1962).

The only attempt at a more or less biological classification prior to the one in this volume is the classification of Lindström (1970). In this first multielement approach to classification, Lindström recognized 2 orders, 8 superfamilies, 21 families, and 4 subfamilies, many of which are used in the present classification. He based his classification on multielement apparatus similarities and skeletal differences among supposed natural groups.

Lindström's approach has been expanded in this volume, and our classification is based on structural and chemical differences of elements (2 orders), grouping of similar apparatus reconstructions (11 superfamilies), and distinctive apparatus or element composition (47 families) for some 180 genera. The apparatus structures for an additional 48 genera have not been interpreted and these genera are listed as "family unknown." Thus, 180 of some 240 valid conodont genera (thought to be complete through at least September, 1975) are here interpreted biologically. The residue of unassigned genera, together with reinterpretations and possible corrections of the assigned genera, outline a fertile field of research for the student of conodonts. In this first attempt at a unified biologic classification, it is important to remember that the authors regard the classification as provisional, especially at the suprageneric level.

OUTLINE OF CLASSIFICATION

The following outline of the phylum Conodonta summarizes taxonomic relationships, geologic occurrence, and numbers of recognized genera and subgenera in each suprageneric group. A single number refers to genera; where two numbers are given, the second indicates subgenera in addition to nominotypical ones.

Class Conodonta, 240; 1. *U.Precam.-U.Trias.*
Order Paraconodontida, 15. *Uppermost Precam.-M. Ord.*
Superfamily Amphigeisinacea, 1. *L.Cam.-M.Cam.*
  Amphigeisinidae, 1. *L.Cam.-M.Cam.*
Superfamily Furnishinacea, 14. *Uppermost Precam.-M.Ord.*
  Furnishinidae, 12. *Uppermost Precam.-L.Ord.*
  Westergaardodinidae, 2. *M.Cam.-M.Ord.*
Order Conodontophorida, 225; 1. *U.Cam.-U.Trias.*
Superfamily Proconodontacea, 11. *U.Cam.-L.Ord.*
  Clavohamulidae, 3. *U.Cam.-L.Ord.*
  Cordylododontidae, 3. *U.Cam.-L.Ord.*
  Oneotodontidae, 5. *U.Cam.-L.Ord.*
Superfamily Fryxellodontacea, 1. *L.Ord.*
  Fryxellodontidae, 1. *L.Ord.*
Superfamily Prioniodontacea, 36; 1. *Ord.-Dev.*
  Balognathidae, 3. *L.Ord.-Sil.*
  Cyrtoniodontidae, 6. *Ord.*
  Icriodontidae, 6. *M.Ord.-Sil.(Llandov.), Sil. (Ludlov.)-U.Dev.*
  Oepikodontidae, 1. *L.Ord., ?M.Ord.*
  Paracordylodontidae, 1. *L.Ord.*
  Periodontidae, 2. *Ord.*
  Phragmodontidae, 1. *M.Ord.-U.Ord.*
  Polyplacognathidae, 2. *L.Ord.-M.Ord.*
  Prioniodontidae, 1; 1. *L.Ord.-M.Ord.*
  Pygodontidae, 1. *M.Ord.*
  Rhipidognathidae, 5. *L.Ord.-Sil.*
  Pterospathodontidae, 6. *Sil.*
  Distomodontidae, 1. *Sil.*
Superfamily Chirognathacea, 8. *Ord.*
  Chirognathidae, 2. *M.Ord.*
  Multioistodontidae, 6. *Ord.*
Superfamily Panderodontacea, 10. *L.Ord.-U.Dev.*
  Panderodontidae, 4. *L.Ord.-M.Dev.*
  Scolopodontidae, 2. *L.Ord.-M.Ord.*
  Belodellidae, 4. *L.Ord.-U.Dev.*
Superfamily Distacodontacea, 18. *U.Cam.-U.Ord.*
  Acanthodontidae, 1. *L.Ord.*
  Drepanoistodontidae, 7. *Ord.*
  Juanognathidae, 1. *L.Ord.-M.Ord.*
  Protopanderodontidae, 2. *Ord.*
  Proconodontidae, 1. *U.Cam., ?L.Ord.*
  Oistodontidae, 3. *L.Ord.-M.Ord.*
  Strachanognathidae, 1. *Ord.*
  Teridontidae, 1. *U.Cam.-L.Ord.*
  Ulrichodinidae, 1. *L.Ord.*
Superfamily Hibbardellacea, 4. *M.Ord.-U.Penn.*
  Hibbardellidae, 4. *M.Ord.-U.Penn.*
Superfamily Gondolellacea, 18. *U.Carb.-U.Trias.*
  Gondolellidae, 1. *M.Penn.-L.Perm.*
  Ellisoniidae, 9. *U.Carb.-U.Trias.*
  Xaniognathidae, 8. *Perm.-Trias.*
Superfamily Polygnathacea, 48. *U.Ord.-L.Trias.*
  Kockelellidae, 2. *Sil.*
  Cryptotaxidae, 2. *M.Dev.-U.Dev.*
  Cavusgnathidae, 7. *U.Dev.-L.Perm.*
  Idiognathodontidae, 7. *U.Dev.-L.Perm.*
  Polygnathidae, 22. *U.Ord.-U.Carb.*
  Anchignathodontidae, 8. *L.Carb.-L.Trias.*
Superfamily Unknown, 12.
  Bactrognathidae, 7. *U.Dev.-L.Miss.*
  Elictognathidae, 4. *U.Dev.-L.Miss.*
  Mestognathidae, 1. *U.Miss.-L.Penn.*
Family Unknown, 48, in Appendix, 11.

RANGES OF TAXA

The stratigraphic distribution of orders, superfamilies, and families of Conodonta recognized in the *Treatise* is indicated graphically in the table that follows (compiled by Jack D. Keim).

TABLE 6. *Stratigraphic Distribution of the Conodonta.*

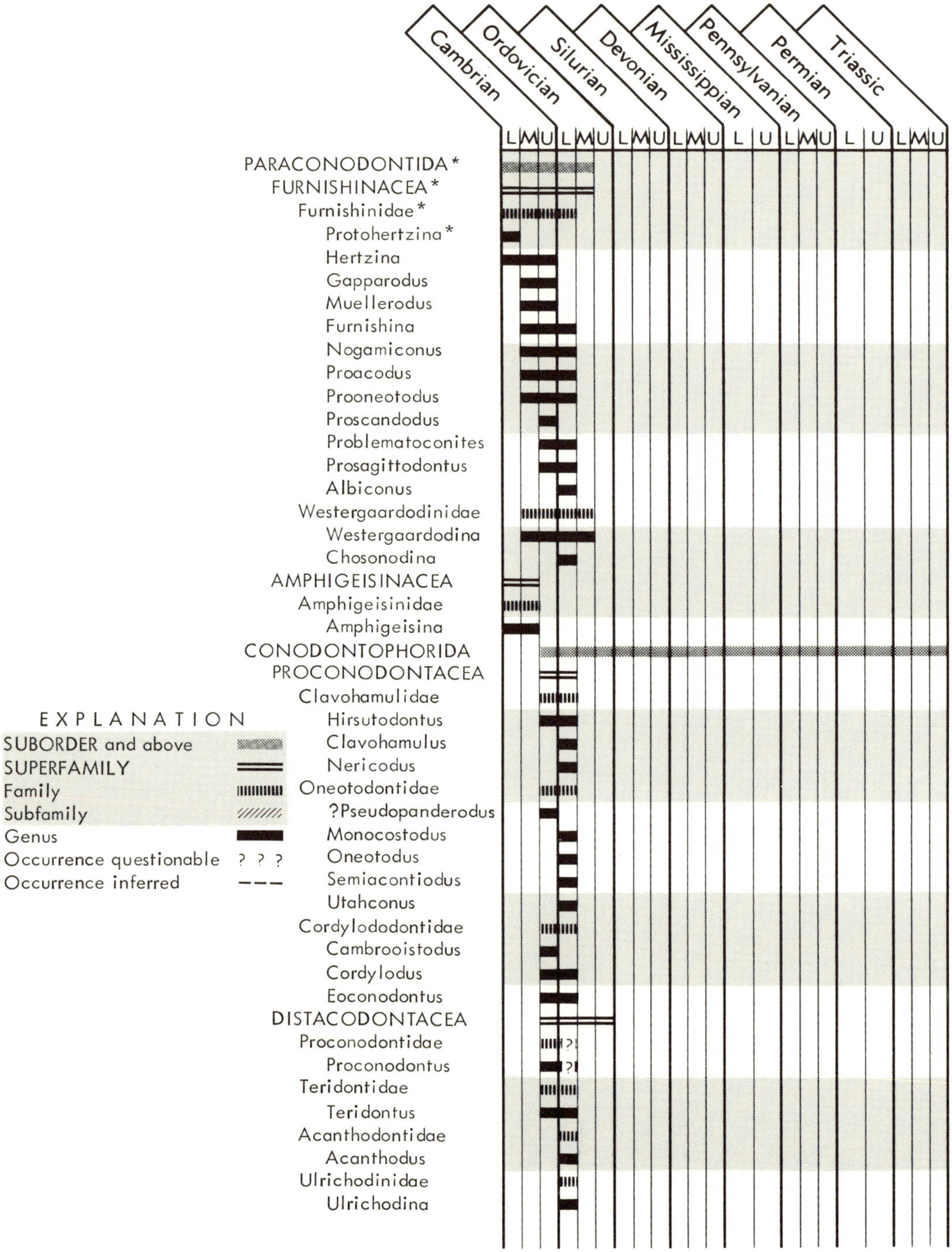

* Range starts at upper Precambrian.

TABLE 6. *(Continued)*

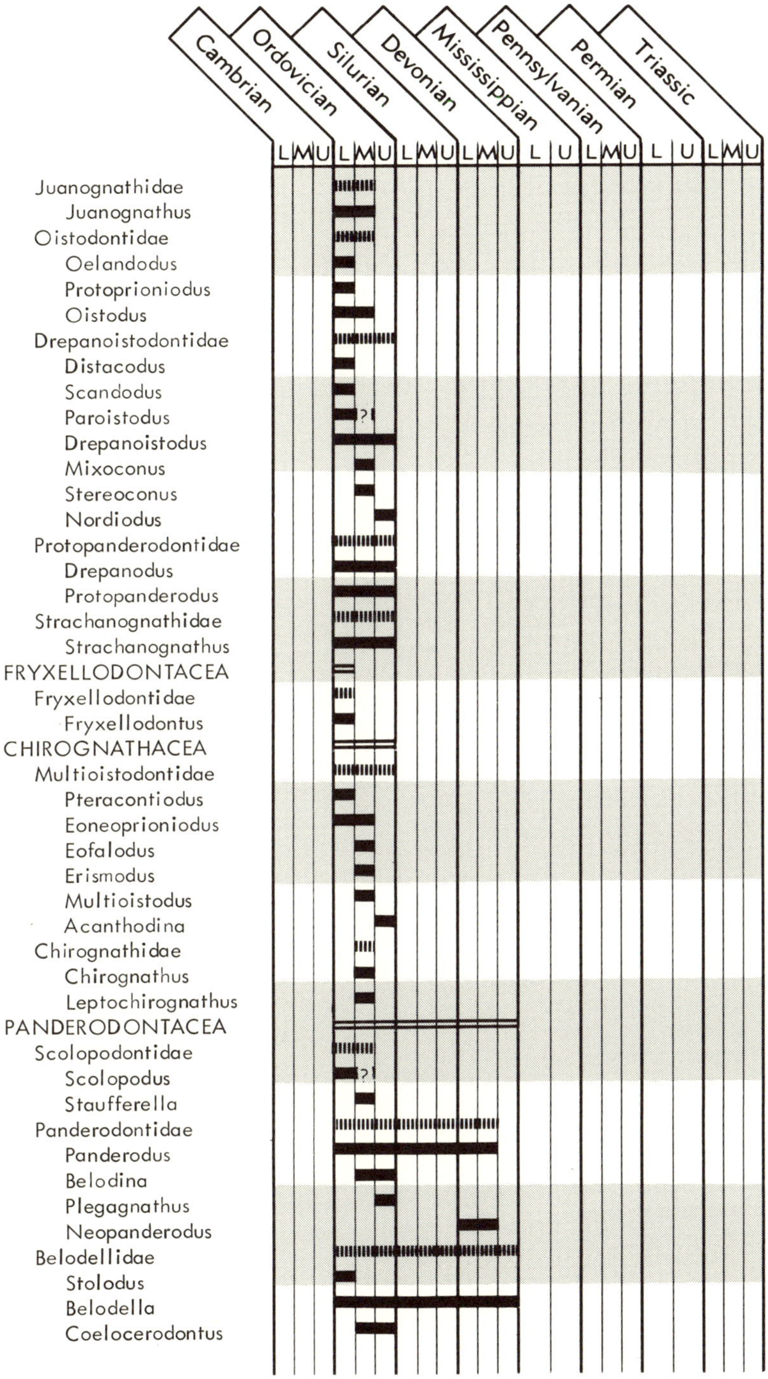

TABLE 6. (Continued)

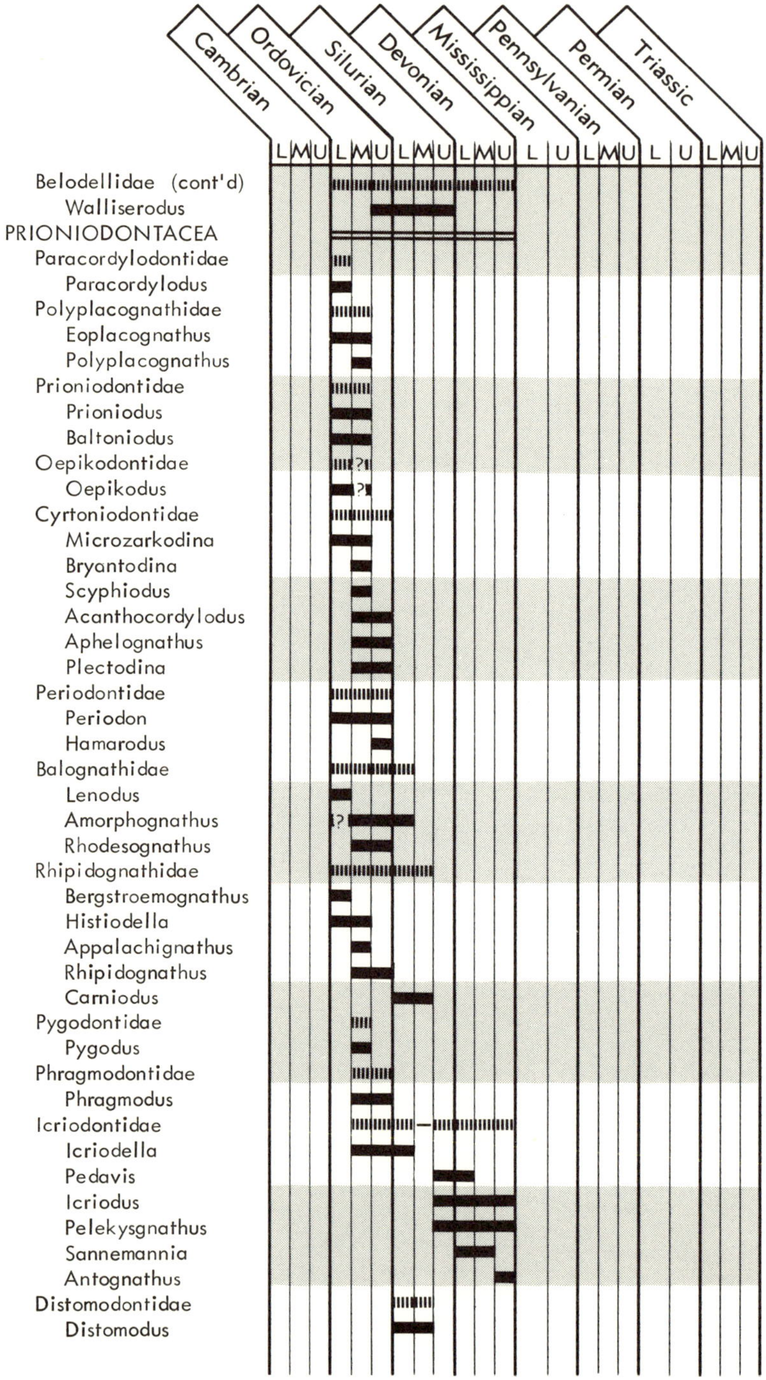

TABLE 6. *(Continued)*

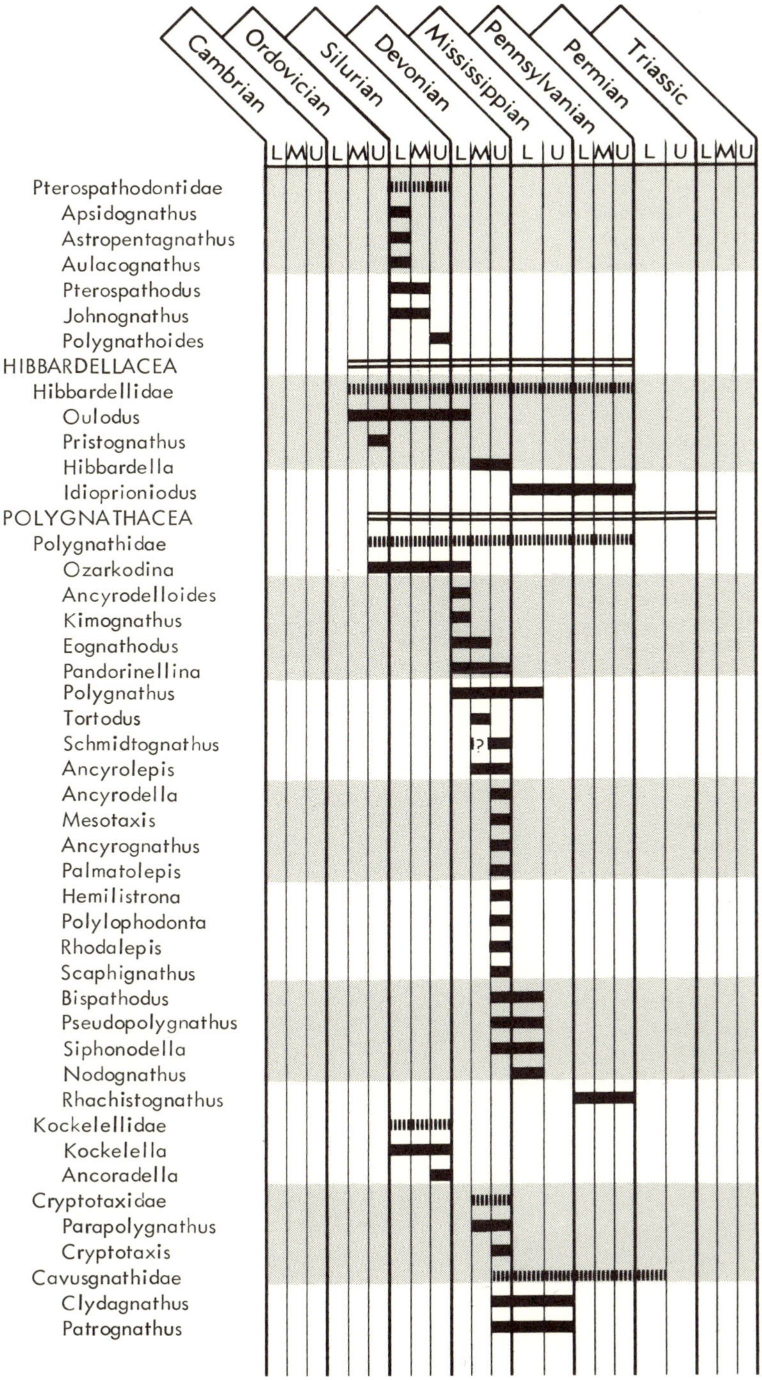

Table 6. *(Continued)*

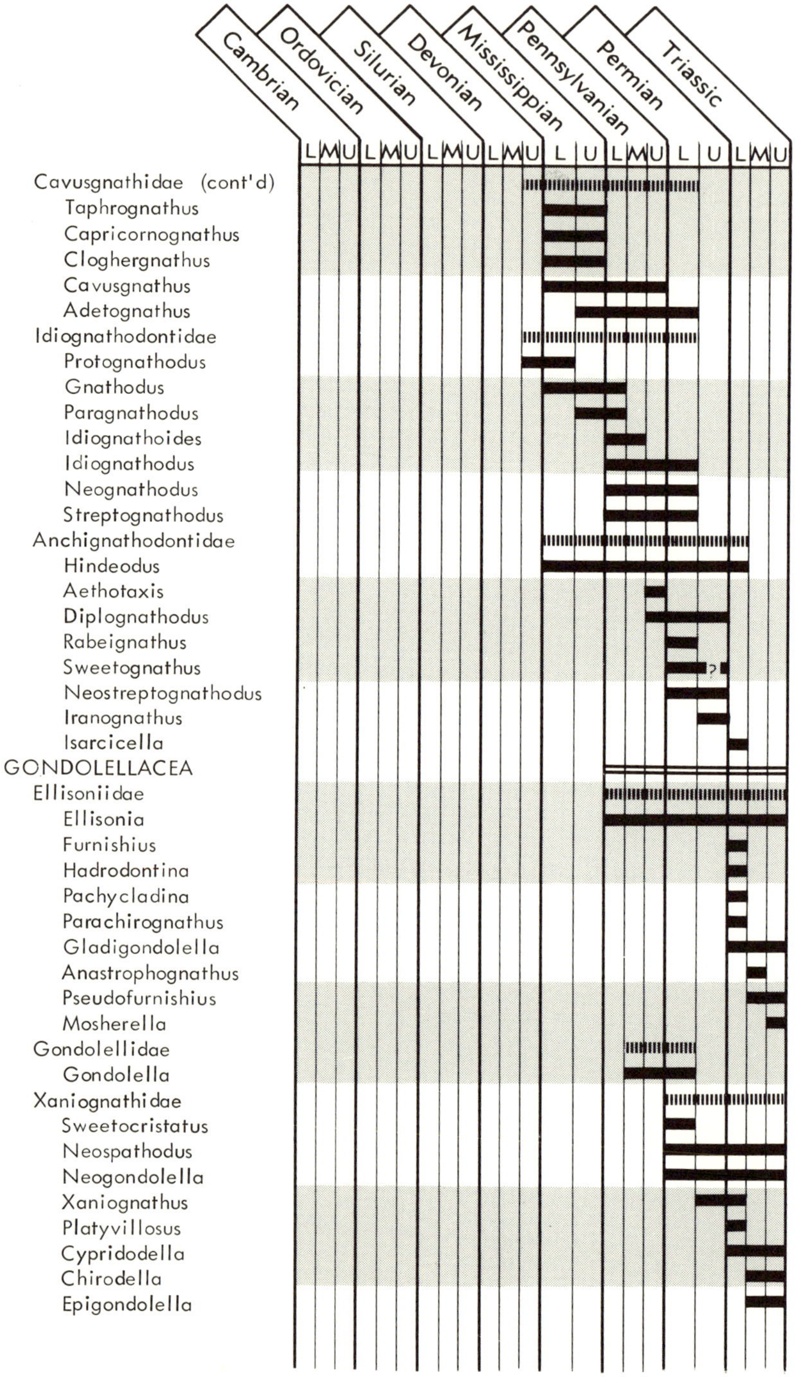

TABLE 6. *(Continued)*

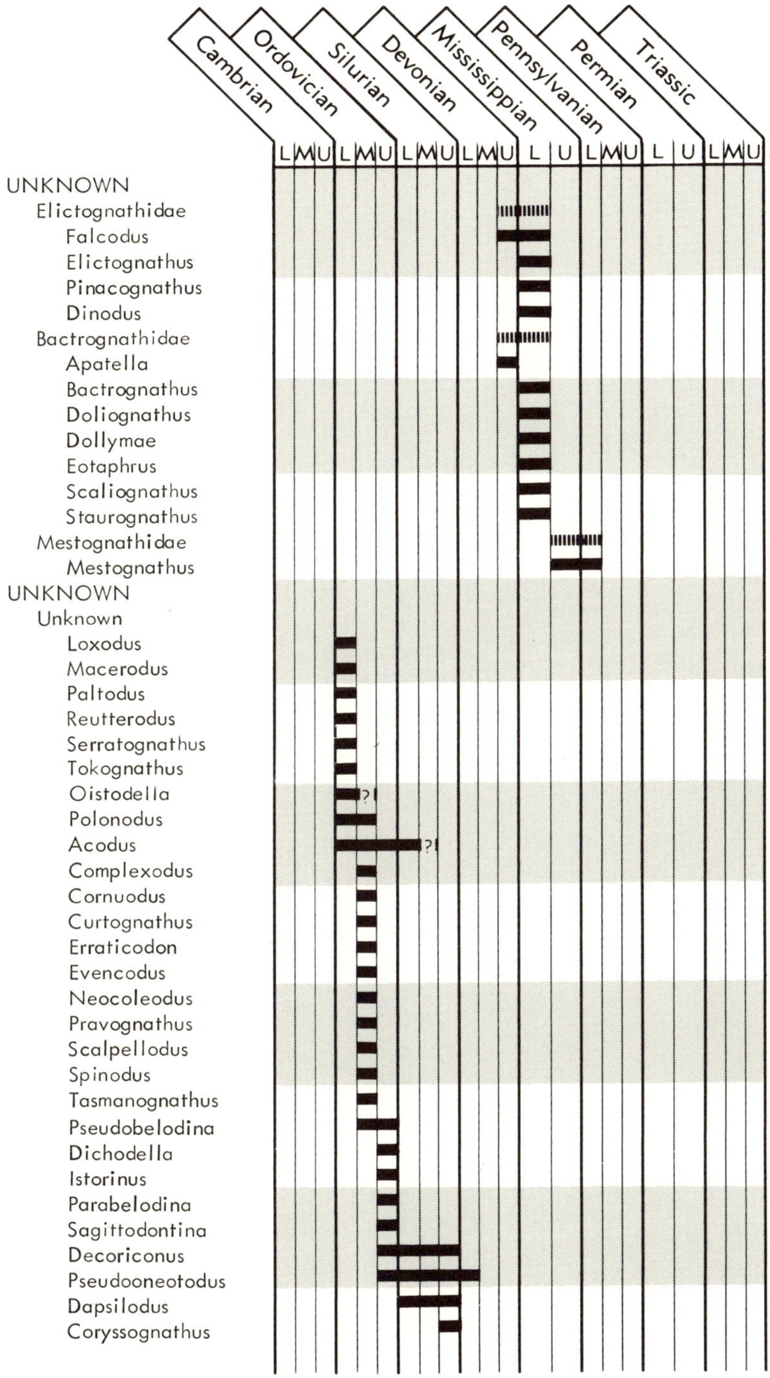

TABLE 6. *(Continued)*

	Cambrian L	Cambrian M	Cambrian U	Ordovician L	Ordovician M	Ordovician U	Silurian L	Silurian M	Silurian U	Devonian L	Devonian M	Devonian U	Mississippian L	Mississippian U	Pennsylvanian L	Pennsylvanian M	Pennsylvanian U	Permian L	Permian U	Triassic L	Triassic M	Triassic U
Unknown (cont'd)																						
Rotundacodina										X												
Angulodus										X	X	X	X	X	X	X	X					
Bryantodus												X										
Euprioniodina												X										
Ligonodina												X										
Lonchodina												X										
Playfordina												X										
Prioniodina												X										
Elsonella												X										
Polygnathellus												X										
Scutula												X										
Diplododella												X										
Synprioniodina												X										
Tripodellus												X										
Branmehla												X	X									
Metaprioniodus												X	X									
Mehlina												X	X									
Subbryantodus												X	X									
Apatognathus												X	X	X								
Hindeodina												X	X	X								
Oligodus													X									
Hindeodelloides													X									
Laterignathus													X	X								
Roundya													X ? X	X	X							
Geniculatus														X								
Magnilateralla														X								
Ptilognathus														X								
Kladognathus														X								
Lambdagnathus														X								
Pavlovites																		X				
Caenodontus																			X			

SYSTEMATIC DESCRIPTIONS

Phylum CONODONTA Eichenberg, 1930

[*nom. transl.* CLARK, herein, *ex* Conodontophorida EICHENBERG, 1930, p. 181, order] [Diagnosis by D. L. CLARK]

Extinct group of marine animals, mostly pelagic during some or all of their lives; most commonly preserved parts microscopic coniform, ramiform, or pectiniform elements occurring in various combinations and apparatus patterns and functioning as internal supports of body. Elements composed principally of carbonate apatite laminae built up by outer accretion; histologically complex. Element apparatuses suggest that most conodonts had bilaterally symmetrical bodies. *U.Precam.-U.Trias.*

Class CONODONTA Eichenberg, 1930

[*nom. transl.* CLARK, herein, *ex* Conodontophorida EICHENBERG, 1930, p. 181, order]

Diagnosis as for phylum. *U.Precam.-U. Trias.*

Order PARACONODONTIDA Müller, 1962

[Paraconodontida MÜLLER, 1962c, p. *W*248] [Diagnosis by J. F. MILLER]

Elements characterized by large, deep, basal cavities and lack of white matter; exterior of some specimens covered with layer of dark organic material (similar layer may line basal cavity); growth lamellae fewer and more widely spaced than in elements of Conodontophorida; only first few growth lamellae continuous on all sides, other lamellae beginning below tip, wrapping around basal margin, extending inside basal cavity, ending below apex of basal cavity (later deposited lamellae beginning successively farther below tip of cusp and ending successively farther below apex of basal cavity); growth pattern from tip downward, rather than from base upward as in elements of Conodontophorida; basal structure (basal funnel) indistinct or absent. All genera probably possessing unimembrate apparatuses. *Uppermost Precam.(Yudom.)-M.Ord.* [Most species of Cambrian age.]

Superfamily AMPHIGEISINACEA Miller, new

[Materials for this superfamily prepared by J. F. MILLER]

Nongeniculate coniform elements distinguished by unusual three-layered wall; inner layer (lining basal cavity) and outer layer (covering element) thin and apparently mostly or entirely organic; middle layer thick and mostly apatite. *L.Cam.-M.Cam. transition.*

Family AMPHIGEISINIDAE Miller, new

Diagnosis as for superfamily. *L.Cam.-M.Cam. transition.*

Amphigeisina BENGTSON, 1976, p. 187 [**Hertzina? danica* POULSEN, 1966, p. 4; OD]. Extremely long (up to 8 mm), slender, proclined, symmetrical coniform elements with basal cavity extending to tip. Anterior face smoothly rounded; posterior face concave; posterolateral edges drawn out into two prominent keels. *L.Cam.-M.Cam. transition,* Eu.(Sweden-?Eng.)-?Asia(Sib.).——FIG. 63,*1*. **A. danica* (POULSEN), Eu.(Sweden); *1a,b,* right oblique view, transv. sec., about ×20 (Bengtson, 1976).

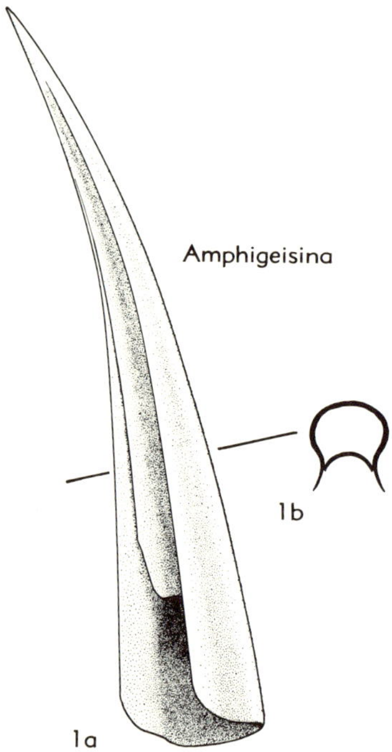

FIG. 63. Amphigeisinidae (p. W111).

Superfamily FURNISHINACEA Miller, new

[Materials for this superfamily prepared by J. F. MILLER]

Nongeniculate coniform and unusual multicuspate elements with two-layered wall structure; outer layer thin, composed mostly of organic matter; inner layer thick, composed mostly of apatite. *Uppermost Precam. (Yudom.)-M.Ord.*

Family FURNISHINIDAE Müller & Nogami, 1971

[Furnishinidae MÜLLER & NOGAMI, 1971, p. 18]

Coniform elements with growth lamellae discontinuous on outside of cusp and on inside of basal cavity. *Uppermost Precam.-L.Ord.*

Furnishina MÜLLER, 1959, p. 451 [**F. furnishi*; OD]. Asymmetrical coniform elements, proclined to erect, some bent laterally; base large and broadly expanded, with distinct cusp above base; basal cavity large and deep; anterior face flat, posterior and anterolateral carinae resulting in triangular cross section, cross section in some modified by secondary lateral carinae. *M.Cam.-L.Ord.*, N.Am. (USA, widespread in W. and SW.)-Eu.(Sweden-Pol.-Ger., glacial erratics)-Asia(Sib.-China-S.Korea-Turkey-Iran)-Australia (Queensl.).——FIG. 64,*7*. **F. furnishi*, holotype, U. Cam. (Gallatin Ls.), USA (Wyo.); *7a-c*, post., lat. views, transv. sec., ×80 (Müller, 1959).

Albiconus MILLER, 1980, p. 8 [**A. postcostatus*; OD]. Symmetrical coniform elements, proclined to erect; base narrow, tapering gently to tip, tip gently bent posteriorly; basal cavity extending to tip; anterior face flat, posterior margin drawn out into prominent costa; cross section roundly triangular. *L.Ord.(Symphysurina Z.)*, N.Am.(Utah-Nev.-Okla.-Texas-Alberta).——FIG. 65,*1*. **A. postcostatus*, USA (Utah); *1a*, lat. view, ×67; *1b,c*, post. view of holotype with transv. sec. at base, ×67 (Miller, 1980).

Gapparodus ABAIMOVA, 1978, *M.Cam.-U.Cam.*, see addendum.

Hertzina MÜLLER, 1959, p. 454 [**H. americana*; OD]. Proclined coniform elements, essentially symmetrical, slender; basal cavity deep, extending nearly to tip, cusp very small; posterior face essentially flat, anterior face rounded, carinae on posterolateral edges. *Cam.*, N.Am.(Nev.-N.Y.)-Eu. (Scand.-Ger., glacial erratics)-Asia(China-Sib.-Kazakh.-S.Korea-Turkey).——FIG. 64,*1*. **H. americana*, U.Cam.(*Elvinia* Z.), USA(Nev.); *1a,b*, lat. view, transv. sec., ×80 (Müller, 1959).

Muellerodus MILLER, 1980, p. 27, *nom. subst. pro Muellerina* SZANIAWSKI, 1971, *non* Bassiouni, 1965, an ostracode [**Distacodus (?) cambricus* MÜLLER, 1959, p. 450; OD]. Proclined to reclined elements, cusp in some bending laterally, tip of cusp in some recurved signoidally; base large, basal cavity deep; anterior and posterior edges rounded, lateral faces each with long, prominent costa. *M.Cam.-U.Cam.*, N.Am.(Nev.-N.Y.)-Eu.(Sweden-Pol.-Ger., glacial erratics)-Asia(China-Turkey).——FIG. 64,*3*. **M. cambricus* (MÜLLER), holotype, U.Cam.(Zone 1), Eu.(Sweden); *3a,b*, lat. view, transv. sec., ×60 (Müller, 1959).

Nogamiconus MILLER, 1980, p. 28 [**Proacodus? sinensis* NOGAMI, 1966, p. 356; OD]. Asymmetrical elements; basal cavity large and deep, cusp, if present, very small; anterior or posterior keel, or both, present, with one or more lateral carinae. *M. Cam.-L.Ord.*, Asia (China-S.Korea-?Turkey)-Australia(Queensl.)-N.Am.(N.Y.-Newf.).——FIG. 64, *6*. **N. sinensis* (NOGAMI), holotype, U.Cam.(Kushan beds), Asia(China); *6a,b*, lat. views, about ×50 (Nogami, 1966).

Proacodus MÜLLER, 1959, p. 458 [**P. obliquus*; OD]. Asymmetrical elements with large base and small cusp; basal cavity large and deep; cusp essentially round except for one lateral costa, costa expanding into carina at base. *M. Cam.-L.Ord.*, N. Am. (Nev.-Utah-Wyo.)-Eu. (Sweden-Ger., glacial erratics)-Asia(China-Sib.).——FIG. 64,*4*. **P. obliquus*, holotype, U.Cam.(Zone 5d), Eu.(Ger.);

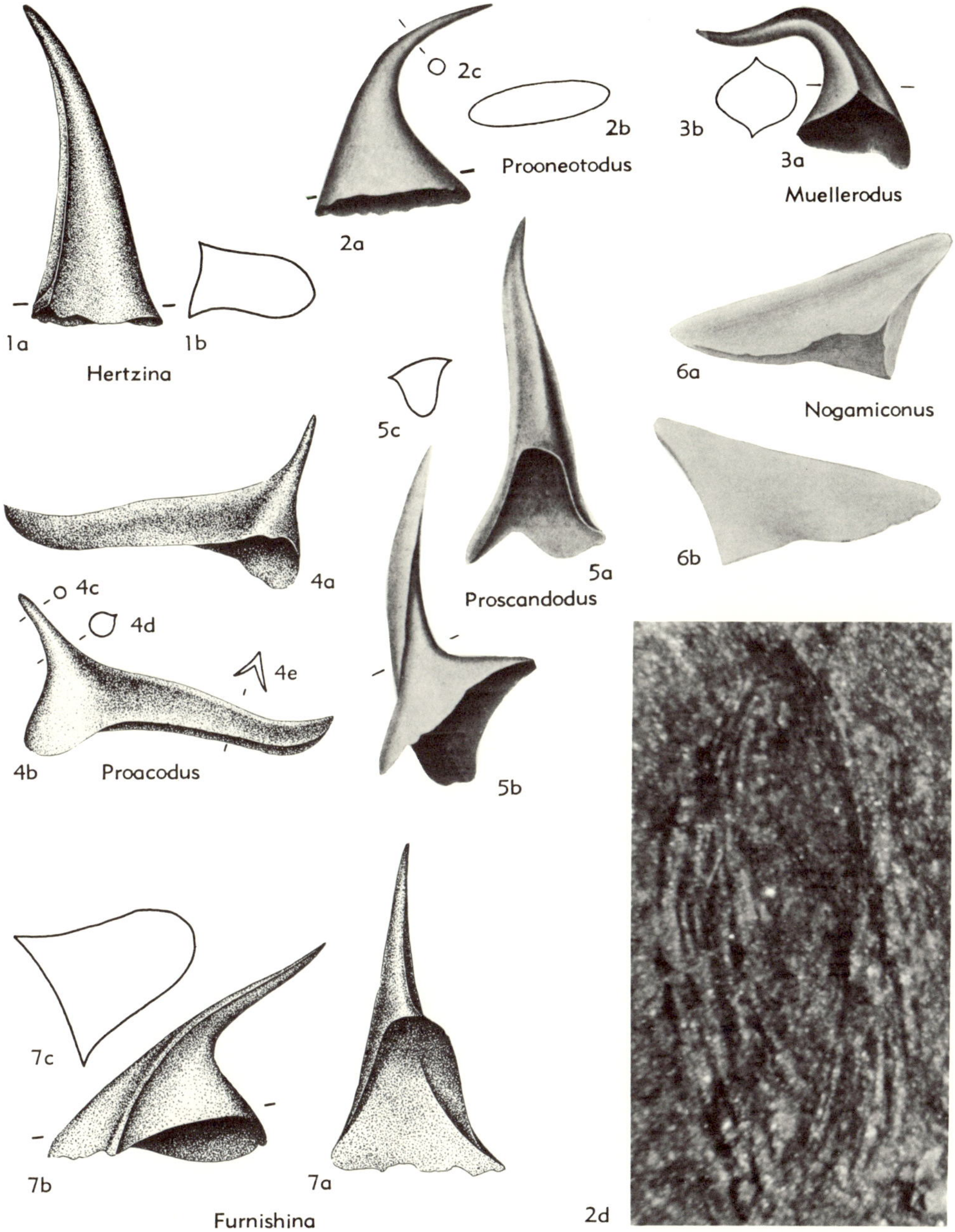

FIG. 64. Furnishinidae (p. W112-W114).

4a, post. view, ×80; *4b-e,* ant. view with transv. secs., ×80 (Müller, 1959).

Problematoconites MÜLLER, 1959, p. 471 [*P. *perforata*; M]. Symmetrical elements, round to oval in cross section; basal cavity large and deep; like *Prooneotodus* but lower part of base perforated by small circular holes. *U.Cam.-L.Ord.,* N.Am.(Utah-Nev.-Okla.-Wyo.)-Eu.(Sweden-Ger., glacial erratics)-Asia(Iran)-Australia(Queensl.).——FIG. 65,*4*. **P. perforatus,* holotype, U.Cam.(Zone 5d), Eu. (Ger.); lat. view, ×54 (Müller, 1959).

Prooneotodus MÜLLER & NOGAMI, 1971, p. 17

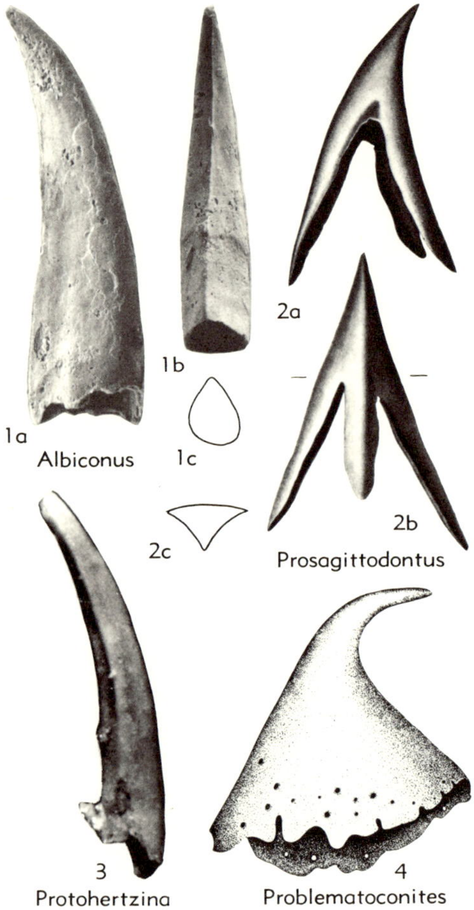

FIG. 65. Furnishinidae (p. W112-W114).

[**Oneotodus gallatini* MÜLLER, 1959, p. 457; OD]. Symmetrical elements lacking ornamentation; proclined; base usually large and in some expanded posteriorly in lower portion; basal cavity large and deep; cross section round to oval. [Natural assemblages of *P. tenuis* (MÜLLER), 1959, are unimembrate, consisting of two oppositely curved sets, each of 4 to 5 coniform elements. Sets are curved like a pair of parentheses. Tips of elements are all close together at the same end of the assemblage and bases are somewhat spread apart. These assemblages are the basis for the interpretation that all Paraconodontida are unimembrate.] *M.Cam.-L.Ord.,* N.Am.(N.Y.-widespread in W. and SW. states of USA-Alberta-Oaxaca)-Eu.(G.Brit.-Sweden-Ger., glacial erratics)-Asia (Sib.-Kazakh.-China-S. Korea-Turkey-Iran)-Australia(Queensl.).——FIG. 64,*2a-c*. **P. gallatini* (MÜLLER), holotype, U.Cam., USA(Wyo.); lat. view with transv. secs. near tip and base, ×60 (Müller, 1959).——FIG. 64,*2d*. *P. tenuis* (MÜLLER), U.Cam. or L.Ord., Eu.(G.Brit.); natural assemblage on black shale, ×27 (Miller, n); for interpretative diagram of *2d* with bases of elements spread apart, see Fig. 53,*5*.

Prosagittodontus MÜLLER & NOGAMI, 1971, p. 17 [**Sagittodontus dahlmani* MÜLLER, 1959, p. 460; OD]. Symmetrical, proclined, elongate, pyramidal elements; cross section triangular due to carinae on posterior and both lateral edges; base occupying nearly entire unit to exclusion of cusp, basal margin in some highly arched between carinae, basal cavity large and deep. *U.Cam.-L.Ord.,* N.Am. (Nev.-Utah-Wyo.)-Eu.(Sweden-Ger., glacial erratics)-Asia(China-Kazakh.-Iran)-Australia(Queensl.). ——FIG. 65,*2*. **P. dahlmani* (MÜLLER), holotype, U.Cam.(Zone 5d), Eu.(Ger.); *2a-c,* lat. and post. views, transv. sec., ×40 (Müller, 1959).

Proscandodus MÜLLER & NOGAMI, 1971, p. 18 [**Scandodus tortilis* MÜLLER, 1959, p. 464; OD]. Asymmetrical, proclined to erect elements; base large and greatly expanded posteriorly; cusp well developed, bent, large basal cavity opening to one side; prominent posterolateral carinae on each side of cusp producing broadly triangular cross section. *U.Cam.,* N.Am.(Nev.-?S.Dak.)-Eu.(Sweden-Ger., glacial erratics)-Asia(China).——FIG. 64,*5*. **P. tortilis* (MÜLLER), holotype, Eu.(Ger.); *5a-c,* post. and lat. views, transv. sec., ×40 (Müller, 1959).

Protohertzina MISSARZHEVSKY, 1973, p. 54 [**P. anabarica*; OD]. Symmetrical, erect, slender, coniform elements with large basal cavity extending nearly to tip of cusp; anterior rounded, posterior with prominent keel, some elements also with posterolateral keels. *Uppermost Precam.(Yudom.)-L.Cam.(Tommot.),* Asia(Sib.-Kazakh.).——FIG. 65,*3*. **P. anabarica,* uppermost Precam., USSR; lat. view, about ×70 (Missarzhevsky, n).

Family WESTERGAARDODINIDAE Müller, 1959

[Westergaardodinidae MÜLLER, 1959, p. 445]

Aberrant paraconodonts with 2 to 5 cusplike projections of subequal size. Basal cavity continuous from side to side with growth lamellae interrupted on under side or divided into two lateral cavities with growth lamellae continuous around base. *M.Cam.-M.Ord.*

Westergaardodina MÜLLER, 1959, p. 465 [**W. bicuspidata*; OD]. Unimembrate elements with 2 or 3 cusplike projections; middle projection, if present, usually smaller; basal cavity large, in symmetrical forms may be replaced by 2 lateral cavities, or in asymmetrical forms by single lateral cavity. Small spheres associated with type species may belong to same taxon. *M.Cam.-M.Ord.,* N. Am.(N.Y.-widespread in W. and SW. states of USA)-Eu.(Pol.-Sweden-Ger., glacial erratics)-Asia

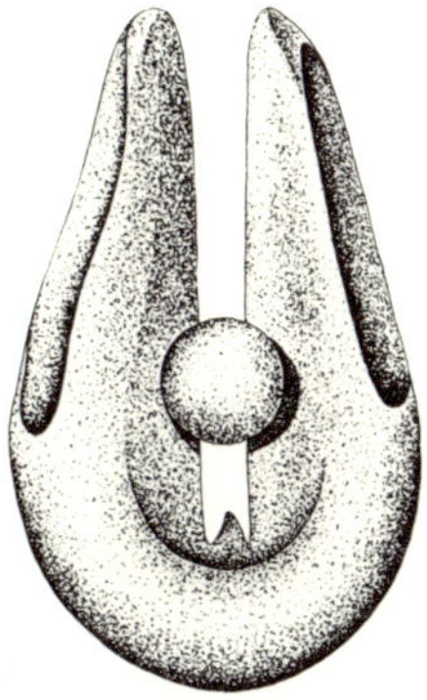

1

Westergaardodina

2

Chosonodina

Fig. 66. Westergaardodinidae (p. *W*114-*W*115).

(Sib.-China-Turkey-Iran)-Australia(Queensl.).——Fig. 66,*1*. **W. bicuspidata,* U.Cam.(Zone 5d), Eu.(Ger.); about ×55 Müller, 1959).

Chosonodina Müller, 1964, p. 99 [**C. herfurthi*; OD]. Elements with 5 to 7 denticles; basal cavity double and present on both sides. *L.Ord.,* Asia (Korea)-N.Am.(Okla.-Utah)-Australia (Queensl.).——Fig. 66,*2*. **C. herfurthi,* Asia(Korea); ×37 (Müller, n).

Order CONODONTOPHORIDA Eichenberg, 1930

[Conodontophorida Eichenberg, 1930, p. 181; =euconodonts of Bengtson, 1976] [Diagnosis by Gilbert Klapper]

Elements characterized by lack of exterior layer of dark organic material; element proper and basal plate or funnel mostly apatite; growth lamellae more numerous and more closely spaced than in elements of Paraconodontida; all growth lamellae continuous around tip (including basal plate or funnel); growth pattern from tip upward and outward in element proper and outward and downward in basal plate or funnel; basal plate or funnel present in best preserved material; white matter usually in cusp and denticles, except in elements of Chirognathacea and some Distacodontacea, in these, white matter greatly reduced or missing and basal cavities shallow. Genera represented by uni- or multimembrate apparatuses. *U.Cam.-U.Trias.*

Superfamily PROCONODONTACEA Miller, new

[Materials for this superfamily prepared by J. F. Miller]

Coniform and rare dolabriform elements mostly with at least some white matter; basal cavities ranging from extremely deep to virtually absent. Apparatuses unimembrate or bimembrate, some with symmetry transitions. Sculpture consisting of keels, costae, spines, nodes, and granules. *U.Cam.-L.Ord.*

Family CLAVOHAMULIDAE Lindström, 1970

[Clavohamulidae Lindström, 1970, p. 430]

Nongeniculate coniform elements with granulose, nodose, or spinose sculpture. *U.Cam.(Corbinia apopsis Subzone)-L.Ord.*

Clavohamulus Furnish, 1938, p. 326 [**C. densus*; OD]. Unusual coniform elements, proclined; basal cavity usually very shallow and in some reduced to flat or convex attachment area; cusp blunt and in some so reduced as to be longer than high; cross section generally round to oval; spine occurring as posterior process in type species, commonly lacking in other species; surface sculpture of fine granular nodes. *L.Ord.,* N.Am.(up. Mississippi River valley-Pa.-widespread in W. and SW. states of USA-Arctic Can.)-NW.Greenl.-Asia(Sib.).——Fig. 67,*3*. **C. densus,* syntype, USA(Minn.); *3a,* basal oblique view; *3b,* post. oblique view; about ×200 (Miller, n).

Hirsutodontus Miller, 1969, p. 431 [**H. hirsutus*; OD] [=*Strigaconus* Druce & Jones, 1971]. Coniform elements, proclined to recurved, rounded in cross section; similar to *Teridontus,* but base with nodes or spines. *U.Cam.(Corbinia apopsis Subzone)-L.Ord.(Symphysurina Zone),* N.Am. (widespread in USA-Alberta)-Greenl.-Asia(Sib.)-Australia(Queensl.).——Fig. 67,*2a*. **H. hirsutus,* L.Ord., USA(Utah); lat. oblique view, ×125

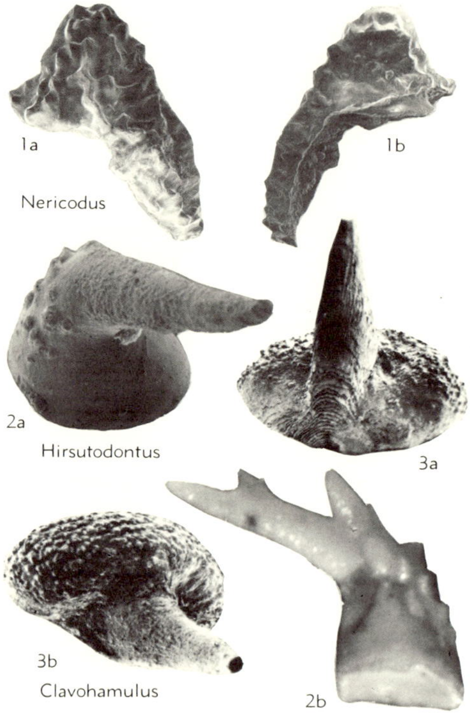

FIG. 67. Clavohamulidae (p. W115-W116).

(Miller, n).——FIG. 67,*2b*. *H. simplex* (DRUCE & JONES), L.Ord., USA(Texas); lat. view, ×100 (Miller, n).

Nericodus LINDSTRÖM, 1955, p. 570 [**N. capillamentum*; M]. Unusual arched coniform elements with surface highly modified by unevenly distributed nodes, in some joining to form irregular ridges. [This genus is known only from a few incomplete specimens.] *L.Ord.*, Eu.(Sweden).——FIG. 67,*1*. **N. capillamentum; 1a,b,* ant. and post. views, ×50 (Miller, 1980).

Family CORDYLODODONTIDAE Lindström, 1970

[*nom. transl.* MILLER, herein, *ex* Cordylodontinae LINDSTRÖM, 1970, p. 429]

Elements with relatively deep basal cavities; forming probable bimembrate apparatuses usually composed of symmetrical elements with rounded edges and asymmetrical elements with sharp edges and a lateral carina. Cusp composed of white matter. Elements either both coniform or both dolabrate. *U.Cam.-L.Ord.*

Cordylodus PANDER, 1856, p. 33 [**C. angulatus*; SD ULRICH & BASSLER, 1926, p. 8]. Inferred bimembrate apparatus of two types of dolabrate ramiform elements; denticles 1 to 5 or more on posterior edge or on posterior process; basal cavity large, moderately deep to shallow, extending beneath posterior process, in some extending into one or more denticles; tip of basal cavity subparallel to sides of cusp or recurved anteriorly. Rounded element symmetrical or nearly so, cusp and denticles round to oval in cross section. Compressed element usually asymmetrical due to lateral bend of cusp and prominent lateral carina, both features lost in some advanced species, such species with symmetrical compressed elements. Cusp and denticles of compressed element strongly compressed laterally, both with sharp edges. Rounded element more distinctive, usually 2 to 3 times as abundant as compressed element; in advanced species, compressed element rare or absent. [The apparatus of the type species probably includes rounded elements described as *C. angulatus* PANDER and compressed elements described as *C. prion* LINDSTRÖM, 1955. Elements described as *Pravognathus aengensis* LINDSTRÖM, 1955, cannot be definitely excluded from the apparatus of the type species. SWEET & BERGSTRÖM (1972) suggested that the apparatus of *C. angulatus* includes elements described as *C. rotundatus* PANDER, but the latter more likely represents the rounded element of a separate species.] *U.Cam.(Corbinia apopsis Subzone)-L.Ord.*, widespread, N.Am.-Eu.-Asia-Australia.——FIG. 68,*3*. **C. angulatus*, L.Ord., USA (Texas); *3a-c*, rounded element, lat. view, transv. sec., and shape of basal cavity; *3d-f*, compressed element, lat. view, transv. sec., and shape of basal cavity; ×100 (Miller, n).

Cambrooistodus MILLER, 1980, p. 9 [**Oistodus cambricus* MILLER, 1969, p. 431; OD]. Inferred bimembrate apparatus; elements coniform, proclined to erect. Asymmetrical geniculate element distinctive, large to small, asymmetrical due to bending of cusp laterally; basal cavity very deep to shallow; cusp correspondingly short to long, composed of white matter, strongly compressed laterally, with prominent anterior and posterior keels; carina on concave side of base. Symmetrical nongeniculate element not distinctive, apparently not distinguishable from symmetrical element of *Eoconodontus*. *U.Cam.(Trempeal.)*, N.Am.(Utah-Nev.-Texas-Okla.-Alberta).——FIG. 68,*2*. **C. cambricus* (MILLER), USA(Texas); *2a,b*, asymmetrical element, lat. view and transv. sec., ×55 (Miller, 1980).

Eoconodontus MILLER, 1980, p. 21 [**Proconodontus notchpeakensis* MILLER, 1969, p. 438; M]. Inferred bimembrate apparatus; elements nongeniculate coniform, proclined to erect; basal cavity large and moderately deep to very deep; cusp very short to long and composed of white matter. Asymmetrical element bent laterally with carina on concave side; cusp strongly compressed laterally, anterior and posterior keels prominent. Symmetrical element lacking carina, in some with prominent anterior and posterior keel, in some keels lost, resulting in oval cross section. Symmetrical element 2 to 3 times as abundant as asymmetrical element. *U.Cam.(Trempeal.)-L.Ord.*, N.Am.(Pa.-Wis.-widespread in W. and SW. states of USA-Alberta-Dist. Mackenzie-Oaxaca)-Asia(?China-Korea-Sib.-Turkey-Iran)-Australia (Queensl.).——FIG. 68,*1*. **E. notchpeakensis* (MILLER), U.Cam., USA (Texas); *1a,b,* symmetrical element, lat. view, transv. sec., ×110; *1c-e,* asymmetrical element, lat. view, transv. secs. near midcusp and base, ×110 (Miller, 1980).

Family ONEOTODONTIDAE Miller, new

Nongeniculate coniform elements, proclined to erect, forming apparent multimembrate apparatuses by symmetry transition. One element lacking costae, other elements with multiple lateral or posterior costae. *U.Cam.-L.Ord.*

Oneotodus LINDSTRÖM, 1955, p. 581 [**Distacodus? simplex* FURNISH, 1938, p. 328; OD]. Nongeniculate coniform elements forming apparent symmetry-transition series. Elements proclined to erect, cusp composed entirely of white matter; basal cavity shallow, triangular, terminating in conical tip, anterior edge of basal cavity close to anterior margin of cusp, anterior margin of cusp smooth, remainder of cusp smooth or bearing multiple low costae beginning above basal margin and extending to near tip; cross section subcircular, ellipsoidal, or flattened to recessed posteriorly, depending on presence and distribution of costae. Apparatus consisting of many symmetry variants, ranging from elements with no costae to those with many. [The holotype of *O. simplex* (FURNISH) is noncostate; slightly younger elements, incorrectly identified as *Scolopodus cornutiformis* BRANSON & MEHL by various authors, possess numerous costae and are believed to be part of this apparatus. ETHINGTON & BRAND (1981) recommended restriction of *Oneotodus* to the type species until restudy of other species in the genus is made. Some of these species are assignable to *Prooneotodus* MÜLLER & NOGAMI, 1971 and *Teridontus* MILLER, 1980.] *L.Ord.*, N.Am.——FIG. 69,*3*. **O. simplex* (FURNISH), lectotype, USA (Iowa); lat. view, ×250 (Clark, n).

Monocostodus MILLER, 1980, p. 26 [**Acodus sevierensis* MILLER, 1969, p. 418; M]. Slender, erect to reclined, coniform elements; cross section round below about bend in cusp; narrow, sharp costa extending from bend in cusp to tip; costa usually on one or the other side, producing dextral and sinistral specimens, but rare symmetrical specimens have posterior costa, forming a symmetry transition. *L.Ord.(Symphysurina Z.)*, N.Am.(Utah-Nev.-

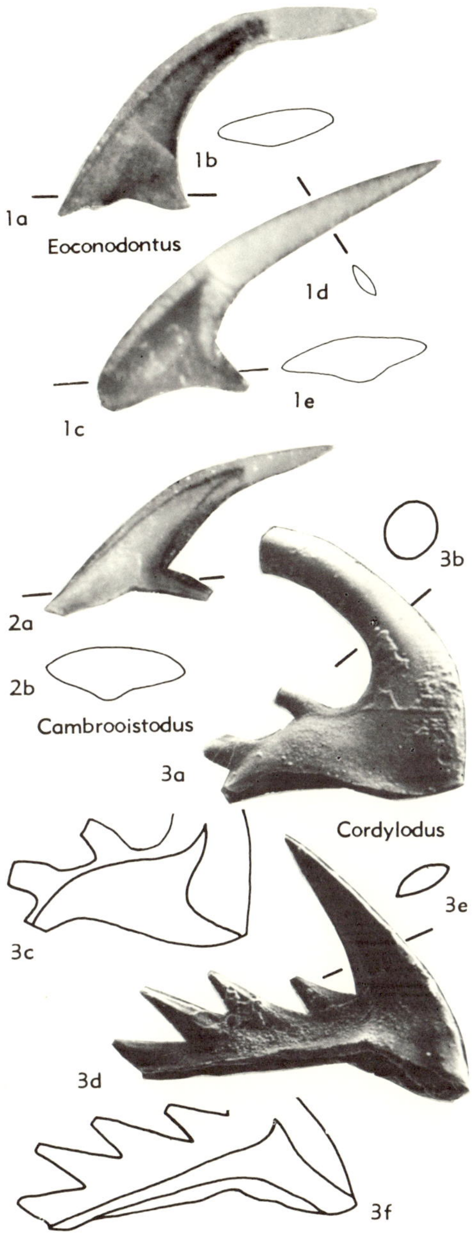

FIG. 68. Cordylododontidae (p. *W*116-*W*117).

Texas-Okla.-Wyo.-Mont.)-Australia(Queensl.).——FIG. 69,*4*. **M. sevierensis* (MILLER), USA(Texas); *4a,* basal oblique view; *4b,c,* lat. view, transv. sec. near midcusp; *4d,* shape of basal cavity; about ×95 (Miller, 1980).

?**Pseudopanderodus** LANDING, 1979, *U.Cam.*, see addendum.

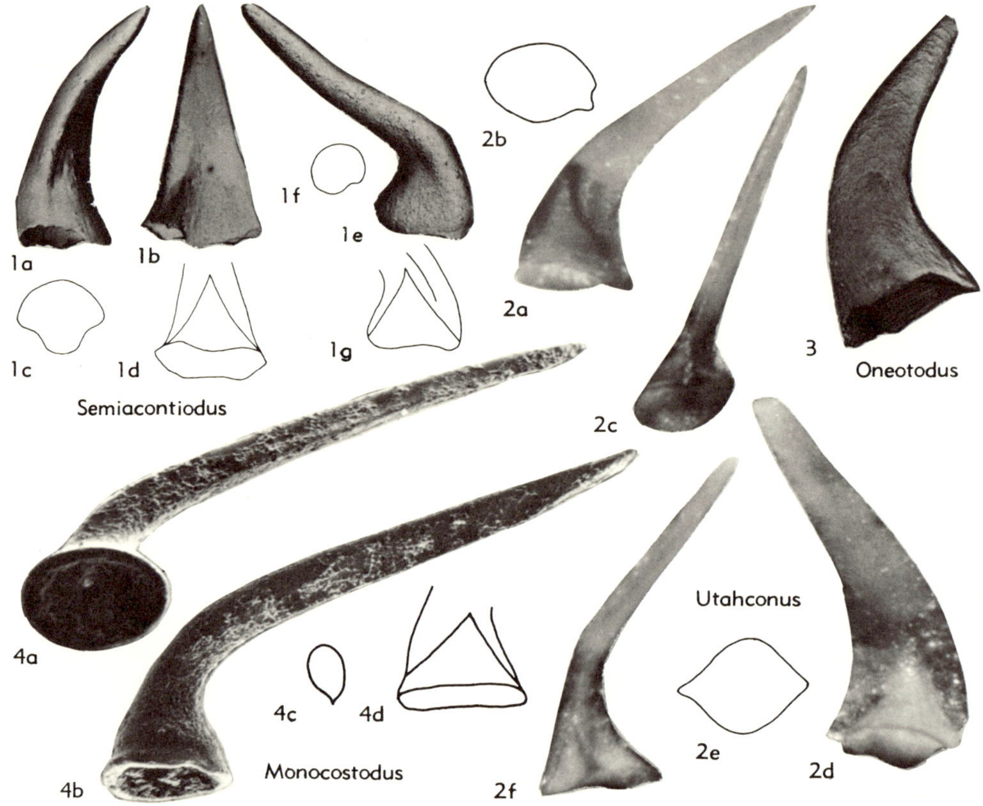

FIG. 69. Oneotodontidae (p. *W*117-*W*118).

Semiacontiodus MILLER, 1969, p. 420 [**Acontiodus (Semiacontiodus) nogamii*; OD]. Erect to reclined coniform elements of two types, arranged in symmetry transition. Symmetrical element much less abundant than asymmetrical element; anterior side lacking costae (also lacking on asymmetrical element); somewhat compressed anteroposteriorly; lateral or posterolateral costa on both sides; posterior costa may be present; fine striae on all sides. Asymmetrical element generally round to oval at base, base in some slightly extended posteriorly; lateral costa on one side or the other, resulting in dextral and sinistral forms; groove often present posterior to costa. *L.Ord.(Missisquoia Z.-Symphysurina Z.),* N.Am.(Utah-Wis.-Texas-Okla.-S.Dak.-Pa.-Alberta)-Asia(Sib.-?China).——FIG. 69,*1*. **S. nogamii* (MILLER), USA(Okla.); *1a-d,* symmetrical element, *1a,* lat. oblique view, *1b-d,* post. view, transv. sec., shape of basal cavity; *1e-g,* asymmetrical element, lat. view, transv. sec., shape of basal cavity; ×100 (Miller, 1980).

Utahconus MILLER, 1980, p. 35 [**Paltodus utahensis* MILLER, 1969, p. 436; OD]. Inferred bimembrate apparatus of coniform elements forming symmetry transition; elements proclined, usually bent to one side; basal cavity round and conical, diameter about equal to height; base a prominent cone modified by one or two large costae extending from basal margin to tip of cusp; cusp large and composed of white matter, costate. Unicostate element usually more abundant; usually asymmetrical with right or left lateral costa, rare symmetrical specimens with posterior costa. Bicostate element usually asymmetrical with lateral and posterolateral costae, rare symmetrical specimens with lateral costa on each side. *L.Ord.,* N.Am. (Pa.-Wis.-widespread in W. and SW. states of USA-Alberta)-Asia (Turkey)-Australia (Queensl.). ——FIG. 69,*2a-f*. **U. utahensis* (MILLER), USA (Utah); *2a-c,* unicostate element, lat. view, transv. sec., post. view; *2d-f,* bicostate element (slightly broken at anterobasal corner), post. view, transv. sec., lat. view; ×110 (Miller, 1980).

Superfamily FRYXELLODONTACEA Miller, new

[Materials for this superfamily prepared by J. F. MILLER]

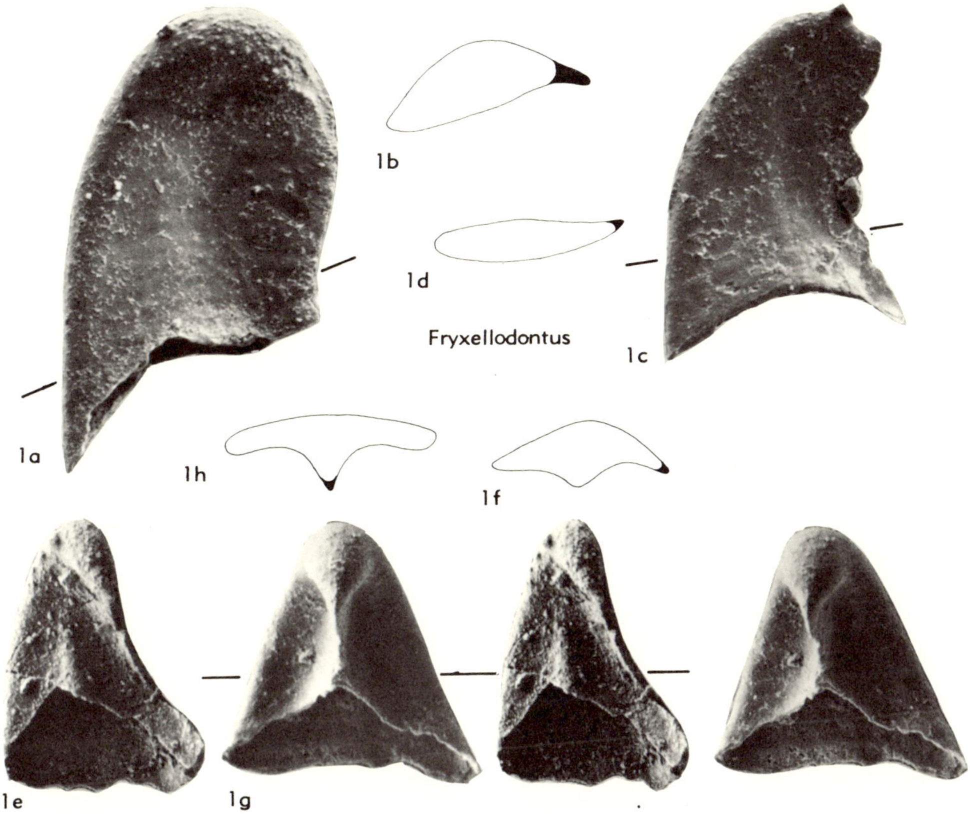

FIG. 70. Fryxellodontidae (p. *W*119).

Nongeniculate coniform elements with deep basal cavities, forming multimembrate apparatuses by symmetry transition. White matter present in minor amounts. *L.Ord.*

Family FRYXELLODONTIDAE Miller, new

Diagnosis as for superfamily. *L.Ord.*

Fryxellodontus MILLER, 1969, p. 426 [**F. inornatus*; OD]. Inferred tri- or quadrimembrate apparatus of unusual coniform elements forming symmetry transition; elements proclined with large, deep, generally compressed basal cavities and little, if any, white matter; posterior edge drawn out into thin flap; small amount of white matter in flap posterior to tip of basal cavity. *Sa* element compressed anteroposteriorly in plane perpendicular to posterior flap; element usually wider than high. *Sb* elements occurring as sinistral and dextral units in about equal numbers; basal cavity compressed in plane forming distinct angle with posterior flap; in some, posterolateral carina corresponding in position with flap. Sc_1 (planar) element laterally compressed in same plane with posterior flap. Sc_2 (serrate) element like Sc_1 but posterior edge of flap serrate, element lacking in one species. *Sb*, Sc_1, Sc_2 elements generally higher than wide, and may be somewhat twisted. *Sb* elements several times more abundant than *Sa; Sc* elements several times more abundant than *Sb.* Upper anterior and lateral faces on all elements may have series of paired ridges. *L.Ord.(Missisquoia Z.-Symphysurina Z.),* N.Am.(N.Y.-Mo.-widespread in W. and SW. states of USA-Arctic Can. islands)-Greenl.-Australia(Queensl.).——FIG. 70,*1*. **F. inornatus,* USA(Okla.); *1a,b,* Sc_1 element, lat. view, transv. sec.; *1c,d,* Sc_2 element, lat. view, transv. sec.; *1e,f, Sb* element, post. view (stereopair), transv. sec.; *1g,h, Sa* element, post. view (stereopair), transv. sec.; black area on transv. secs. position of prominent ridge; ×110 (Miller, 1980).

Superfamily PRIONIODONTACEA Bassler, 1925

[*nom. transl.* LINDSTRÖM, 1970, p. 434, *ex* Prioniodontidae BASSLER, *nom. correct.* MOORE & SYLVESTER-BRADLEY, 1957b, p. 28, *pro* Prioniodidae BASSLER, 1925, p. 218] [Unless noted otherwise, materials for this superfamily prepared by S. M. BERGSTRÖM]

Apparatus in most, if not all forms multimembrate, characteristically sexi- or septimembrate, and composed of *P, M,* and *S* elements, but variously reduced or modified in some forms; most elements ramiform or pectiniform, multidenticulate, with basal cavity along most of length of processes; white matter abundant in cusps and denticles, only exceptionally *(Bergstroemognathus)* missing; many forms with prominent surface microsculpture on lateral and upper surfaces. *Ord.-Dev.*

Family BALOGNATHIDAE Hass, 1959

[*nom. transl.* LINDSTRÖM, 1970, p. 435, *ex* Balognathinae HASS, 1959, p. 379]

Apparatus septimembrate in at least some forms; *P* element robust, more or less platformlike, with one or several lateral processes and well-developed, wide basal cavity, central rows of denticles on upper surface and characteristic microsculpture laterally (best shown by SEM); *M* and *S* elements more delicate than *P* elements, ramiform, multidenticulate; four main types of *S* elements form transition series. *L.Ord.-Sil. (Llandov.).*

Amorphognathus BRANSON & MEHL, 1933b, p. 126 [**A. ordovicica*; OD]. [=*Ambalodus* BRANSON & MEHL, 1933b; *Balognathus* RHODES, 1953a; *Holodontus* RHODES, 1953a; *Keislognathus* RHODES, 1955; *Rosagnathus* RHODES, 1955; *Goniodontus* ETHINGTON, 1959a; *Tvaerenognathus* BERGSTRÖM, 1962; *?Tripodontus* KNÜPFER, 1967]. Apparatus septimembrate, with morphologically different left and right elements in *P* positions. *Pa* elements pastiniscaphate, greatly expanded laterally, with simple or bifid lateral processes and wide basal cavity; denticulation of upper surface of all processes restricted to central row without lateral ribs or nodes. *Pb* elements pastinate with well-developed posterior and lateral processes and short anterior process. *M* elements tertiopedate with short, weakly denticulate posterior process and longer, denticulate anterior and lateral processes. *Sa* elements alate with long posterior process and shorter lateral process. *Sb* elements tertiopedate but otherwise similar to *Sa* elements. *Sc* elements bipennate with long, laterally denticulate anterior process. *Sd* elements quadriramate with long posterior and shorter anterior and lateral processes. In most species, processes of all ramiform elements laterally compressed and narrow basal cavity extending along entire length; denticulation on posterior process of "hindeodelloid" type. [Apparatus reconstruction: BERGSTRÖM, 1971.] *?L.Ord., M.Ord.-Sil.(Llandov.),* Eu.-N.Am.——FIG. 71,*1*. **A. ordovicicus,* U.Ord.(Maravillas F.), USA (Texas); *1a,b,* sinistral *Pa* element, upper and lower views, ×32, ×35; *1c,* dextral *Pa* element, upper view, ×32; *1d,e,* sinistral *Pb* element, ant.-lat. and lower views, ×62; *1f,* dextral *Pb* element, ant.-lat. view, ×62; *1g, M* element, lat. view, ×60; *1h, Sc* element, lat. view, ×62; *1i, Sa* element, lat. view, ×62; *1j, Sb* element, lat. view, ×62; *1k, Sd* element, lat. view, ×62 (Bergström, n).

Lenodus SERGEEVA, 1963, p. 138 [**L. clarus*; OD]. Apparatus unknown. Genus based on modified tertiopedate ramiform element with anterior and posterior processes of subequal length and much shorter lateral process; anterior process directed downward, provided with numerous, subequal-sized, confluent denticles, topmost denticle forming short, suberect cusp; posterior process adenticulate; lateral process with upper edge developed into series of nodes; basal cavity deep and wide, occupying under side of processes. [Elements of the type species are basically similar to *M* elements of *Amorphognathus* and *Rhodesognathus*; however, no elements similar to others in apparatuses of these genera have been reported from the type strata of *L. clarus,* and it is unlikely that *Lenodus* is synonymous with either of these genera.] *L. Ord.(Arenig.),* Eu.(USSR).——FIG. 71,*3*. **L. clarus,* USSR(Baltic); *3a-c,* post-lat. view, lat. views, ×71; *3d,e,* post.-lat. and lower views, diagr., ×76 (Sergeeva, 1963).

Rhodesognathus BERGSTRÖM & SWEET, 1966, p. 392 [**Ambalodus elegans* RHODES, 1953a; OD]. Apparatus probably septimembrate; *M* and *S* elements closely similar to, and at present indistinguishable from, those of *Amorphognathus*. *Pa* and *Pb* elements pastinate, lamellar, with short lateral process, large basal cavity beneath processes, and distinct cusp; inner side of posterior process in *Pa* element may have platformlike flange. [*Rhodesognathus* is distinguished from *Amorphognathus* by its lack of pastiniscaphate elements in *Pa* position.] *M.Ord.-U.Ord.,* N.Am.-Eu.——FIG. 71,*2*. **R.* sp. cf. *R. elegans* (RHODES), M.Ord., Eu. (Sweden); *2a,b, Pa* element, lat. views, ×92; *2c,d, Pb* element, lat. views, ×92 (Bergström, n).

Family CYRTONIODONTIDAE Hass, 1959

[*nom. transl.* BERGSTRÖM, herein, *pro* Cyrtoniodontinae HASS, 1959, p. 378]

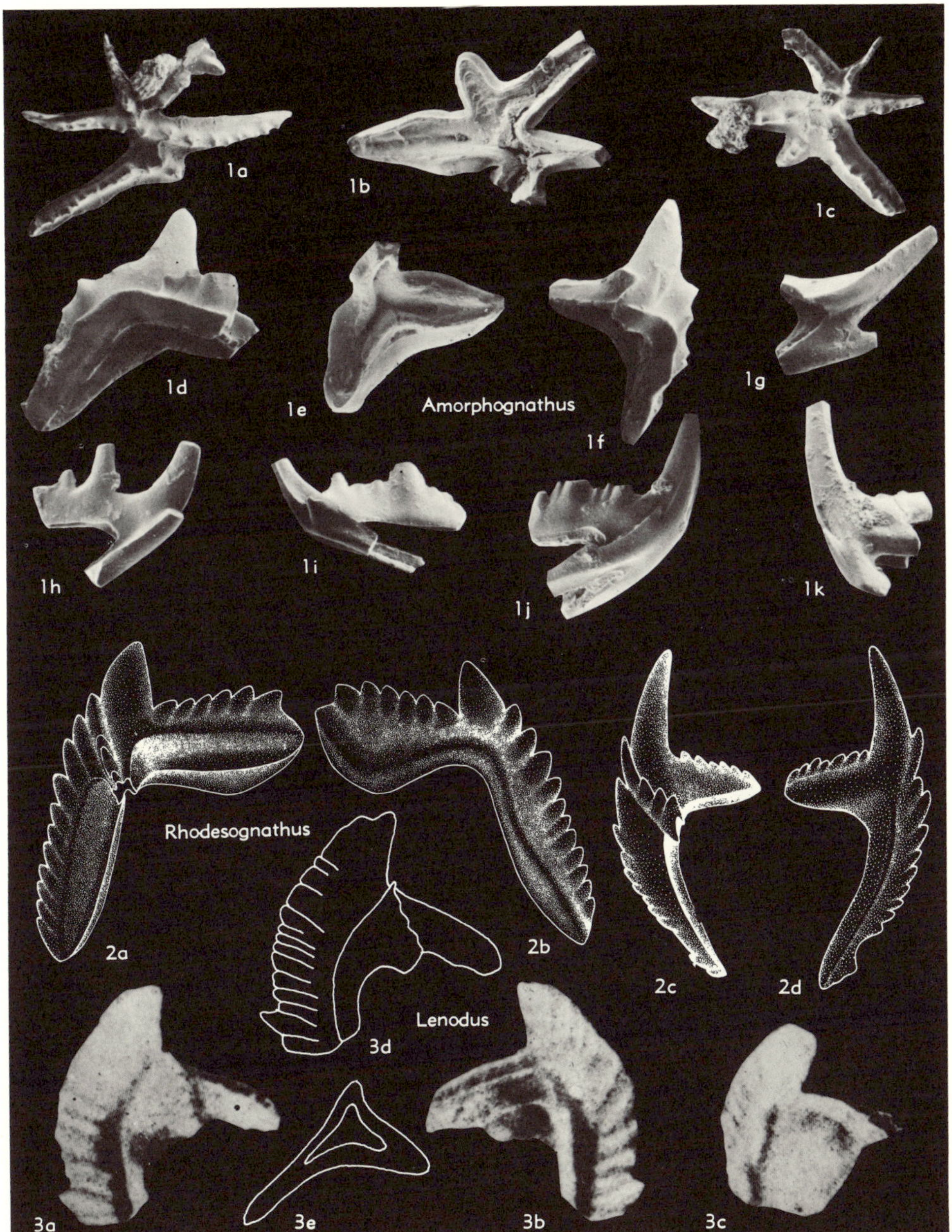

FIG. 71. Balognathidae (p. W120).

Apparatus sexi- or septimembrate in most forms, possibly reduced in some genera; *P* elements carminate, angulate, digyrate, segminate, or pastinate; *M* elements dolabrate in most forms but in some digyrate or geniculate, coniform; *S* elements ramiform, multidenticulate, forming transition series from dolabrate to digyrate or tertio-

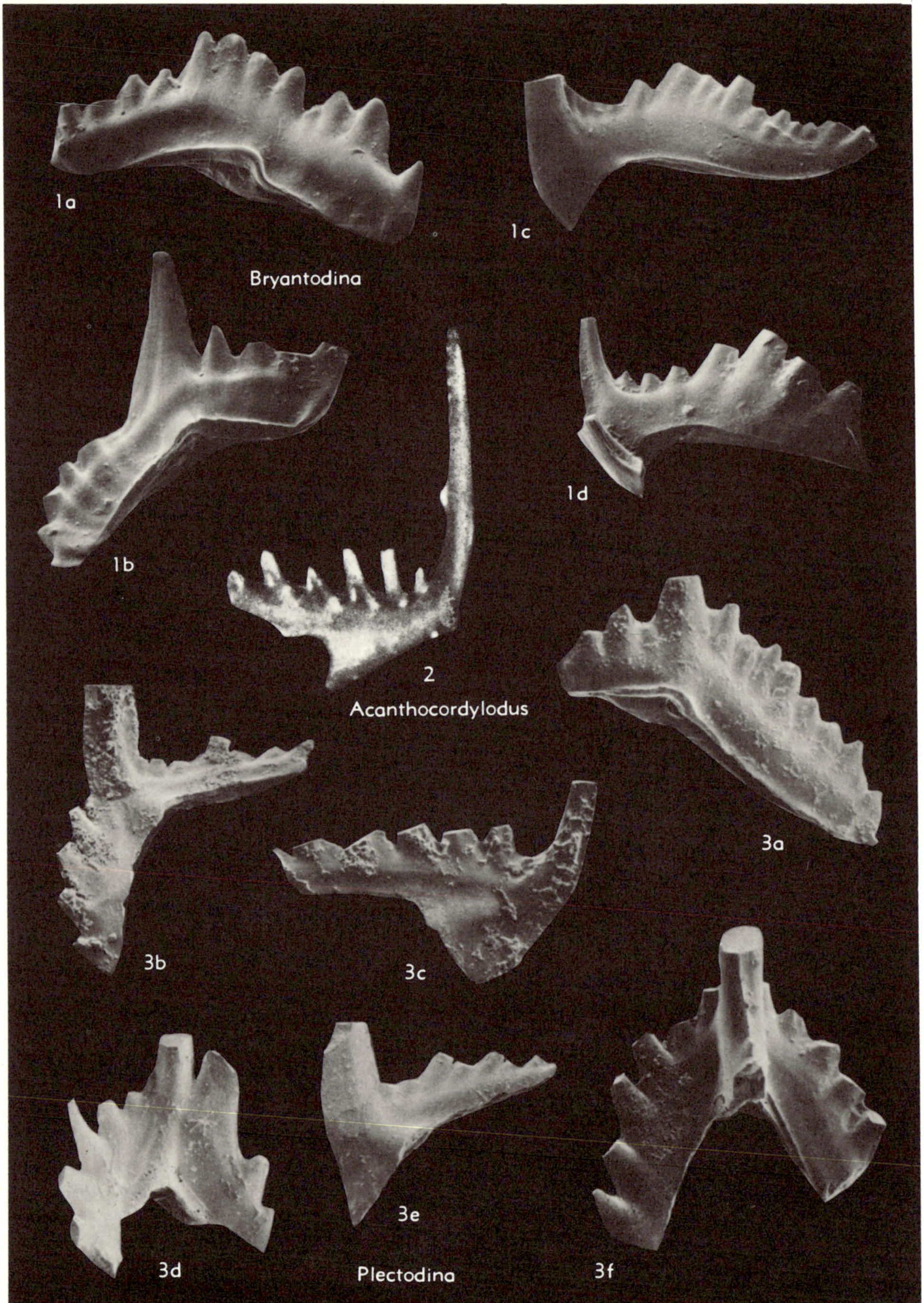

FIG. 72. Cyrtoniodontidae (p. W124).

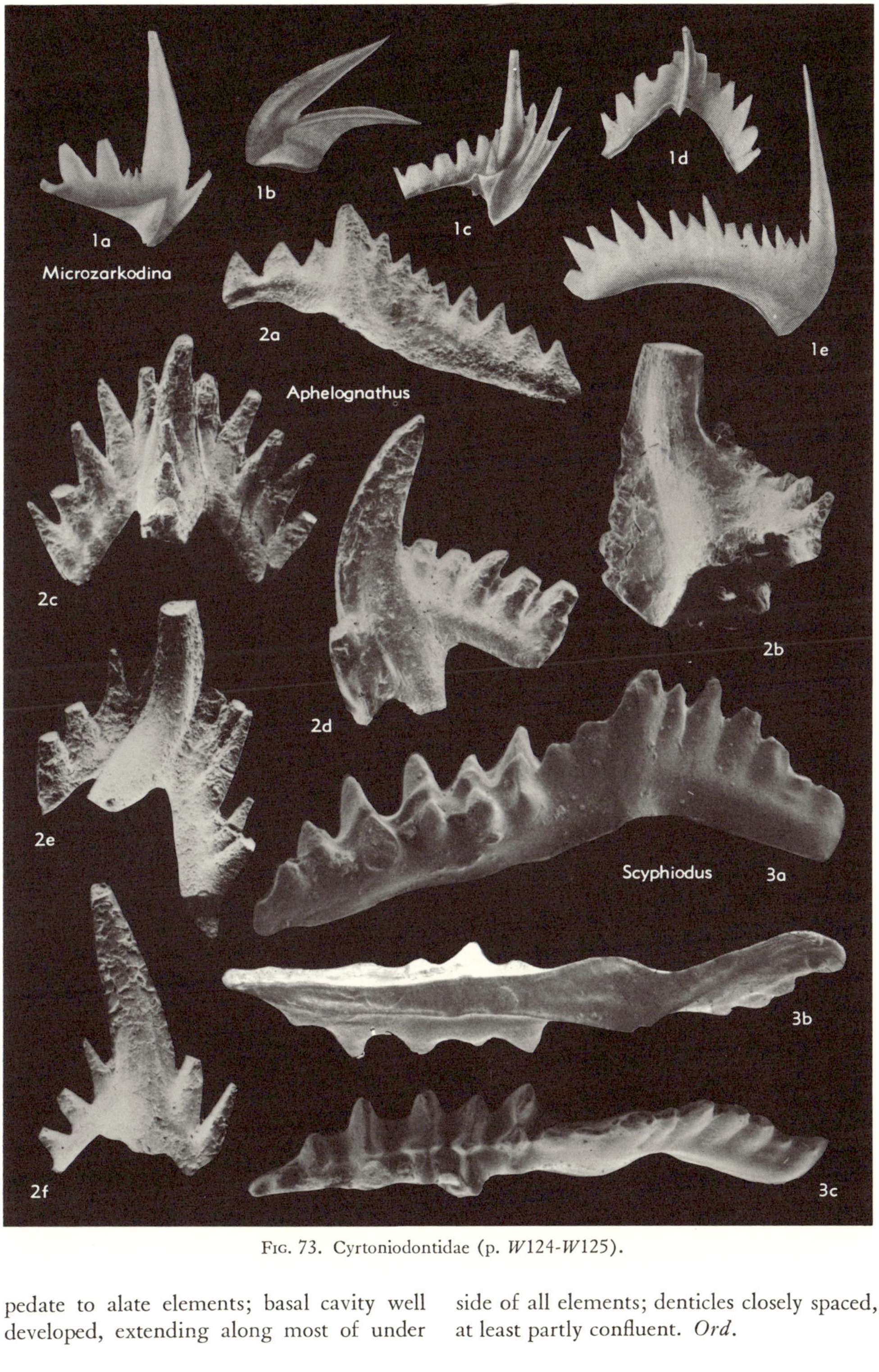

Fig. 73. Cyrtoniodontidae (p. *W*124-*W*125).

pedate to alate elements; basal cavity well developed, extending along most of under side of all elements; denticles closely spaced, at least partly confluent. *Ord.*

Acanthocordylodus MOSKALENKO, 1973, p. 49 [**A. fidelis*; OD]. Apparatus unknown; elements dolabrate with discrete denticles of subequal size on posterior process and long, slender, slightly reclined cusp, posterior or anterior edge, or both, with series of short but distinct, nodelike denticles. *M.Ord.-U.Ord.*, Asia(Sib.).——FIG. 72,*2*. **A. fidelis*, U.Ord.(Dolborsky F.); lat. view, ×65 (Moskalenko, 1973).

Aphelognathus BRANSON, MEHL, & BRANSON, 1951, p. 9 [**A. grandis*; OD]. Apparatus probably septimembrate, including lamellar elements. *Pa* elements angulate, in some species with distinct gap in denticle row immediately anterior to subcentral cusp; denticles robust, confluent along most of length; basal cavity flaring laterally, particularly large beneath cusp, but extending along entire under side of element. *Pb* elements pastinate in some species, bipennate in others, with denticles similar to those in *Pa* element. *M* elements dolabrate with laterally flaring base; bipennate in some species. *S* elements forming transition series from alate through digyrate to dolabrate or bipennate units, each with prominent cusp, in cross section robust denticles more or less rounded, basal cavity well-developed, extending along entire length of processes. [*Aphelognathus* is similar in many respects to *Plectodina*, but differs in form of the *Pa* element. Apparatus reconstruction: SWEET, THOMPSON, & SATTERFIELD, 1975.] *M.Ord.-U.Ord.*, N.Am.-Asia(Sib.).——FIG. 73,*2*. **A. grandis*, U. Ord., USA(Colo.); *2a, Pb* element, lat. view, ×70; *2b, M* element, lat. view, ×61; *2c, Sa* element, post. view, ×79; *2d, Sc* element, lat. view, ×61; *2e, Sb* element, post.-lat. view, ×70; *2f, Pa* element, lat. view, ×79 (Shatzer, 1976).

Bryantodina STAUFFER, 1935a, p. 131 [**B. typicalis*; OD] [=*Tortoniodus* STAUFFER, 1935a]. Apparatus probably seximembrate although only five components have been identified so far. *Pa* element carminate, slightly angulate, with basal cavity consisting of subapical pit and narrow slits along processes; denticles of somewhat variable size, confluent along at least half of length; cusp not appreciably larger than many denticles. *Pb* elements digyrate, with prominent cusp and multidenticulate processes. *M* element not identified. *S* elements forming transition series. *Sa* elements alate, with long, sinous posterior process and very short, straight, laterally directed, lateral processes with one or two denticles. *Sb* elements tertiopedate, otherwise similar to *Sa* elements except for asymmetrically developed lateral processes, one lateral process carrying a large denticle rivaling cusp in size. *Sc* elements dolabrate with long, sinous, multidenticulate posterior process and adenticulate anterior process. Basal cavity in ramiform elements narrow and shallow but extending along entire length of processes. [Apparatus reconstruction: WEBERS, 1966.] *M.Ord.*, N.Am.-Asia(Sib.).——FIG. 72,*1*. **B. typicalis*, Glenwood F., USA (Minn.); *1a, Pa* element, lat. view, ×57; *1b, Pb* element, lat. view, ×57; *1c, Sc* element, lat. view, ×65; *1d, Sa* element, lat. view, ×64 (Bergström, n).

Microzarkodina LINDSTRÖM, 1971, p. 57 [**Prioniodina flabellum* LINDSTRÖM, 1955; OD]. Apparatus seximembrate, composed of lamellar elements. *P* elements carminate to segminate, with prominent cusp and relatively few denticles, especially on the short anterior process. *M* elements geniculate, coniform, with base extended posteriorly. *S* elements with long, slender cusp, relatively discrete denticles, and shallow basal cavity. *Sa* elements alate, without posterior process. *Sb* elements digyrate. *Sc* elements dolabrate. *Sd* elements quadriramate, with poorly developed anterior and lateral processes. [Apparatus reconstruction: LINDSTRÖM, 1971.] *L.Ord.-M.Ord.*, Eu.-N.Am.——FIG. 73,*1*. **M. flabellum* (LINDSTRÖM), L.Ord.(low. Arenig.), Eu.(Sweden); *1a, P* element, lat. view; *1b, M* element, lat. view; *1c*, digyrate *Sb* element, post.-lat. view; *1d*, alate *Sa* element, post. view; *1e*, dolabrate *Sc* element, lat. view; all ×70 (Van Wamel, 1974).

Plectodina STAUFFER, 1935a, p. 152 [**Prioniodus aculeatus* STAUFFER, 1930, p. 126; OD; =*Plectodina dilata* STAUFFER, 1935a] [=*Trichonodella* BRANSON & MEHL, 1948, *nom. subst. pro Trichognathus* BRANSON & MEHL, 1933a, *non* GEMMIGER & HAROLD, 1868 *nec* BERTHOLD, 1927, both beetles; *Eoligonodina* BRANSON, MEHL, & BRANSON, 1951; ?*Zygognathus* BRANSON, MEHL, & BRANSON, 1951]. Apparatus basically seximembrate, elements lamellar. *P* elements carminate to angulate or pastinate with laterally partly confluent denticles; in some species, *Pa* elements with short lateral process or costa; *M* elements dolabrate or digyrate, with basal cavity wall flaring laterally; *S* elements forming symmetry transition series from dolabrate or bipennate through digyrate to alate elements; cusp of *S* elements long, slender; denticles shorter than cusp, discrete along part of length, of relatively uniform size; basal cavity shallow and narrow in all elements, extending along all processes. [Apparatus reconstruction: BERGSTRÖM & SWEET, 1966; SWEET & BERGSTRÖM, 1972.] *M.Ord.-U.Ord.*, N. Am.-Eu.-Australia-Asia(Sib.).——FIG. 72,*3*. *Plectodina* sp., U.Ord.(Cobourg F.), Can.(Ont.); *3a, Pa* element, lat. view; *3b, Pb* element, lat. view; *3c, M* element, lat. view; *3d, Sc* element, lat. view; *3e, Sb* element, lat. view; *3f, Sa* element, post. view; ×55 (Bergström, n).

Scyphiodus STAUFFER, 1935b, p. 617 [**S. primus*; OD]. Apparatus unknown, may be unimembrate; genus based on lamellar, anguloplanate elements with platformlike anterior process bearing 3 rows of denticles and relatively short, subcentral, reclined cusp. Posterior process bladelike with single row of laterally compressed denticles. Basal cavity pitlike beneath cusp, rather deep and wide beneath anterior process, groovelike beneath posterior process. [As noted by WEBERS (1966, p. 46), elements of *Scyphiodus* show, apart from development of the platformlike anterior process, close similarity to angulate elements described as *Bryantodina maxima* STAUFFER, 1935a. *Scyphiodus* is probably more closely related to *Bryantodina, Plectodina,* and *Phragmodus* than to other Middle and Upper Ordovician genera with platformed pectiniform elements.] *M.Ord.,* N.Am.(W. midcontinent).——FIG. 73,*3*. **S. primus,* Platteville F., USA(Minn.); *3a-c,* lat., lower, and upper views, ×102 (Bergström, n).

Family ICRIODONTIDAE Müller & Müller, 1957

[*nom. transl.* LINDSTRÖM, 1970, p. 436, *ex* Icriodontinae, *nom. transl. et correct.* HASS, 1959, p. 379, *pro* Icriodidae MÜLLER & MÜLLER, 1957, p. 1105] [Materials for this family prepared by GILBERT KLAPPER and S. M. BERGSTRÖM]

Essentially trimembrate apparatus characterized by scaphate *Pa* elements. *Pb* and *S* elements are simple cones or modifications thereof. Apparatus of Ordovician species probably pentamembrate. *M.Ord.-Sil.(Llandov.), Sil.(Ludlov.)-U.Dev.(Famenn.).*

Icriodus BRANSON & MEHL, 1938, p. 159 [**I. expansus*; OD; for discussion of validation of *Icriodus* and type species, see ZIEGLER, 1975, p. 68] [=*Acodina* STAUFFER, 1940, p. 418; *Latericriodus* MÜLLER, 1962a, p. 114; *Caudicriodus* BULTYNCK, 1976, p. 19; *Praelatericriodus* BULTYNCK, 1976, p. 40]. Main process of *Pa* element characteristically consisting of three longitudinal rows of nodes but including forms with transverse ridges; denticulate lateral processes developed near posterior end in some. *Pb* element nongeniculate, laterally compressed coniform element with sharp anterior and posterior keels. *S* elements relatively smooth, simple cones of circular to elliptical cross section. [Reconstruction: KLAPPER & PHILIP, 1971; modified herein.] *Sil.(Pridol.)-U. Dev. (Famenn.),* Eu.-N. Afr.-Asia-N. Am.-Australia.——FIG. 74,*4a,b*. **I. expansus,* disputed M.Dev.-U.Dev. boundary interval (*Schmidtognathus hermanni–Polygnathus cristatus* Z., Rapid Mbr., Cedar Valley Ls.), USA(Iowa); *Pa* element, upper and lower views, ×37 (Klapper, n).——FIG. 74,*4c-e*. *I. steinachensis* AL-RAWI, L.Dev. (McMonnigal Ls.), USA(Nev.); *4c, Pa* element, upper view; *4d,e, Pb* element, inner lat. and outer lat. views, ×37 (Klapper & Philip, 1971).

Antognathus LIPNYAGOV in KOZITSKAYA *et al.,* 1978, *U.Dev.,* see addendum.

Icriodella RHODES, 1953a, p. 285 [**I. superba*; OD] [=*Rhynchognathodus* ETHINGTON, 1959b, p. 1128, *nom. subst. pro Rhynchognathus* ETHINGTON, 1959a, p. 286, *non* JAEKEL, 1929, a fish; *Sagittodontus* RHODES, 1953a]. Apparatus probably pentamembrate, including lamellar elements. *Pa* elements pastinoscaphate with long anterior process bearing double row of denticles, short cusp, short adenticulate lateral process, and bladelike posterior process with one row of confluent denticles. *Pb* elements tertiopedate, more or less pyramidal, with short cusp and adenticulate or weakly denticulate processes. *M* elements bipennate or dolabrate, flaring laterally, with adenticulate posterior process and weakly denticulate anterior process. *S* elements tertiopedate, subpyramidal; of two types, one with 3 denticulate processes, the other similar but with anterior process adenticulate. Basal cavity large in all elements, especially in *P* elements. *M.Ord.-Sil.(Llandov.),* N.Am.-Eu.——FIG. 74,*1*. **I. superba,* M.Ord.(Lexington Ls.), USA(Ky.); *1a-c, Pa* element, lat., upper, lower views, ×33; *1d,e, M* element, lat. views, ×82; *1f-g, Pb* element, upper and lat. views, ×26; *1h, S* element with 2 denticulate processes, lat. view, ×47; *1i,j, S* element with 3 denticulate processes, lat. and post.-upper views, ×47 (Bergström & Sweet, 1966).

Pedavis KLAPPER & PHILIP, 1971, p. 446 [**Icriodus pesavis* BISCHOFF & SANNEMANN, 1958, p. 96; OD]. Main process of *Pa* element like that of *Icriodus* but additionally with two anteriorly directed lateral processes and a posterior process. *Pb* element as in *Icriodella* but denticulation on processes much better developed. *S* elements strongly costate cones that may develop 1 or 2 accessory posterior denticles. [Reconstruction: KLAPPER & PHILIP, 1971.] *Sil.(Ludlov.)-L.Dev.,* Eu.-Asia-N.Am.-Australia.——FIG. 74,*3*. **P. pesavis* (BISCHOFF & SANNEMANN), L.Dev.(McMonnigal Ls.), USA(Nev.); *3a, Pa* element, upper view; *3b, S* element, lat. view; *3c, Pb* element, lat. view; ×37 (Klapper & Philip, 1971).

Pelekysgnathus THOMAS, 1949, p. 424 [**P. inclinata*; OD] [=*Drepanodina* MOUND, 1968, p. 480]. *Pa* element like that of *Icriodus* but main process characteristically with only a single longitudinal row of nodes. *Pb* element nongeniculate, laterally compressed coniform element with sharp anterior and posterior keels. *S* elements varying from smooth to costate, unkeeled simple cones of circular to elliptical cross section. [Reconstruction: KLAPPER & PHILIP, 1972.] *Sil. (Pridol.)-U.Dev.(Famenn.),* Eu.-N.Am.-Australia.——FIG. 74,*2*. **P. inclinatus,* U.Dev. (*Scaphignathus subserratus-Pelekysgnathus inclinatus* Fauna, Maple Mill Sh.), USA(Iowa); *2a, Pa* element, lat. view; *2b, Pb* element, lat. view; *2c, S* element, lat. view; ×37 (*2a,* Klapper, n; *2b,c,* Klapper & Philip, 1972).

Sannemannia AL-RAWI, 1977, *L.Dev.-M.Dev.,* see addendum.

Family OEPIKODONTIDAE Bergström, new

Apparatus basically trimembrate, consisting of lamellar pastinate and modified

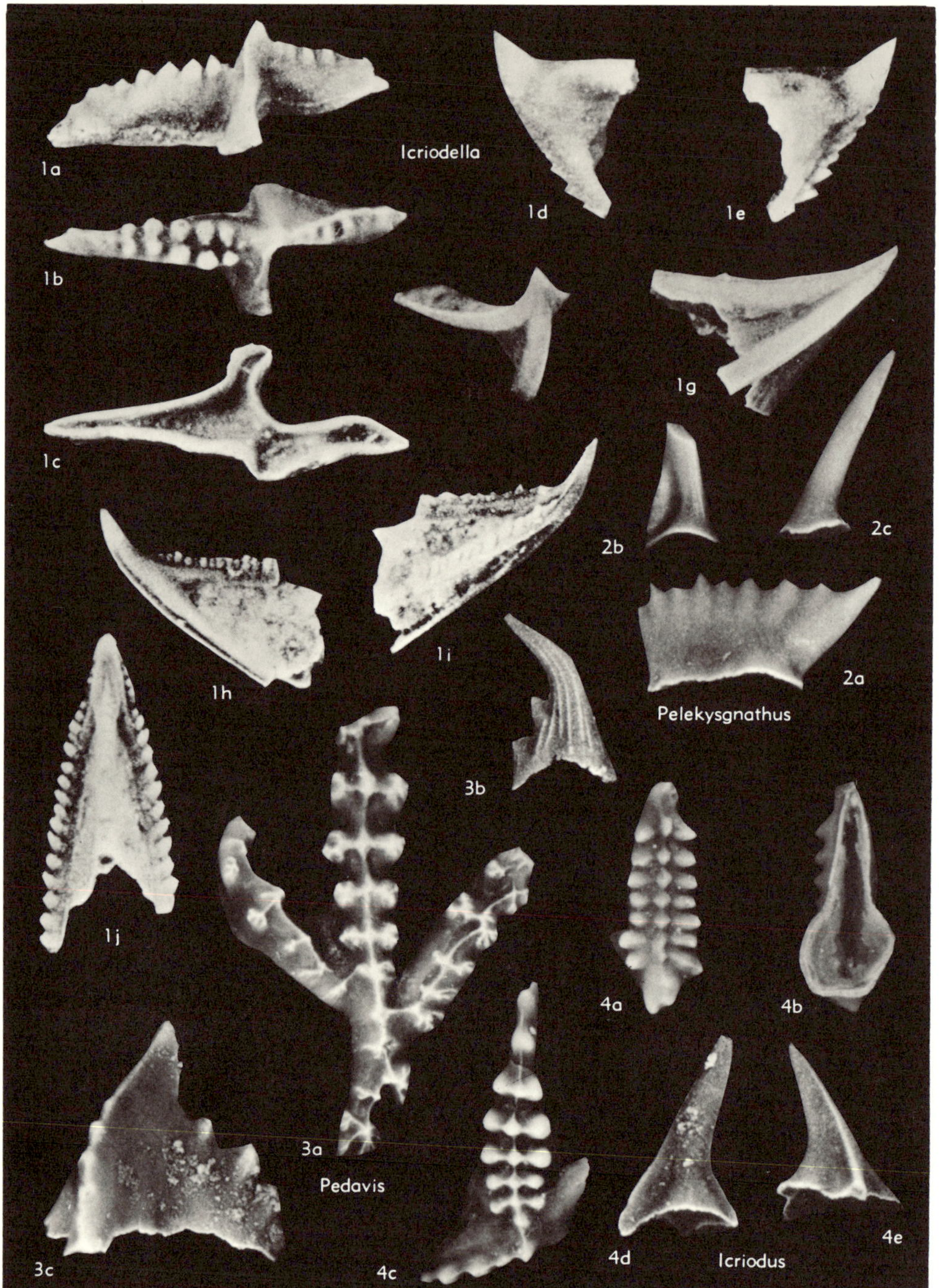

FIG. 74. Icriodontidae (p. W125).

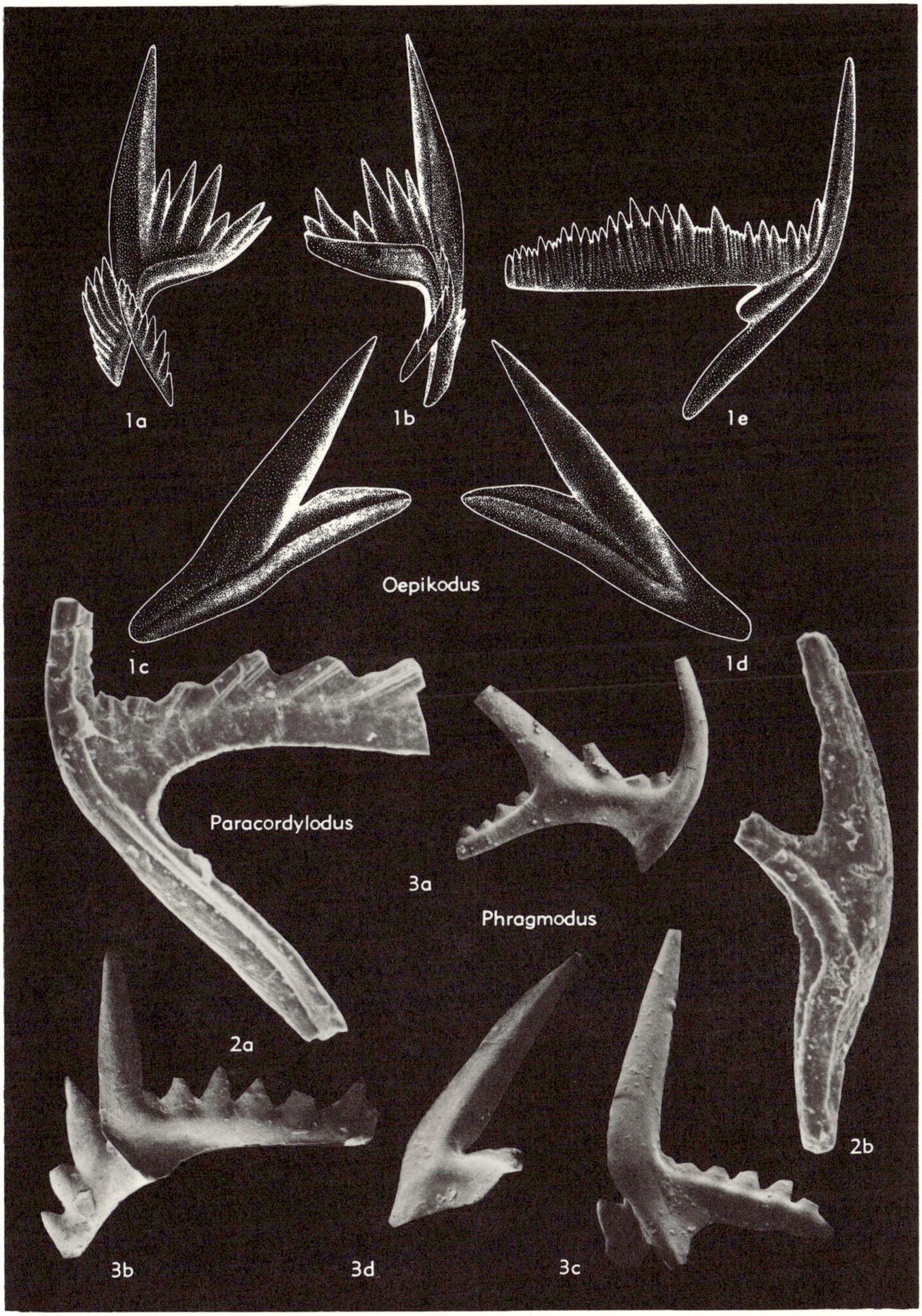

FIG. 75. Oepikodontidae, Paracordylodontidae, Phragmodontidae (p. *W*128-*W*129).

quadriramate, denticulate, ramiform elements and geniculate, adenticulate, coniform elements. Pastinate elements with well-developed, laterally compressed cusp, diverging downward from cusp one anterior and one posterior process and a lateral process, lateral process in some twisted posteriorly; all processes, or at least the posterior one, with denticles confluent along at least part of their length. Quadriramate elements with slender cusp, long adenticulate anterior process, long, laterally compressed multidenticulate posterior process, and rudimentary, adenticulate lateral processes. Basal cavity in both ramiform elements narrow but extending along entire length of processes. Geniculate element with rather long, reclined cusp and long base considerably extended anteriorly and posteriorly, base in some flaring slightly toward one side; basal cavity small, developed as subapical central pit, narrow and shallow grooves extending from pit along most of base. *L. Ord., ?M.Ord.*

Oepikodus LINDSTRÖM, 1955, p. 570 [**O. smithensis*; OD]. Diagnosis as for family. [*Oepikodus* is similar to *Prioniodus*, but is distinguished by the presence of only two basic types of ramiform elements.] *L.Ord., ?M.Ord.,* Eu.-N.Am.-S.Am.-Australia-Asia(China).——FIG. 75,*1*. **O. smithensis,* L.Ord.(*Dalecarlicus* Z.), Eu.(Sweden); *1a,b,* pastinate element, lat. views; *1c,d,* geniculate element, lat. views; *1e,* quadriramate element, lat. view; ×50 (Bergström, n).

Family PARACORDYLODONTIDAE Bergström, new

Apparatus probably bimembrate but may contain additional elements; elements lamellar, strongly laterally compressed, geniculate and dolabrate with recurved carinate cusp. Geniculate elements with arched, anteroposteriorly extended base. Dolabrate elements with adenticulate anterior process directed posteriorly and downward and with an essentially straight denticulate posterior process of similar length as anterior process; denticles reclined, strongly laterally compressed, discrete, highest at midlength of process. Basal cavity shallow and small in both types of elements. Cusp, denticles, and anterior process with conspicuous microstriae. *L.Ord.*

Paracordylodus LINDSTRÖM, 1955, p. 584 [**P. gracilis*; OD]. Diagnosis as for family. [VAN WAMEL (1974) proposed that the apparatus of *Paracordylodus gracilis* includes an additional element, but study of available collections of that species from Europe and North and South America has not yet confirmed the correctness of that view. Apparatus reconstruction: BERGSTRÖM, EPSTEIN, & EPSTEIN, 1972; SWEET & BERGSTRÖM, 1972.] *L.Ord.,* Eu.-N.Am.-S.Am.——FIG. 75,*2*. **P. gracilis,* USA (Pa.); *2a,* dolabrate element; *2b,* geniculate element; ×100 (Bergström, Epstein, & Epstein, 1972).

Family PERIODONTIDAE Lindström, 1970

[Periodontidae LINDSTRÖM, 1970, p. 435]

Apparatus multimembrate, in at least some forms septimembrate; similar to that of representatives of Phragmodontidae but with different type of *P* elements and most of *S* elements in transition series having short denticulate processes, rather than adenticulate anterior costae. *Ord.*

Periodon HADDING, 1913, p. 33 [**P. aculeatus*; M] [=*Loxognathus* GRAVES & ELLISON, 1941; *Falodus* LINDSTRÖM, 1955]. Apparatus multimembrate, composed of lamellar elements. *P* elements of two types; one angulate to bipennate, the other digyrate with posterior process strongly twisted out of plane of anterior process; both types with distinct cusp, multidenticulate processes, and basal cavity extending along entire length of processes. *M* elements geniculate, coniform, with anteriorly denticulate cusp and posteriorly extended base. *S* elements forming transition series from dolabrate through bipennate and tertiopedate to alate elements. All *S* elements with long, denticulate posterior process and basal cavity with prominent zone of recessive basal margin a short distance behind cusp. Denticles of variable size, laterally compressed, more or less confluent; those on posterior process a short distance behind cusp may rival cusp in size. *Ord.,* Eu.-N.Am.-S.Am.-Australia(N.Z.).——FIG. 76,*1*. *P. grandis* (ETHINGTON), M.Ord.(Prosser F.), USA(Minn.); *1a,c,* two types of *P* elements, lat. views, ×61; *1b, M* element, lat. view, ×61; *1d-f,* three types of *S* elements, lat. views, ×77, ×102, ×110 (Bergström, n).

Hamarodus VIIRA, 1975, p. 87 [**Distomodus europaeus* SERPAGLI, 1967, p. 64; OD]. Apparatus type unknown; described elements basically nongeniculate, laterally compressed, lamellar, coniform with proclined to suberect cusp and wide, more

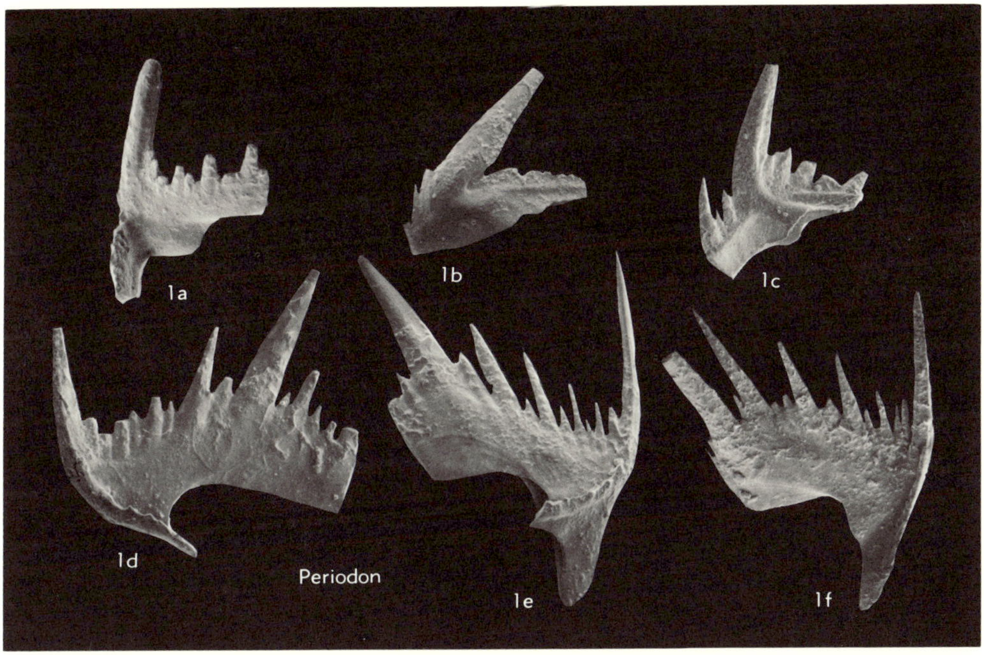

FIG. 76. Periodontidae (p. *W*128).

or less triangular base, base flaring toward one side. In some, both anterior and posterior edges of base with small number of short denticles. Basal cavity deep, extending in some elements halfway to tip of cusp. [*Hamarodus* elements exhibit similarity to those of *Distomodus*; however, it seems unlikely that these genera are closely related.] *U.Ord.*, Eu.(Baltoscandia-Italy-Eng.-Nor.)-?Asia(Sib.).——FIG. 77,*1*. **H. europaeus* (SERPAGLI), Slandrom Ls., Eu.(Sweden); *1a,b*, lat. views, same specimen, about ×100 (Bergström, n).

Family PHRAGMODONTIDAE Bergström, new

Apparatus unknown in type species of *Phragmodus,* in others septimembrate. *P* elements pastinate, or angulate and pastinate, with distinct cusp and multidenticulate anterior and posterior processes and lateral process, if present, adenticulate or weakly denticulate. *M* elements geniculate, coniform or dolabrate, with well-developed cusp and base flaring laterally. *S* elements superficially similar to each other but forming transition series, expressed in disposition of adenticulate costae on cusp, from dolabrate through tertiopedate and alate to quadriramate units; each of these elements with long, multidenticulate, in some species sinuous posterior process and shallow basal cavity extending along entire length of processes. *M.Ord.-U.Ord.*

Phragmodus BRANSON & MEHL, 1933b, p. 98 [**P. primus*; OD] [?=*Dichognathus* BRANSON & MEHL, 1933a; ?*Cyrtoniodus* STAUFFER, 1935a; ?*Subcordylodus* STAUFFER, 1935a]. Diagnosis as for family. *M.Ord.-U.Ord.*, N.Am.-Eu.-Asia-Australia.——FIG. 75,*3*. *P. undatus* BRANSON & MEHL, M.Ord.(Lexington Ls.), USA(Ky.); *3a, S* element, lat. view; *3b,c,* two types of *P* element, lat. views; *3d, M* element, lat. view; ×73 (Bergström, n).

Family POLYPLACOGNATHIDAE Bergström, new

Apparatus apparently reduced to bimembrate type by loss of *S* and *M* elements; *P* elements stelliplanate and pastiniplanate, lamellar, with well-developed platforms; basal cavity restricted to central narrow groove on under side of processes. *L.Ord.-M.Ord.*

Polyplacognathus STAUFFER, 1935b, p. 615 [**P. ramosus*; OD] [=*Petalognathus* DRYGANT, 1974, *non* DUMÉRIL & BIBRON, 1854, a reptile]. Apparatus composed of paired elements; stelliplanate

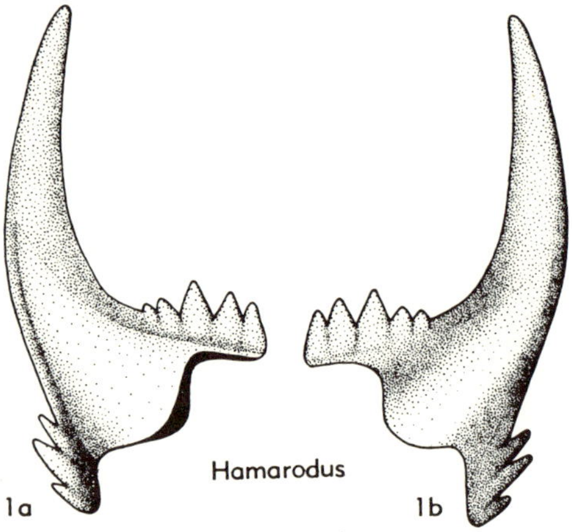

FIG. 77. Periodontidae (p. *W*128-*W*129).

elements with one anterior, one posterior, and four lateral processes; pastiniplanate elements crudely Y-shaped with subcentral cusp and one anterior, one posterior, and one lateral process; all elements with central denticle row and abundant nodes, ridges, and small denticles scattered over entire upper surface of processes. [For distinguishing characters between *Polyplacognathus* and *Eoplacognathus* see latter genus. *Polyplacognathus* is separated from *Amorphognathus* by lack of ramiform elements in the apparatus and by different types of basal cavity, denticulation, and process arrangement. Apparatus reconstruction: BERGSTRÖM & SWEET, 1966; SCHOPF, 1966; WEBERS, 1966.] *M.Ord.*, N.Am.-Eu.-Asia(Sib.).——FIG. 78,*1*. **P. ramosus*, Platteville F., USA(Minn.); *1a,c*, stelliplanate element, upper, lower views; *1b*, pastiniplanate element, upper view; ×45 (Bergström, n).

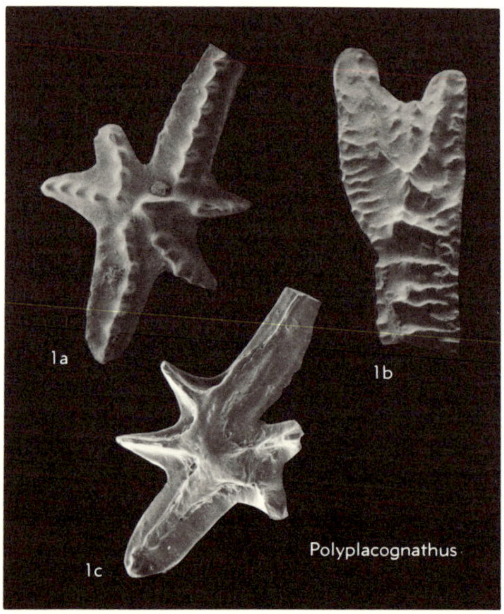

FIG. 78. Polyplacognathidae (p. *W*129-*W*131).

Eoplacognathus HAMAR, 1966, p. 52 [**Ambalodus lindstroemi* HAMAR, 1964, p. 258; OD] [=*Priomorphognathus* KNÜPFER, 1967]. Apparatus composed of *Pa* and *Pb* elements but left and right elements in each pair not mirror images, morphologically distinct. *Pa* elements stelliplanate with an anterior process, in some bladelike, a posterior process, and 2 to 3 lateral processes, one lateral process bifid in some. *Pb* elements pastiniplanate, Y-shaped, with long anterior process bladelike in some distally, short posterior process forming angle with anterior process, and short lateral process de-

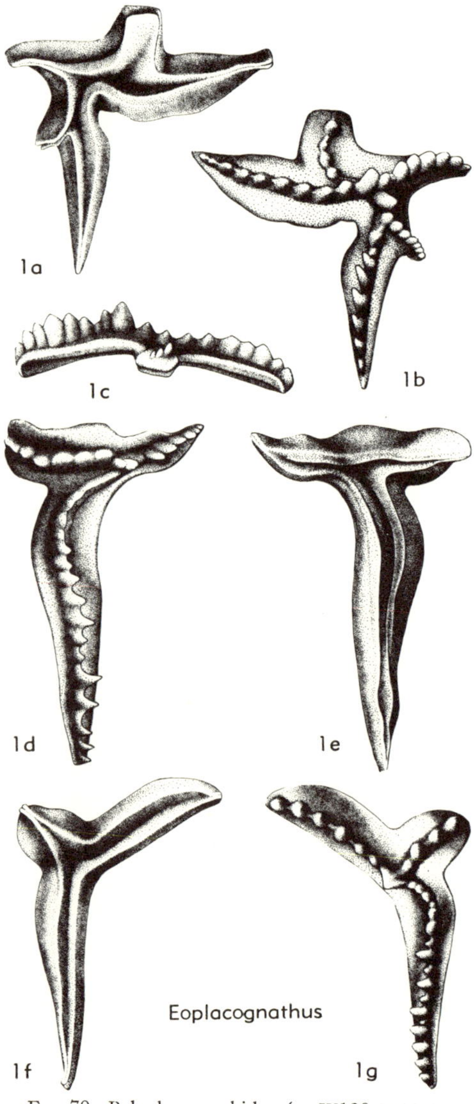

FIG. 79. Polyplacognathidae (p. *W*130-*W*131).

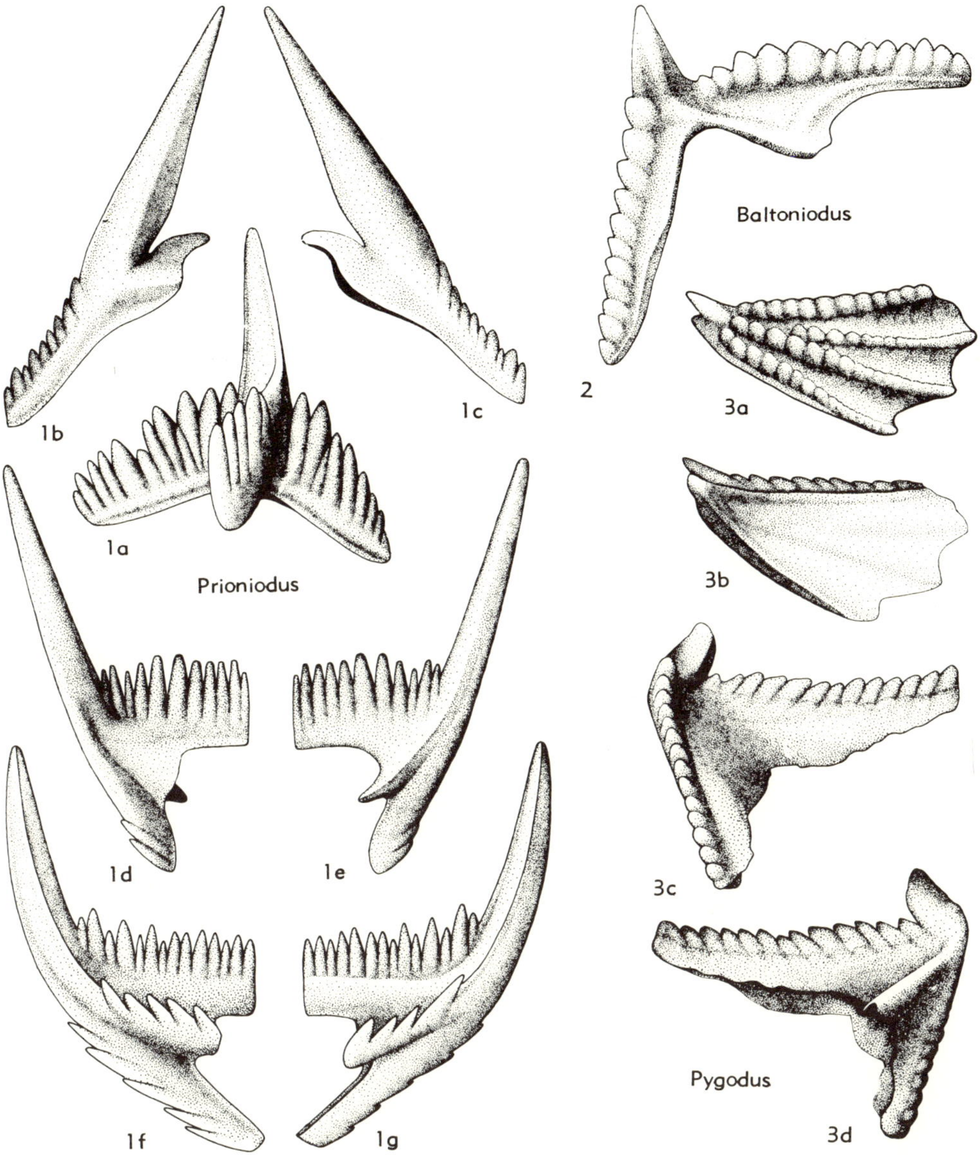

FIG. 80. Prioniodontidae, Pygodontidae (p. *W*132).

parting from point of junction between anterior and posterior processes, some elements with short cusp at this junction. Single row of relatively low denticles centrally located on each process of all elements; no additional denticle rows or nodes present laterally. [Elements of *Eoplacognathus* are distinguished from corresponding ones in *Polyplacognathus*, with which they have been confused frequently, by their lack of accessory denticulation lateral to the central denticle rows and by process configuration; further, right and left elements in each pair in *Polyplacognathus* are near mirror images and not different as in *Eoplacognathus*. Apparatus reconstruction: BERGSTRÖM, 1971, 1973.] *L.Ord.-M.Ord.*, Eu.(Baltoscandia-Pol.-G. Brit.-USSR-Ire.)-N. Am. (Appalachian Mts.-Great Basin-Okla.-Ark.).——FIG. 79,*1*. *E. foliaceus* (FÅHRAEUS), M.Ord.(Folkeslunda Ls.), Eu.(Sweden); *1a-c*, stelliplanate element, lower, upper, lat. views; *1d,e*, sinistral pastiniplanate element, upper, lower views; *1f,g*, dextral pastiniplanate element, lower, upper views; ×50 (Bergström, n).

Family PRIONIODONTIDAE Bassler, 1925

[*nom. correct.* MOORE & SYLVESTER-BRADLEY, 1957b, p. 28, *pro* Prioniodidae BASSLER, 1925, p. 218]

Apparatus sexi- or septimembrate; *P* elements pastinate, with 3 denticulate processes; *M* elements geniculate, in some species denticulate above anterobasal corner; *S* elements forming transition series from bipennate through tertiopedate and alate to quadriramate units, all with denticulate processes, distinct cusp, and basal cavity extending along entire length of processes. *L.Ord.-M.Ord.*

Prioniodus PANDER, 1856, p. 29 [**P. elegans*; SD MILLER, 1889, p. 315]. Diagnosis as for family. *L.Ord.-M.Ord.,* Eu.-N.Am.-S.Am.-Australia.

P. (Prioniodus) PANDER, 1856, p. 29 [=*Belodus* PANDER, 1856; *Gothodus* LINDSTRÖM, 1955; *Tetraprioniodus* LINDSTRÖM, 1955]. Apparatus seximembrate with only one type of *P* element. Basal cavity narrow, shallow, but extending along entire length of all processes. *M* and *Sc* elements with distinctly denticulate anterior process. [Precise phylogenetic relations between the various forms here included in *Prioniodus* are poorly known and the genus may possibly not be a sound taxonomic unit; however, because transitional forms between *Prioniodus (Baltoniodus)* and *P. (Prioniodus)* have been reported, it appears justified to group these forms pending further study. Apparatus reconstruction: BERGSTRÖM, 1968, 1971.] *L.Ord., ?M.Ord.,* Eu.-N. Am.-S.Am.-Australia.——FIG. 80,*1*. **P. elegans,* L.Ord.(*Didymograptus balticus* Subzone), Eu. (Sweden); *1a, P* element, ant.-lat. view; *1b,c, M* element, lat. views; *1d,e,* bipennate *S* element, lat. views; *1f,g,* quadriramate *S* element, lat. views; ×50 (Bergström, n).

P. (Baltoniodus) LINDSTRÖM, 1971, p. 55 [**Prioniodus navis* LINDSTRÖM, 1955, p. 590; OD] [?=*Trapezognathus* LINDSTRÖM, 1955; *Volchodina* SERGEEVA, 1974]. Apparatus in most species septimembrate, including two types of *P* elements; one with inner side of posterior process expanded laterally, forming wide basal cavity; the other with small basal cavity; basal cavity in all ramiform elements relatively deep and wide. *M* and *Sc* elements with weakly, if at all, denticulate anterior process. [Apparatus reconstruction: BERGSTRÖM, 1971; LINDSTRÖM, 1971.] *L.Ord.-M.Ord.,* Eu.-N.Am.-S.Am.-Australia.——FIG. 80,*2*. *P. (Baltoniodus) variabilis* BERGSTRÖM, M.Ord.(Dalby Ls.), Eu.(Sweden); *Pb* element, lat. view, ×45 (Bergström, n).

Family PYGODONTIDAE Bergström, new

Apparatus bimembrate (may possibly be tetramembrate, see below), composed of modified tertiopedate and stelliscaphate, lamellar elements. Tertiopedate elements with short central cusp, well-developed anterior and posterior processes, and short, laterally directed, adenticulate or weakly denticulate, lateral process; denticles of subequal size, reclined, laterally compressed, confluent along most of their length; basal cavity large and deep, extending over entire under side of processes. Stelliscaphate elements triangular in outline, flat to arched, with short cusp at posterior corner of unit, 3 to 4 rows of low, equal-sized denticles with characteristic surface microsculpture diverging from cusp, two of these rows lateral, forming margins of element, other rows central; whole under side of element occupied by wide but shallow basal cavity, in some with basal funnel. *M.Ord.*

Pygodus LAMONT & LINDSTRÖM, 1957, p. 67 [**P. anserinus*; OD] [=*Haddingodus* SWEET & BERGSTRÖM, 1962]. Diagnosis as for family. [Associated as a rule with the tertiopedate and stelliscaphate elements of several species of *Pygodus,* but always in much lower numbers, are elements originally described as *Tetraprioniodus lindstroemi* and *Hibbardella pyramidalis* by SWEET & BERGSTRÖM (1962). They have some characters in common with the *Pygodus* elements, but their low frequency is difficult to explain if they are a part of the *Pygodus* apparatus. Apparatus reconstruction: BERGSTRÖM, 1971.] *M.Ord.,* Eu.-N.Am.-Asia(China)-Australia.——FIG. 80,*3*. **P. anserinus,* Dalby Ls., Eu.(Sweden); *3a,b,* stelliscaphate element, upper and lower views; *3c,d,* tertiopedate element, lat. views; ×40 (Bergström, n).

Family RHIPIDOGNATHIDAE Lindström, 1970

[Rhipidognathidae LINDSTRÖM, 1970, p. 432]

Apparatus multimembrate, in some forms trimembrate, in others with up to seven types of elements; *P?* elements in most forms multidenticulate, carminate, angulate, or segminate; *S* elements ramiform or modified ramiform, forming transition series that, when complete, includes four types of alate,

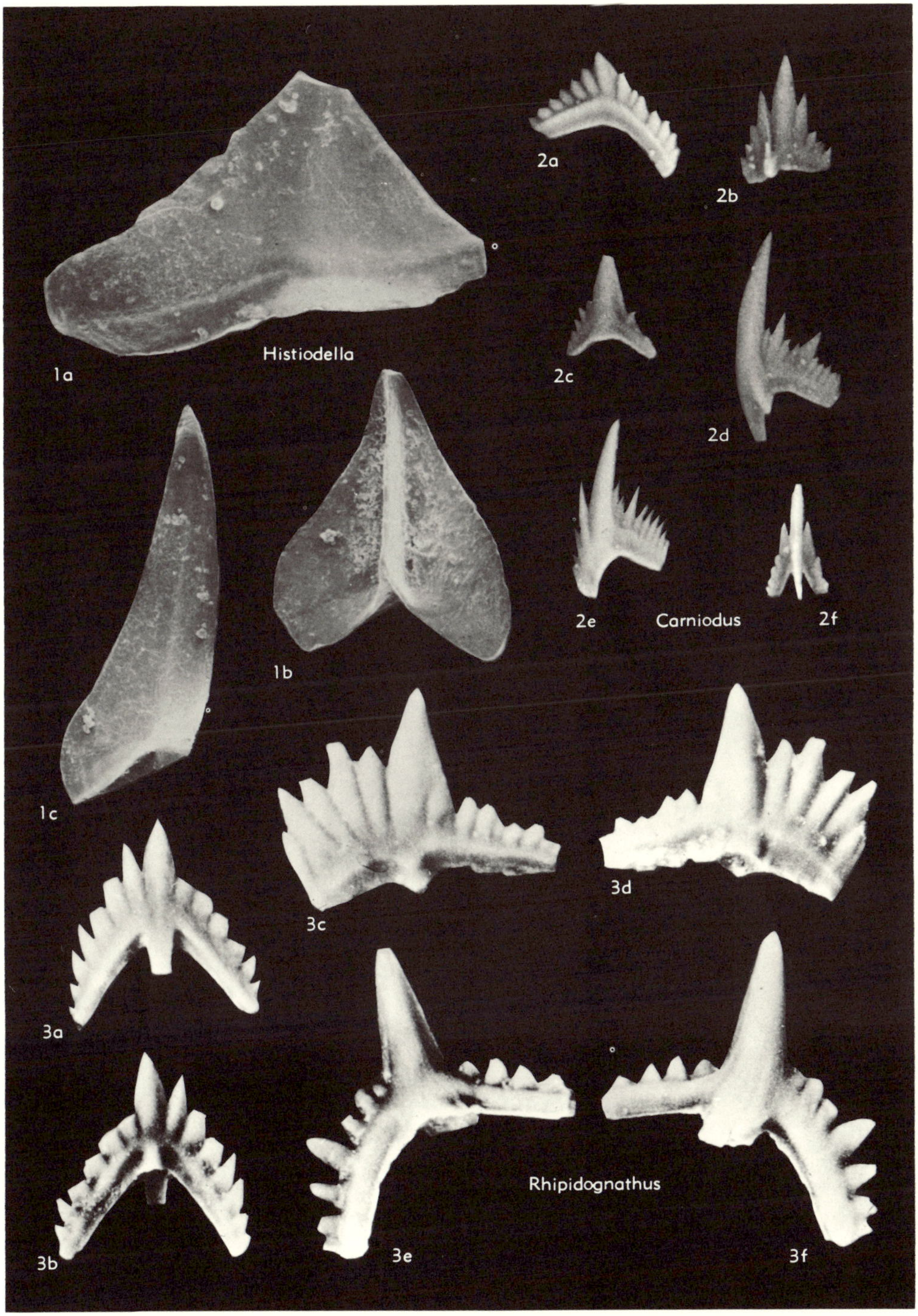

FIG. 81. Rhipidognathidae (p. *W*134-*W*135).

digyrate, and dolabrate elements but may be reduced to include only two types of alate and angulate, dolabrate, digyrate, or modified tertiopedate elements. *L.Ord.-Sil. (Wenlock.).*

Rhipidognathus Branson, Mehl, & Branson, 1951, p. 10 [**R. symmetrica*; OD]. Apparatus trimembrate, including alate and two types of angulate elements. Alate elements bilaterally symmetrical to slightly asymmetrical, anteroposteriorly compressed, convex anteriorly and concave posteriorly, without posterior process but with anterobasal boss. One type of angulate elements arched to almost straight, slightly bowed; other type distinctly arched and bowed; both types with stout cusp. All elements with numerous subequal denticles along upper margins of processes; in some forms denticles laterally compressed and partially confluent laterally; in others denticles rounded and discrete. Basal cavity developed as subapical pit and shallow grooves along processes. [Apparatus reconstruction: Bergström & Sweet, 1966.] *M.Ord.-U.Ord.,* N.Am.——Fig. 81,*3*. *R. discretus* Bergström & Sweet, M.Ord.(Catheys F.), USA (Tenn.); *3a,b,* alate element, ant., post. views; *3c-f,* two types of angulate elements, lat. views; ×31 (Bergström & Sweet, 1966).

Appalachignathus Bergström & others, 1974, p. 227 [**A. delicatulus*; OD]. Apparatus apparently septimembrate; *Pa* elements long, segminate, laterally compressed, with numerous subequal-sized denticles and slitlike basal cavity, cavity conspicuously expanded at posterior end. *Pb* elements long, bowed, angulate, with numerous subequal-sized denticles and slitlike basal cavity. *M?* elements laterally compressed, with one convex and one concave side and short cusp flanked by several confluent denticles. *S* elements of four types forming transition series from alate through digyrate to modified dolabrate, all with more or less confluent denticles, cusp, and relatively narrow basal cavity. [Similar in some respects to *Bergstroemognathus, Appalachignathus* is distinguished by being clearly lamellar, by having *Pb* elements, and by the appearance of denticulation in the ramiform elements. Further, available collections suggest that *Bergstroemognathus* had fewer types of elements in the apparatus than *Appalachignathus* (see Serpagli, 1974). Apparatus reconstruction: Bergström & others, 1974.] *M.Ord.,* N.Am. ——Fig. 82,*1*. **A. delicatulus,* USA; *1a,b, Pa* element, lat. and lower views; *1c,d, Pb* element, lat. and lower views; *1e-j,* three *S* elements each in post. and lat. views, forming transition series; ×23 (Bergström & others, 1974).

Bergstroemognathus Serpagli, 1974, p. 39 [**Oistodus extensus* Graves & Ellison, 1941, p. 13; OD]. Apparatus apparently trimembrate, consisting of alate, segminate, and tertiopedate hyaline elements. Alate elements with very short, indistinct, adenticulate posterior process, nearly bilaterally symmetrical, with multidenticulate lateral processes and basal cavity developed as subapical pit beneath cusp and narrow, shallow grooves along processes. Segminate elements basically coniform, geniculate, with long denticulate anterior process and reclined posterior cusp; basal cavity relatively wide beneath cusp but developed as shallow, narrow groove along entire anterior process. Tertiopedate elements asymmetrical to nearly symmetrical with prominent cusp, short denticulate or adenticulate posterior process, and short denticulate anterolateral process. Denticles of all elements laterally compressed and more or less confluent, without white matter. [For differences between *Bergstroemognathus* and *Appalachignathus,* see *Appalachignathus.* Apparatus reconstruction: Serpagli, 1974.] *L.Ord.,* N.Am.(N.Y.-Texas-Pa.-Newf.)-S.Am.(Arg.).——Fig. 82,*2*. **B.*

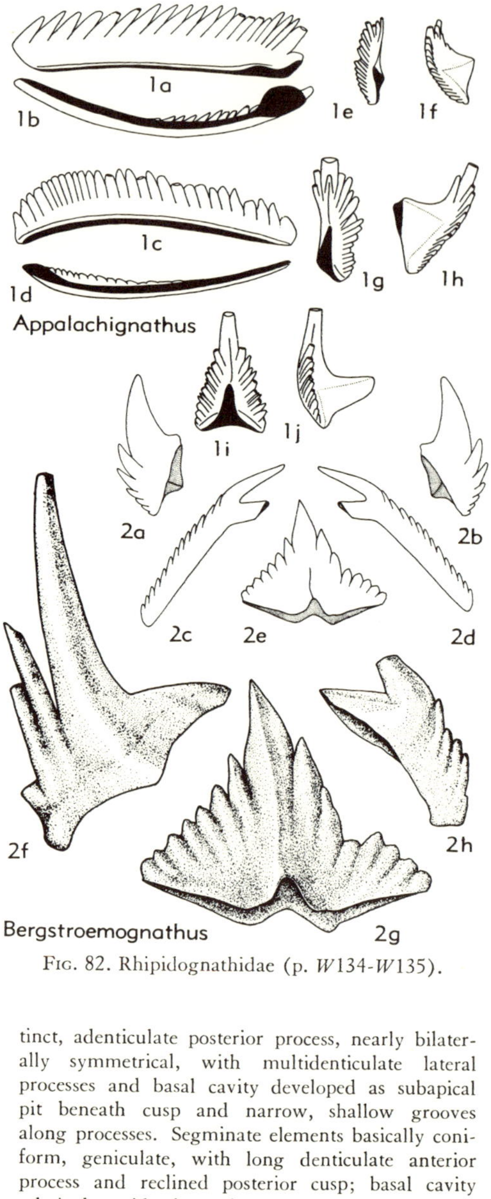

Fig. 82. Rhipidognathidae (p. W134-W135).

extensus (Graves & Ellison), San Juan F., S.Am. (Arg.); *2a,b,* lat. views of tertiopedate elements, diagr., ×11; *2c,d,* lat. views of segminate element, diagr., ×15; *2e,* post. view of alate element, diagr., ×13; *2f,* tertiopedate element, lat. view, ×54; *2g,* alate element, post. view, ×25; *2h,* segminate element, lat. view, ×67 (Serpagli, 1974).

Carniodus Walliser, 1964, p. 30 [**C. carnulus*; OD]. *Pa* element characterized by short, slightly arched, compressed blade with prominent central cusp; short, denticulate lateral process extending from cusp in some. *Pb* element with arched, thick blade, small central cusp, and well-developed, narrow, marginal ledges. *M* element with low cusp, short, weakly denticulate anticusp, and short posterior process. *S* symmetry-transition elements characterized by high, slender cusp and long, arched posterior process bearing compressed denticles and secondary cusp generally developed at mid-length. [Reconstruction: Walliser, 1964; Barrick & Klapper, 1976.] *Sil.(Llandov.-low.Wenlock),* Eu.-N.Am.——Fig. 81,*2.* **C. carnulus,* up. Llandov.-low. Wenlock. (*Pterospathodus amorphognathoides* Z., Clarita F.), USA(Okla.); *2a, Pb* element, lat. view; *2b, Pa* element, lat. view; *2c, M* element, lat. view; *2d, Sc* element, lat. view; *2e, Sb* element, lat. view; *2f, Sa* element, post. view; ×36 (Barrick & Klapper, 1976).

Histiodella Harris, 1962, p. 207 [**H. altifrons*; OD]. Apparatus in at least some species tetramembrate, including carminate, modified alate, and digyrate lamellar elements. Carminate elements strongly compressed laterally, in some species subtriangular in lateral view, with straight basal margin and shallow basal cavity extending along most of under side; upper margin smooth, serrated, or developed into series of laterally confluent denticles. Alate and digyrate elements forming transition series; these elements conspicuously compressed anteroposteriorly, lacking posterior process but with winglike lateral processes that may have small serrations or short, confluent denticles; basal cavity shallow and narrow, developed mainly as subapical pit. [Reconstruction of apparatus by McHargue (1974) includes a geniculate element; however, such an element is missing in some large collections of *Histiodella* and is not included in the apparatus herein.] *L.Ord.-M.Ord.,* N.Am.-Eu.-Asia(Sib.)-Australia.——Fig. 81,*1.* **H. altifrons,* M.Ord.(Joins F.),USA(Okla.); *1a,* carminate element, lat. view; *1b,* alate element, post. view; *1c,* modified digyrate element, post. view; ×123 (Bergström, n).

Family PTEROSPATHODONTIDAE Cooper, 1977

[Pterospathodontidae Cooper, 1977, p. 1062] [Materials for this family prepared by Gilbert Klapper]

Apparatus at least bimembrate; *Pa* element either pastiniscaphate or carminiplanate; *Pb* element either angulate, anguliplanate, or pastiniscaphate. *Sil.(Llandov.-Ludlov.).*

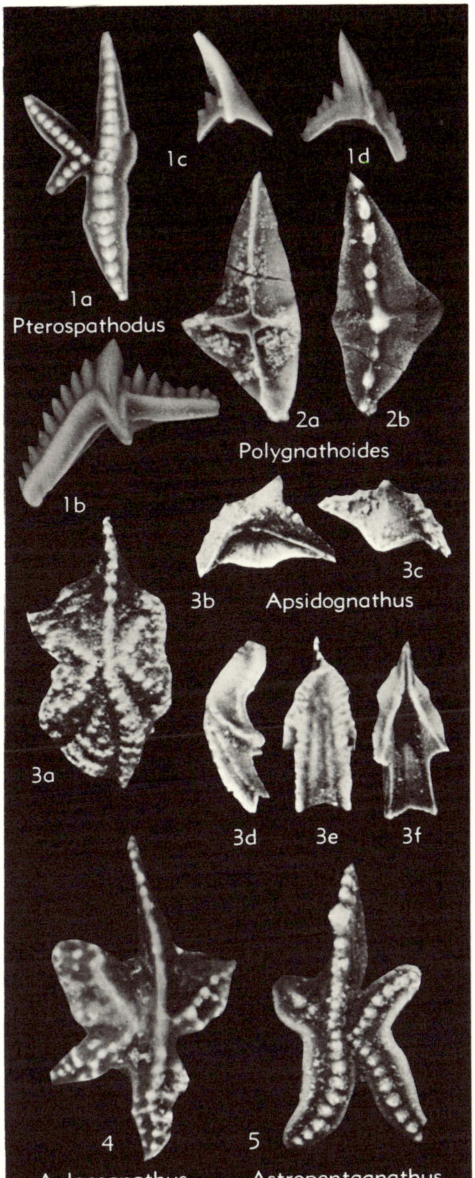

Fig. 83. Pterospathodontidae (p. *W*135-*W*136).

Pterospathodus Walliser, 1964, p. 66 [**P. amorphognathoides*; OD] [=*Llandoverygnathus* Walliser, 1972, p. 76]. *Pa* element either pastiniscaphate like that of *Astropentagnathus* but with cavity much more restricted and inner lateral

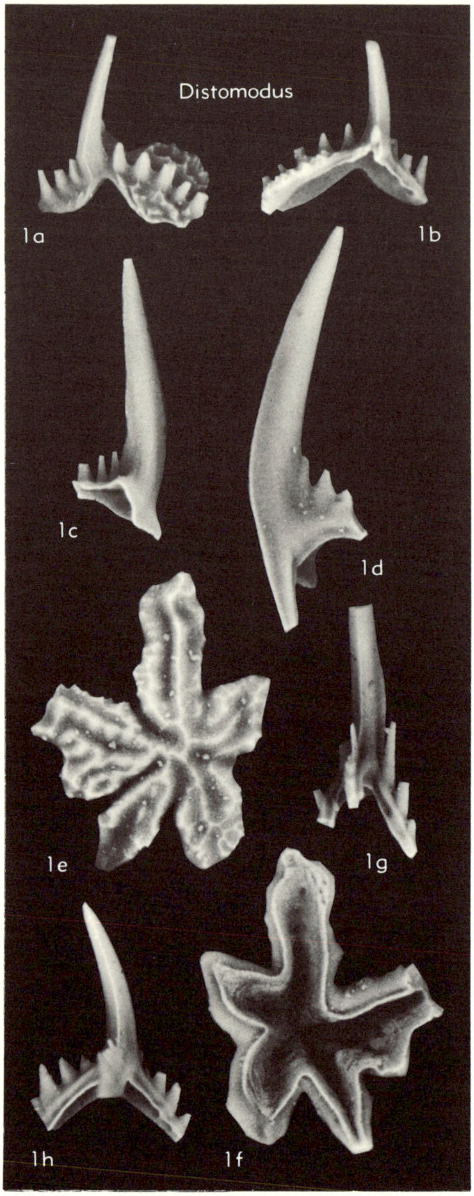

FIG. 84. Distomodontidae (p. W137).

process not well developed, or carminiscaphate with offset lateral lobes at midlength. *Pb* element either anguliplanate with narrow platform ledges and downward projecting apical lips or angulate with offset lobes of basal cavity. [Reconstruction: WALLISER, 1964; BARRICK & KLAPPER, 1976.] *Sil.(Llandov.-low.Wenlock.)*, Eu.-Asia-N.Am.——FIG. 83,*1*. **P. amorphognathoides*, up. Llandov.-low. Wenlock.(*P. amorphognathoides* Z., Clarita F.), USA(Okla.); *1a, Pa* element, upper view; *1b, Pb* element, lat. view; *1c, M* element, lat. view; *1d, S* element, lat. view; ×27 (Barrick & Klapper, 1976).

Apsidognathus WALLISER, 1964, p. 29 [**A. tuberculatus*; OD] [?=*Astrognathus* WALLISER, 1964, p. 30]. Pastiniscaphate *Pa* element like that of *Astropentagnathus* but with bifurcate lateral processes on both sides and more fully developed platform between processes. *Pb* element anguliplanate with well-developed platform having concentric ridges like that of *Pa* element. Homology of arched, scaphate element with two lateral carinae unclear; originally described as *Pygodus lyra* by WALLISER, 1964. [Reconstruction: WALLISER, 1972, p. 76; ALDRIDGE, 1974, p. 299, suggested inclusion of the *Pb* element, *Ambalodus galerus* WALLISER. An additional element may be the type species of *Astrognathus, A. tetractis*.] *Sil.(Llandov.)*, Eu.-Asia-N.Am.——FIG. 83,*3*. **A. tuberculatus, Pterospathodus celloni* Z., Eu.(Aus., Carnic Alps); *3a*, holotype, *Pa* element, upper view; *3b,c, Pb* element, lat. and oblique-upper views; *3d-f*, arched scaphate element, lat., upper, and lower views; ×27 (Walliser, 1964).

Astropentagnathus MOSTLER, 1967, p. 298 [**A. irregularis*; OD]. *Pa* element pastiniscaphate with bifurcate outer lateral process and anteriorly directed inner lateral process. *Pb* element pastiniscaphate with essentially unarched main process and long inner lateral process. [Reconstruction: SCHÖNLAUB, 1971, p. 42; KLAPPER & MURPHY, 1975, p. 24.] *Sil.(Llandov.)*, Eu.-N.Am.——FIG. 83,*5*. **A. irregularis, Pterospathodus celloni* Z. (Roberts Mts. F.), USA(Nev.); *Pa* element, upper view, ×27 (Klapper & Murphy, 1975).

Aulacognathus MOSTLER, 1967, p. 300 [**A. kuehni*; OD] [=*Neospathognathodus* NICOLL & REXROAD, 1969, p. 42]. Pastiniscaphate *Pa* element like that of *Astropentagnathus*, but *Pb* element with prominent cusp and processes on both sides, unlike that of *Astropentagnathus*. [Reconstruction tentatively indicated by KLAPPER & MURPHY, 1975, p. 24-25; *Aulacognathus ceratoides* (NICOLL & REXROAD) is the probable *Pb* element.] *Sil.(Llandov.)*, Eu.-N.Am.——FIG. 83,*4*. **A. kuehni, Pterospathodus celloni* Z.(Roberts Mts. F.), USA(Nev.); *Pa* element, upper view, ×27 (Klapper & Murphy, 1975).

Johnognathus MASHKOVA, 1977, *Sil.(up.Llandov.-low.Wenlock.)*, see addendum.

Polygnathoides BRANSON & MEHL, 1933a, p. 50 [**P. siluricus*; OD]. *Pa* element carminiplanate with short secondary keels extending from transversely extended pit. *Pb* element anguliplanate with well-developed platform on inner side and prominent cusp. [Reconstruction: COOPER, 1974b, p. 187, text-fig. 8D.] *Sil.(Ludlov.)*, Eu.-N.Am.-Australia.——FIG. 83,*2*. **P. siluricus, P. siluricus* Z.(Roberts Mts. F.), USA(Nev.); *2a,b, Pa* element, lower and upper views, ×27 (Klapper & Murphy, 1975).

Family DISTOMODONTIDAE Klapper, new

[Materials for this family prepared by GILBERT KLAPPER]

Apparatus characteristically seximembrate; *Pa* element scaphate with 4 to 6 processes centrally joined; *Pb* element with large cusp, large basal cavity, and expanded base that may develop into platform; dolabrate *M* element and *S* symmetry-transition series with large cusp, large cavity, and discrete denticles. *Sil.(Llandov.-low.Wenlock.).*

Distomodus BRANSON & BRANSON, 1947, p. 553 [**D. kentuckyensis*; OD] [=*Hadrognathus* WALLISER, 1964, p. 35; *Exochognathus* POLLOCK, REXROAD, & NICOLL, 1970, p. 751]. *Pa* element either with 4 processes joined in cross or 5 to 6 radiating processes; *Pb* element with anterior process on expanded base, or 3 to 4 processes in elements with platform. [Reconstruction: COOPER, 1974b; BARRICK & KLAPPER, 1976.] *Sil.(Llandov.-low.Wenlock.)*, Eu.-Asia-N.Am.——FIG. 84,*1*. *D. staurognathoides* (WALLISER), up.Llandov.-low.Wenlock. (*Pterospathodus amorphognathoides* Z., Clarita F.), USA(Okla.); *1a,b, Pb* element, oblique ant.-lat. and post. views; *1c, M* element, lat. view; *1d, Sc* element, lat. view; *1e,f, Pa* element, upper and lower views; *1g, Sa* element, post. view; *1h, Sb* element, post. view; ×27 (Barrick & Klapper, 1976).

Superfamily CHIROGNATHACEA Branson & Mehl, 1944

[*nom. transl.* LINDSTRÖM, 1970, p. 431, *ex* Chirognathidae BRANSON & MEHL, 1944, p. 237] [Materials for this superfamily prepared by S. M. BERGSTRÖM]

White matter greatly reduced or absent; basal cavity very shallow, in many forms developed only as flat surface; elements mostly ramiform, commonly forming transition series; denticles discrete or free along most of element. *Ord.*

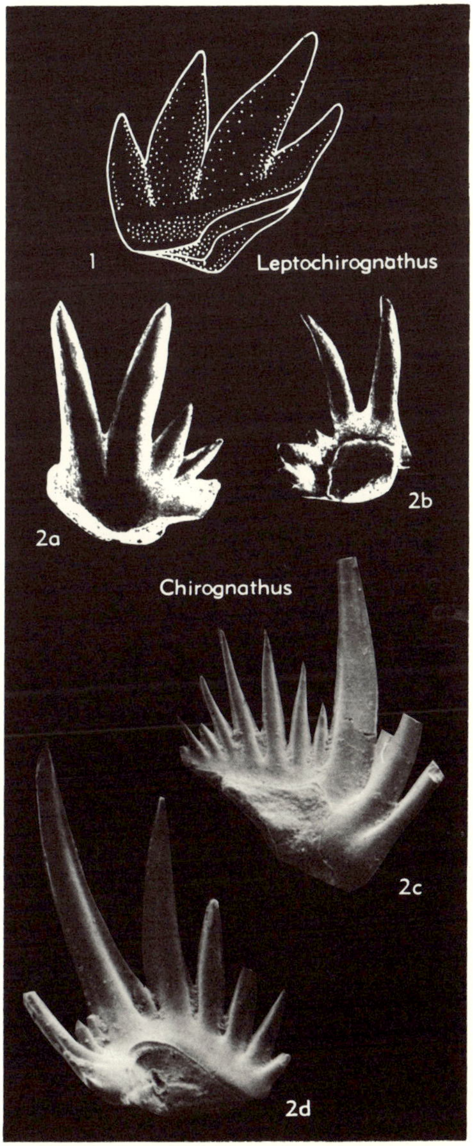

FIG. 85. Chirognathidae (p. *W*137-*W*138).

Family CHIROGNATHIDAE Branson & Mehl, 1944

[Chirognathidae BRANSON & MEHL, 1944, p. 237]

Elements more or less palmate, without prominent cusp, apparently forming simple transition series. *M.Ord.*

Chirognathus BRANSON & MEHL, 1933a, p. 28 [**C. duodactyla*; OD]. Apparatus unknown but probably multimembrate. Elements fibrous (hyaline), palmate, in some more or less angulate, with shallow basal cavity; denticles few, of varying length, discrete along most of element, with more or less rounded cross section. *M.Ord.*, N.Am.-Eu.-Asia. ——FIG. 85,*2a,b*. **C. duodactylus*, Harding Ss., USA(Colo.); ant. and post. views, ×27 (Branson & Mehl, 1933).——FIG. 85,*2c,d*. *C. delicatulus* STAUFFER, Glenwood Sh., USA(Minn.); lat. and post. views, ×68 (Bergström, n).

Leptochirognathus BRANSON & MEHL, 1943, p. 377 [**L. quadrata*; OD]. Apparatus unknown; genus

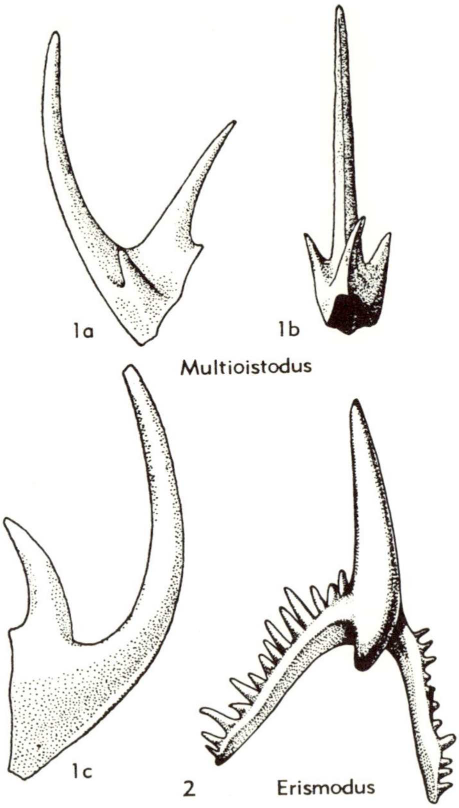

FIG. 86. Multioistodontidae (p. W138-W139).

based on pectiniform, mostly palmate, asymmetrical units consisting of shallowly excavated base with a few denticles along upper margin. Denticles characteristically showing strong lateral compression with sharp edges, wide at base, confluent basally but free apically; usually no distinct cusp. Denticle orientation varying from suberect to almost parallel to base. [*Leptochirognathus* differs from *Chirognathus* in its compressed denticles.] *M.Ord.,* N.Am.-?Asia(Sib.).——FIG. 85,*1*. **L. quadratus,* McLish F., USA(Okla.); ×37 (Hass, 1962).

Family MULTIOISTODONTIDAE Bergström, new

Elements of variable shape, with cusp considerably larger than denticles; apparatus relatively complex, in at least some forms septimembrate with *P, M,* and *S* elements. *Ord.*

Multioistodus CULLISON, 1938, p. 226 [**M. subdentatus*; OD] [=*Trirhadicodus* HARRIS, 1964; *Dirhadicodus* HARRIS, 1964; *Neomultioistodus* HARRIS & HARRIS, 1965; *Tricladiodus* MOUND, 1965a]. Apparatus similar to that of *Eoneoprioniodus* but elements with single, prominent denticle at basal end of each lateral costa as well as on short posterior process. *M.Ord.,* N.Am.——FIG. 86,*1*. **M. subdentatus,* Dutchtown F., USA(Mo.); *1a,b,* alate element, lat. and post. views, ×52; *1c,* dolabrate element, lat. view, ×52 (Lindström, 1964).

Acanthodina MOSKALENKO, 1973, p. 52 [**A. nobilis*; OD]. Apparatus unknown; elements similar to those of *Ptiloconus* but posterior edge of cusp has series of small denticles. *U.Ord.,* Asia(Sib.).——FIG. 87,*2*. **A. nobilis,* USSR(Ketsky F.); post.-lat. view, ×31 (Moskalenko, 1973).

Eofalodus HARRIS, 1962, p. 204 [**E. brevis*; OD]. Apparatus unknown; genus based on geniculate, coniform, mostly hyaline elements with anteroposteriorly extended, laterally flaring base, reclined to recurved cusp with single denticle at anterobasal corner, and relatively shallow, in lateral view subtriangular, basal cavity. [As noted by HARRIS (1962, p. 205), elements of *Eofalodus* show similarity to those of *Falodus* and *Oistodus,* but it is unlikely that *Eofalodus* is closely related to either of these genera. As suggested by MCHARGUE (1974), it is probably more closely related to *Scandodus* and *Eoneoprioniodus.*] *M.Ord.,* N.Am. (Okla.).——FIG. 87,*1*. **E. brevis,* Joins F., USA (Okla.); *1a,d,* lat. views; *1b,c,* upper and lower views; ×38 (Harris, 1962).

Eoneoprioniodus MOUND, 1965b, p. 195 [**E. cryptodens*; OD] [=*Trigonodus* NIEPER, 1969; *Triangulodus* VAN WAMEL, 1974]. Apparatus pentamembrate in at least some species, consisting of four types of nongeniculate and one type of geniculate coniform elements, all hyaline. Elements with long, slender, proclined to suberect cusp and short base with shallow basal cavity; by arrangement of conspicuous lateral costae four nongeniculate element types can be distinguished, forming transition series from dolabrate through digyrate and alate to quadriramate types, all modified and lacking distinct denticles, but with costae extended basally into short processes in some forms. [Apparatus reconstruction: VAN WAMEL, 1974.] *L.Ord.-M.Ord.,* N.Am.-Eu.-Asia(Sib.)-Australia.——FIG. 87,*3*. **E. cryptodens,* M.Ord.(Joins F.), USA(Okla.); *3a,* holotype, lat. view, ×42 (Mound, 1965b); *3b-f,* five types of elements forming transition series, ×36 (McHargue, 1974).

Erismodus BRANSON & MEHL, 1933a, p. 25 [**E. typus*; OD] [?=*Microcoelodus* BRANSON & MEHL, 1933b; *Ptiloconus* SWEET, 1955, *nom. subst. pro Pteroconus* BRANSON & MEHL, 1933b, *non* HINDE in Fox, 1900, a pteropod; *Multicornis* MOSKALENKO, 1970]. Apparatus of type species unknown but septimembrate in closely similar forms. All elements fibrous, most relatively robust, with peglike, discrete denticles, shallow basal cavity, and distinct

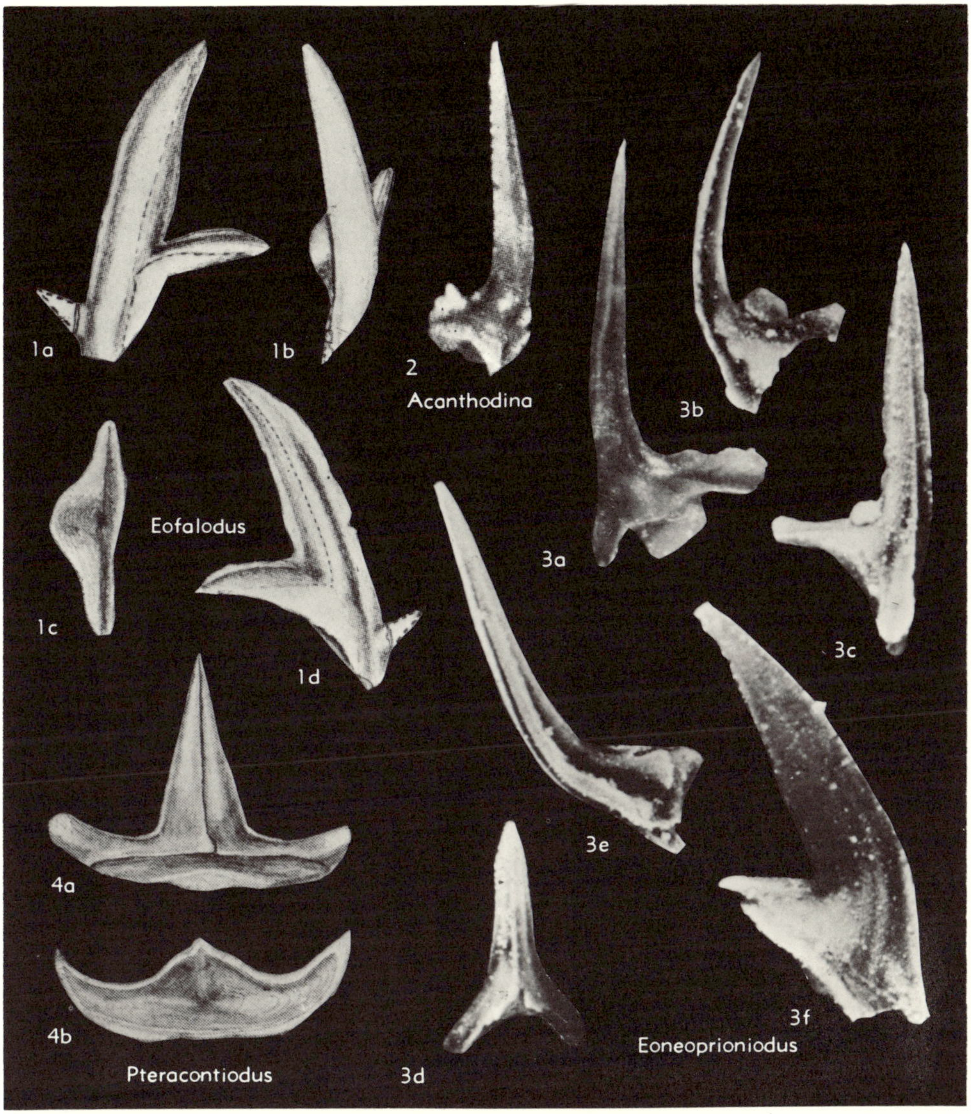

FIG. 87. Multioistodontidae (p. *W*138-*W*140).

cusp. *Pa* elements angulate, laterally bowed, with relatively long posterior and anterior processes forming wide angle with each other. *Pb* elements similar to *Pa* elements but with larger process angle and with abruptly arched posterior process behind cusp. *M* elements modified bipennate with denticulate anterior cusp margin deflected laterally and upper edge of base provided with one or a few denticles near distal end. *S* elements forming transition series from alate through digyrate to bipennate, all modified. *Sa* elements of two types, one symmetrical, one slightly asymmetrical with process angle from 90° to 180°, no posterior process, and anterobasal portion of cusp produced into tonguelike structure in some species. *Sb* elements markedly asymmetrical with process angle of about 90°. *Sc* elements also markedly asymmetrical with short, laterally denticulate anterior process and short, denticulate posterior process. [Apparatus reconstruction: CARNES, 1975.] *M. Ord.,* N.Am.-Eu.-Asia(Sib.).——FIG. 86,*2*. *E. radicans* (HINDE), Chazy., Can.(Que.); *Pa* element, ant. view, ×20 (Hass, 1962).

Pteracontiodus HARRIS & HARRIS, 1965, p. 41 [**P. aquilatus*; OD]. Apparatus unknown; only described element symmetrical, consisting of promi-

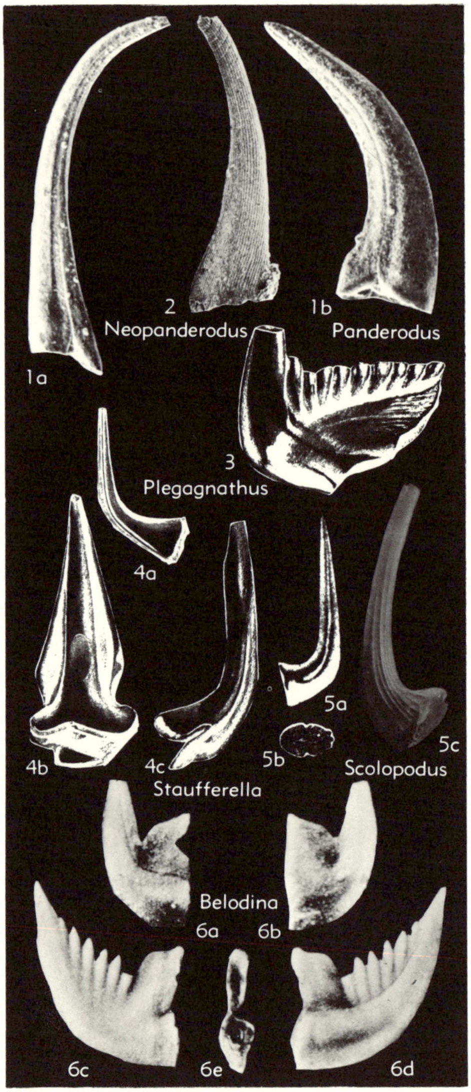

FIG. 88. Panderodontidae, Scolopodontidae (p. *W*140-*W*141).

nent, suberect to recurved, in cross section subtriangular to subrhomboidal cusp with short, low, denticlelike lateral process on each side of unit; basal cavity deepest beneath cusp but extending along under side of processes. *L.Ord.*, N.Am.——FIG. 87,*4*. **P. aquilatus*, West Spring Creek F., USA(Okla.); *4a,b*, post. and lower views, ×35 (Harris & Harris, 1965).

Superfamily PANDERODONTACEA Lindström, 1970

[*nom. transl.* BERGSTRÖM & KLAPPER herein, *ex* Panderodontidae LINDSTRÖM, 1970, p. 433] [Materials for this superfamily prepared by GILBERT KLAPPER and S. M. BERGSTRÖM]

Apparatus uni- or multimembrate, composed of lamellar and, with few exceptions, nongeniculate elements with distinct cusp and more or less prominent longitudinal costae or striations, or both; some elements with denticles on upper margin of cusp or along lateral costae, or both; transition series, if present, mostly expressed in disposition of surface sculpture and cross section of element. *L.Ord.-U.Dev.(Frasn.)*.

Family PANDERODONTIDAE Lindström, 1970

[Panderodontidae LINDSTRÖM, 1970, p. 433]

Apparatus uni- or multimembrate, composed of lamellar coniform elements with relatively deep basal cavity, thick walls, and lateral surfaces with fine longitudinal striations developed especially near basal margin. Most species with only nongeniculate elements, which may form symmetry-transition series expressed by disposition of costae on cusp; others with both geniculate and nongeniculate elements. Upper margin of base may carry single row of denticles or serrations. *L.Ord.-M.Dev.*

Panderodus ETHINGTON, 1959a, p. 284 [**Paltodus unicostatus* BRANSON & MEHL, 1933a, p. 42; OD]. Apparatus apparently basically bimembrate, consisting of one mostly noncostate element and an array of costate elements forming transition series. Elements characteristically adenticulate but some serrated along upper margin of base, with longitudinal groove(s) and costae on lateral faces and wrinkle zone of striations near basal margin. Lateral faces either smooth or having fine longitudinal striations, especially near groove at midheight (best observable with SEM). [Apparatus reconstruction: BERGSTRÖM & SWEET, 1966, p. 355; COOPER, 1975, p. 993; CARNES, 1975, p. 163; BARRICK, 1977.] *L.Ord.-M.Dev.*, Eu.-N.Afr.-Asia-N.Am.-Australia.——FIG. 88,*1*. **P. unicostatus* (BRANSON & MEHL), Sil.(up.Llandov.-low.Wenlock., *Pterospathodus amorphognathoides* Z., Clarita F.), USA(Okla.); *1a*, *Sa* element, obverse lat. view; *1b*, *M* element, obverse lat. view; ×44 (Barrick, 1977).

Belodina ETHINGTON, 1959a, p. 271 [**Belodus grandis* STAUFFER, 1935b, p. 603; OD; =*Belodus compressus* BRANSON & MEHL, 1933b, p. 114] [=*Eobelodina* SWEET & others, 1959, p. 1050;

?*Culumbodina* MOSKALENKO, 1973, p. 64]. Apparatus uni- or multimembrate; elements robust, laterally compressed, basically of coniform type, with expanded base. Some elements with prominent, more or less laterally compressed and confluent denticles on posterior margin of cusp, other elements adenticulate. Some species with geniculate elements. Basal cavity narrow, may be subdivided into two compartments. [Apparatus reconstruction: BERGSTRÖM & SWEET, 1966; MOSKALENKO, 1972.] *M.Ord.-U.Ord.*, N.Am.-Eu.-Asia(Sib.).——FIG. 88,*6*. **B. compressa* (BRANSON & MEHL), U.Ord.(Galena Gr.), USA(Iowa); *6a,b*, adenticulate elements, lat. views; *6c,d*, denticulate element, lat. views; *6e*, basal cavity of denticulate element; ×33 (Bergström & Sweet, 1966).

Neopanderodus ZIEGLER & LINDSTRÖM, 1971, p. 633 [**N. perlineatus*; OD] [=*Parallelocostata* KHODALEVICH & TSCHERNICH, 1973a, p. 28]. Elements like those of *Panderodus* but lateral faces bearing uniform, coarse, longitudinal striations, wrinkle zone less well developed. [Apparatus reconstruction: ZIEGLER, 1975, p. 230.] *L.Dev.(Ems.)-M.Dev.(Givet.)*, Eu.-Asia-N.Am.——FIG. 88,2. **N. perlineatus*, holotype, M.Dev.(Eifel.), Eu.(Ger., Rhenish Slate Mts.); reverse lat. view, ×22 (Ziegler & Lindström, 1971).

Plegagnathus ETHINGTON & FURNISH, 1959, p. 544 [**P. nelsoni*; OD]. Apparatus unknown; elements of type species lamellar, dolabrate, laterally compressed with relatively short, denticulate posterior process and large basal cavity; all denticles and cusp reclined, denticles of about same size, laterally confluent. *U.Ord.*, N.Am.——FIG. 88,*3*. **P. nelsoni*, Stony Mt. F., Can.(Manit.); lat. view, ×50 (Ethington & Furnish, 1960).

Family SCOLOPODONTIDAE Bergström, new

Apparatus apparently multimembrate, composed of coniform nongeniculate elements with rounded cross section and shallow basal cavity; distinct surface sculpture of costae and longitudinal striations. *L.Ord.-M.Ord.*

Scolopodus PANDER, 1856, p. 25 [**S. sublaevis*; SD ULRICH & BASSLER, 1926, p. 7]. Apparatus of type species unknown but apparently including an array of nongeniculate, dominantly hyaline, coniform elements with more or less circular cross section, small base, shallow basal cavity, and proclined to recurved cusp; sides of cusp in most species with numerous lateral costae arranged symmetrically or asymmetrically. [No find of the type species of the genus has been reported since PANDER's time and, as discussed by LINDSTRÖM (1971), there are problems in interpreting the morphology of this species; however, it seems obvious that *S. sublaevis* is morphologically close to, and congeneric with, the well-known species *S. rex* LINDSTRÖM, 1955.] *L.Ord., ?M.Ord.*, Eu.-N.Am.-S.Am.-Asia.——FIG. 88,*5a,b*. **S. sublaevis*, L.Ord., Eu.(Baltic); lat. view and cross section, magnification unknown (Pander, 1856).——FIG. 88,*5c*. *S. rex* LINDSTRÖM, L.Ord.(*Didymograptus balticus* Subzone), Eu.(Sweden); lat. view, ×28 (Bergström, n).

Staufferella SWEET, THOMPSON, & SATTERFIELD, 1975, p. 43 [**Distacodus falcatus* STAUFFER, 1935a, p. 142; OD]. Apparatus basically bimembrate, including two principal types of nongeniculate, longitudinally finely striated, coniform lamellar elements with deep basal cavity. One type bilaterally symmetrical, basally depressed, with unicostate or bicarinate posterior face and prominent anterolateral costae with flaring basal alae. Other type slightly to markedly asymmetrical, with one or two lateral costae and unicostate or bicarinate posterior face of cusp. *M.Ord.*, N.Am.——FIG. 88,*4*. **S. falcatus* (STAUFFER), Galena Gr., USA(Iowa); *4a*, asymmetrical element, lat. view, ×25; *4b,c*, symmetrical element, post. and lat. views, ×38 (Ethington, 1959a).

Family BELODELLIDAE Khodalevich & Tschernich, 1973

[*nom. transl.* BERGSTRÖM & KLAPPER herein, *ex* Belodellinae KHODALEVICH & TSCHERNICH, 1973a, p. 31]

Apparatus uni- or multimembrate, composed of lamellar coniform elements with extremely deep basal cavity and thin wall. Symmetry-transition series developed in at least some genera. *L.Ord.-U.Dev.(Frasn.)*.

Belodella ETHINGTON, 1959a, p. 271 [**Belodus devonicus* STAUFFER, 1940, p. 420; OD] [=*Haplobelodella* KHODALEVICH & TSCHERNICH, 1973b, p. 43]. Apparatus multimembrate, in Ordovician species consisting of an adenticulate nongeniculate element, an adenticulate geniculate element, and a symmetry-transition series of denticulate elements; post-Ordovician species lacking geniculate elements. Elements with short cusp and long base. Denticulate elements with numerous relatively small denticles along upper margin of base; shape of basal cross section varying from triangular to elliptical. [Apparatus reconstruction: BARRICK, 1977; CARNES, 1975, p. 110-111; *Haplobelodella* is a name applied to adenticulate nongeniculate elements associated with other elements of the *Belodella* apparatus.] *L.Ord.-U.Dev.(Frasn.)*, Eu.-N.Afr.-Asia-N.Am.-Australia-S.Am.——FIG. 89,2. *B. silurica* BARRICK, Sil.(Wenlock., *Kockelella stauros* Z., Clarita F.), USA(Okla.); *2a*, *Sa* element; *2b*, *Sc* element; *2c*, *M* element; all lat. views, ×53 (Barrick, 1977).

Coelocerodontus ETHINGTON, 1959a, p. 273 [**C. trigonius*; OD]. Apparatus unknown but may be bimembrate in type species, including elements

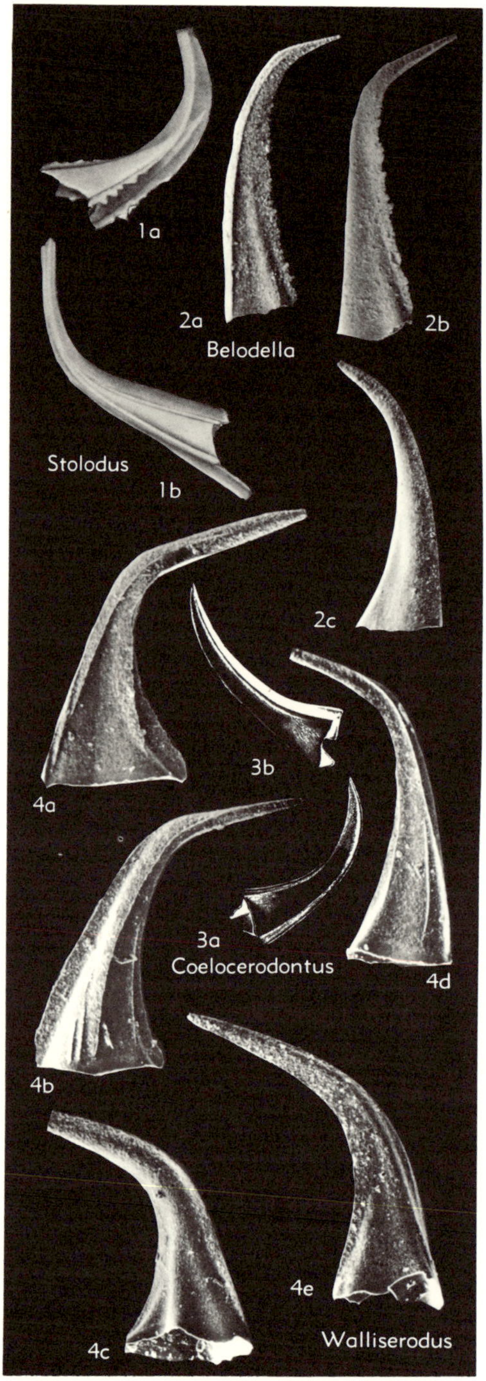

FIG. 89. Belodellidae (p. W141-W142).

with triangular or tetragonal cross section, recurved cusp, keeled edges, and basal cavity extending to tip of element. *M.Ord.-U.Ord.*, N.Am.-Eu.——FIG. 89,*3*. **C. trigonius*, Galena Gr., USA (Iowa); *3a,b*, post. and post.-lat. views, ×48 (Ethington, 1959a).

Stolodus LINDSTRÖM, 1971, p. 51 [**Distacodus stola* LINDSTRÖM, 1955, p. 556; OD]. Apparatus basically unimembrate but elements showing symmetry transition in arrangement of lateral costae. Elements not strongly compressed laterally, with long and relatively wide base, small proclined to recurved cusp, large basal cavity; prominent lateral costae with, in some forms, short denticles. *L.Ord.*, Eu.-N.Am.-S.Am.-Asia(Sib.).——FIG. 89,*1*. **S. stola* (LINDSTRÖM), Eu.(Sweden); *1a,b*, lat. views, ×50 (Van Wamel, 1974).

Walliserodus SERPAGLI, 1967, p. 104 [**Paltodus debolti* REXROAD, 1967, p. 41; OD; =*Acodus curvatus* BRANSON & BRANSON, 1947, p. 554]. Apparatus multimembrate, consisting of an asymmetrical, nondenticulate element and a symmetry-transition series of characteristically multicostate, adenticulate elements with great variation in cross section and in number and position of costae. Elements nongeniculate and strongly costate. [Apparatus reconstruction: COOPER, 1975; BARRICK, 1977.] *U.Ord.(Ashgill.)-Sil.(Ludlov.)*, Eu.-N.Am.——FIG. 89,*4*. *W. sancticlairi* COOPER, Sil. (up. Llandov.-low. Wenlock., *Pterospathodus amorphognathoides* Z., Clarita Fm.), USA(Okla.); *4a, Sa* element; *4b, Sd* element; *4c, Sc* element; *4d, Sb* element; *4e, M* element; all lat. views, approx. ×48 (Barrick, 1977).

Superfamily DISTACODONTACEA Bassler, 1925

[*nom. transl.* LINDSTRÖM, 1970, p. 429, *ex* Distacodontidae BASSLER, *nom. correct.* HASS, 1958, p. 141, *pro* Distacodidae BASSLER, 1925, p. 218] [Materials for this superfamily prepared by S. M. BERGSTRÖM unless noted otherwise]

Apparatus uni- or multimembrate, in many forms bi- or trimembrate, composed of geniculate or nongeniculate coniform elements, or both, with distinct cusp and well-developed basal cavity; much white matter usually present in cusp but restricted or missing in some forms; cusp with prominent costae in many forms; with few exceptions, no denticles on cusp or base. *U.Cam.-U.Ord.*

Family ACANTHODONTIDAE Lindström, 1970

[Acanthodontidae LINDSTRÖM, 1970, p. 433]

Apparatus apparently composed only of nongeniculate coniform lamellar elements with reclined, laterally compressed cusp,

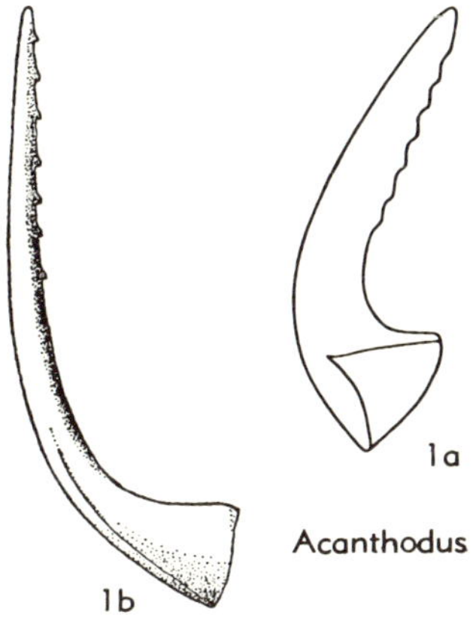

FIG. 90. Acanthodontidae (p. *W*143).

portion of posterior margin of cusp serrate; basal cavity shallow. *L.Ord.*

Acanthodus FURNISH, 1938, p. 336 [**A. uncinatus*; OD]. Diagnosis as for family. *L.Ord.*, N.Am.-Asia(Sib.)-Australia.——FIG. 90,*1a*. **A. uncinatus,* Oneota F., USA(Minn.), diagr. lat. view, ×23 (Furnish, 1938).——FIG. 90,*1b*. *A.* cf. *A. uncinatus,* Dry Creek Sh., USA(Mont.); lat. view, ×52 (Lindström, 1964).

Family DREPANOISTODONTIDAE Bergström, new

Apparatus bi- or trimembrate, composed of laterally compressed geniculate and nongeniculate elements; each cusp laterally smooth or with single longitudinal costa or carina; no conspicuous symmetry transition present based on costae arrangement; white matter abundant or reduced to ropelike growth axis. *Ord.*

Drepanoistodus LINDSTRÖM, 1971, p. 42 [**Oistodus forceps* LINDSTRÖM, 1955, p. 574; OD]. Apparatus basically bimembrate but some forms with two or more types of nongeniculate elements. Geniculate elements with long reclined cusp and posteriorly extended base. Nongeniculate elements with long, recurved to suberect cusp, one lateral face of which may be carinate. Basal cavity restricted to base. [VAN WAMEL (1974) regarded both *Paltodus* PANDER, 1856 and *Scandodus* LINDSTRÖM, 1955 as synonyms of *Drepanoistodus*; however, he erroneously used *Drepanoistodus* for this group of conodonts despite the fact that the former generic names have many years' priority over *Drepanoistodus*. Apparatus reconstruction: BERGSTRÖM & SWEET, 1966; LINDSTRÖM, 1971.] *Ord.*, Eu.-N.Am.-Asia-Australia.——FIG. 91,*1*. **D. forceps* (LINDSTRÖM), L.Ord., Eu.(Sweden); *1a-c,* three types of element, lat. views, ×52 (Van Wamel, 1974).

Distacodus HINDE, 1879, p. 357, *nom. subst. pro Machairodus* PANDER, 1856, *non* KAUP, 1833; a mammal [**Machairodus incurvus* PANDER, 1856; SD MILLER, 1889, p. 313] [=*Machairodia* SMITH, 1907]. Apparatus unknown; elements of type species slender, nongeniculate, coniform, bilaterally symmetrical, with sharp anterior and posterior margins and prominent longitudinal carina on each side of cusp. [No additional specimens of *D. incurvus* have been reported from Estonia or elsewhere since PANDER's time and the species is poorly known.] *L.Ord.(Arenig.),* Eu.(Est.).——FIG. 92,*1*. **D. incurvus* (PANDER), Eu.(Baltic); *1a,b,* lat. views; *1c,* cross section of cusp; magnification unknown (Pander, 1856).

Mixoconus SWEET, 1955, p. 244 [**M. primus*; OD]. Apparatus unknown. Genus based on suberect to reclined, nongeniculate, hyaline, fibrous coniform elements with very shallow basal cavity. Anterior and posterior faces of cusp rounded, lateral faces bicarinate. Anterior and posterior margins, as well as principal lateral carinae, ba-

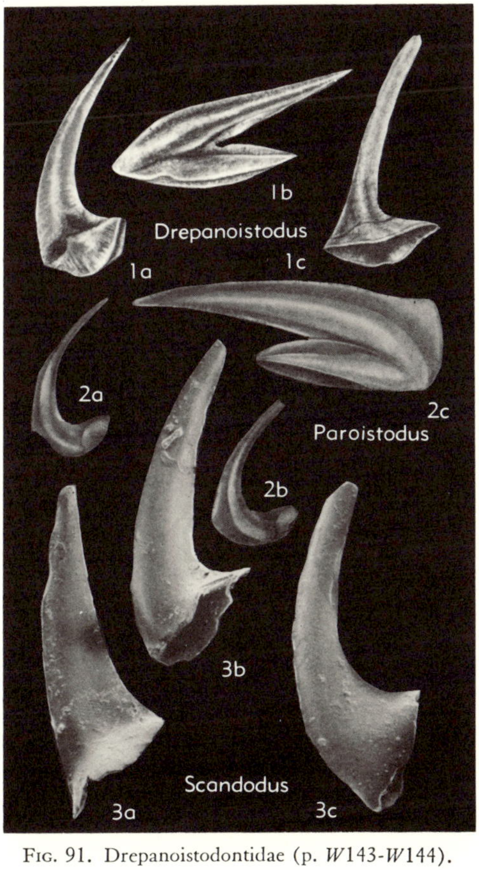

FIG. 91. Drepanoistodontidae (p. *W*143-*W*144).

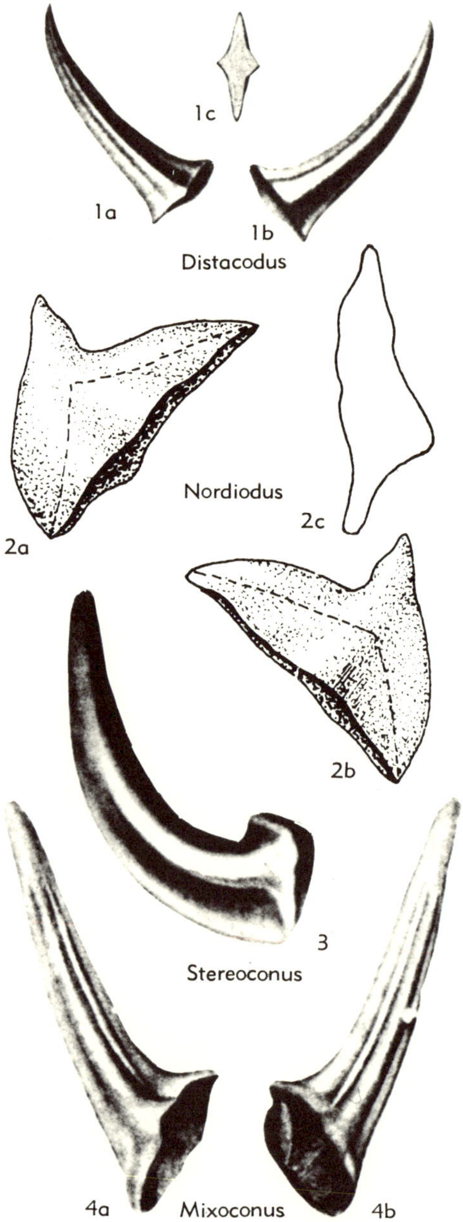

FIG. 92. Drepanoistodontidae (p. W143-W145).

sally extended into very short lobes. *M.Ord.*, N.Am.——FIG. 92,*4*. **M. primus*, Harding Ss., USA(Colo.); *4a,b*, lat. views, ×40 (Sweet, 1955).

Nordiodus SERPAGLI, 1967, p. 77 [**N. italicus*; OD]. Apparatus trimembrate, consisting of one geniculate and two types of nongeniculate, lamellar, laterally compressed, bilaterally asymmetrical, coniform elements. Geniculate elements (*Oistodus rhodesi* SERPAGLI) with reclined to recurved, carinate or costate cusp and posteriorly extended base flaring toward one side. One type of nongeniculate element (including holotype of *N. italicus*) very robust with unusually short, stubby, suberect cusp and very large base. Other type of nongeniculate element (*N. proclinatus* SERPAGLI) also robust with short, proclined cusp and relatively high and wide base. Basal cavity in all elements large, especially in nongeniculate elements, but restricted to base. [As interpreted here, *Nordiodus* has an apparatus reminiscent of *Drepanoistodus* as represented by, for instance, *D. suberectus* BRANSON & MEHL, 1934a. Representatives of *Nordiodus* are, however, distinguished by their short cusp and very large base.] *U.Ord.(Ashgill.)*, Eu.(Italy).——FIG. 92,*2*. **N. italicus*, Italy(Carnic Alps); *2a,b*, nongeniculate element, lat. views; *2c*, same, lower view; ×106 (Serpagli, 1967).

Paroistodus LINDSTRÖM, 1971, p. 46 [**Oistodus parallelus* PANDER, 1856, p. 27; OD]. Apparatus bimembrate. Geniculate elements with reclined cusp and anteroposteriorly extended base tending to be square in lateral view. Nongeniculate elements with recurved cusp, cusp in some elements more or less strongly costate. Prominent zone of recessive basal margin commonly developed at anterobasal corner. Basal cavity relatively shallow. [*Paroistodus* is similar to *Drepanoistodus* and *Paltodus*; for discussion of distinguishing characteristics, see LINDSTRÖM, 1971. Apparatus reconstruction: LINDSTRÖM, 1971.] *L.Ord., ?M. Ord.*, Eu.-N.Am.-S.Am.——FIG. 91,*2*. **P. parallelus* (PANDER), L.Ord.(low.Arenig.), Eu.(Sweden); *2a,b*, nongeniculate elements, lat. views; *2c*, geniculate element, lat. view; ×49 (Van Wamel, 1974).

Scandodus LINDSTRÖM, 1955, p. 592 [**S. furnishi*; OD]. Apparatus trimembrate, composed of largely hyaline elements forming a transition series. Geniculate elements slightly recurved to reclined, with somewhat twisted cusp and short base; basal cavity opening toward one side. One type of nongeniculate element with suberect cusp and short base; another type with recurved cusp and relatively longer base. All elements without lateral costae and with basal cavity restricted to base; white matter in cusp usually present only as thin ropelike growth axis. [*Scandodus* is distinguished from *Oistodus* by lack of prominent lateral costae and dominance of nongeniculate elements in apparatus; from *Drepanoistodus* by being dominantly hyaline and by appearance of geniculate elements; and from *Paltodus* by lack of distinct lateral costae. Apparatus reconstruction: LINDSTRÖM, 1971.] *L. Ord.*, Eu.-?N.Am.——FIG. 91,*3*. **S. furnishi, Didymograptus balticus* Subzone, Eu.(Sweden); *3a*, nongeniculate element, lat. view; *3b,c*, geniculate elements, lat. views; ×48 (Bergström, n).

Stereoconus BRANSON & MEHL, 1933a, p. 27 [**S. gracilis*; OD]. Apparatus unknown; described ele-

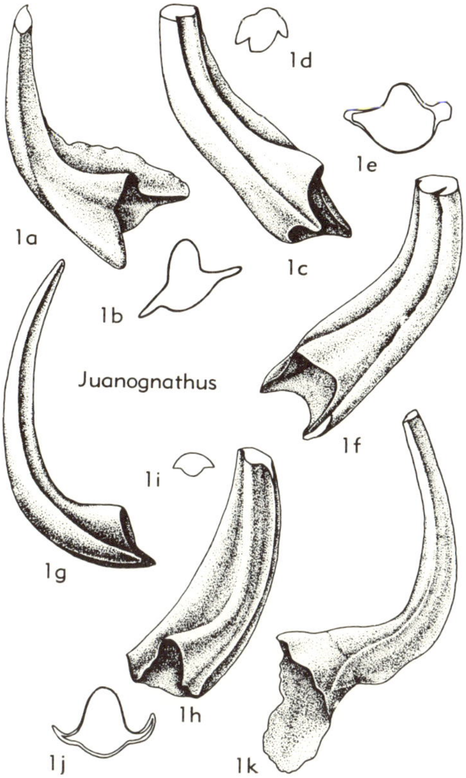

FIG. 93. Juanognathidae (p. *W*145).

ments hyaline, fibrous, nongeniculate, bilaterally symmetrical, laterally carinate, with rounded posterior margin, short base, and shallow basal cavity. [Affinity to *Drepanoistodus* and related genera is currently obscure.] *M.Ord.,* N.Am.-Asia(Sib.).——FIG. 92,*3*. **S. gracilis,* Harding Ss., USA (Colo.); lat. view, ×35 (Sweet, 1955).

Family JUANOGNATHIDAE
Bergström, new

Apparatus unknown but includes nongeniculate and possibly geniculate, coniform, lamellar elements forming symmetry-transition series. Elements elongate, recurved to reclined, with rounded, nonkeeled anterior and posterior margins and more or less oval cross section. Cusp usually twisted with prominent lateral costa on each face. Basal cavity shallow. *L.Ord.-M.Ord.*

Juanognathus SERPAGLI, 1974, p. 49 [**J. variabilis*; OD]. Diagnosis as for family. *L.Ord.-M.Ord.,* S.Am.(Arg.)-N.Am.-Asia(Malaya).——FIG. 93,*1*. **J. variabilis,* L.Ord.(San Juan F.), S.Am.(Arg.); *1a,c,f-h,k,* elements of transition series, post.-lat. views; *1b,d,e,i,j,* cross sections; *a,b,* ×50; *c-f,k,* ×66; *g,* ×52; *h-j,* ×34 (Serpagli, 1974).

Family PROTOPANDERODONTIDAE
Lindström, 1970

[*nom. transl.* BERGSTRÖM herein, *ex* Protopanderodontinae LINDSTRÖM, 1970, p. 433]

Apparatus bimembrate, composed of laterally compressed, lamellar, nongeniculate elements; some forms with cusp smooth laterally, others with prominent longitudinal costae and furrows arranged symmetrically or asymmetrically; cusp more or less twisted in one type of element; white matter abundant in cusp. *Ord.*

Protopanderodus LINDSTRÖM, 1971, p. 50 [**Acontiodus rectus* LINDSTRÖM, 1955; OD]. Apparatus composed of symmetrical and asymmetrical elements forming transition series. Elements with

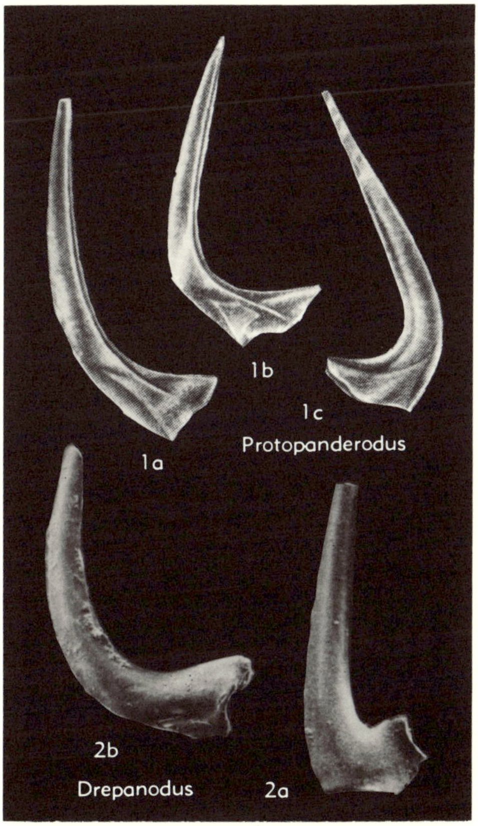

FIG. 94. Protopanderodontidae (p. *W*145-*W*146).

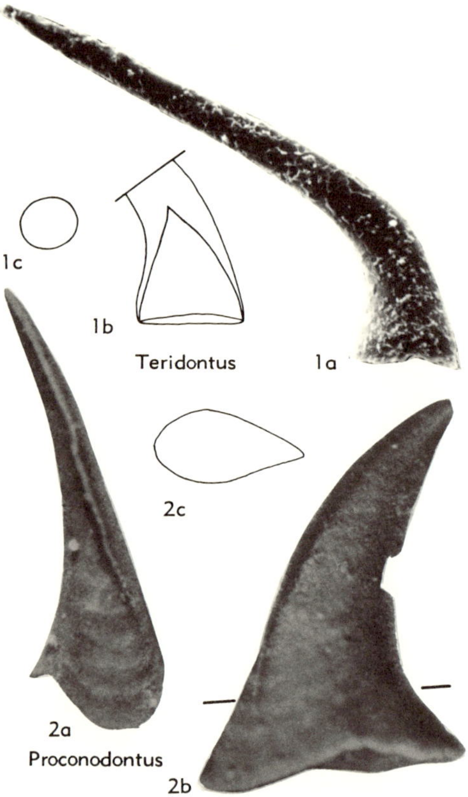

FIG. 95. Proconodontidae, Teridontidae (p. W146, W148).

recurved cusp considerably longer than base of unit. One type of element with strong, symmetrical or asymmetrical, lateral costae; other type with more or less twisted, carinate or grooved cusp; basal cavity of moderate size, restricted to base. [Absence of geniculate elements in the apparatus distinguishes *Protopanderodus* from *Drepanoistodus, Paltodus,* and *Paroistodus.* Apparatus reconstruction: LINDSTRÖM, 1971.] *Ord.,* Eu.-N.Am.-S.Am.-Asia.——FIG. 94,*1.* **P. rectus* (LINDSTRÖM), L.Ord., Eu.(Sweden); *1a,b,* elements with prominent costae, lat. views, ×52, ×37; *1c,* element with noncostate, twisted cusp, lat. view, ×44 (Van Wamel, 1974).

Drepanodus PANDER, 1856, p. 20, *non* MENGE, 1869, Arachnida [**D. arcuatus*; SD MILLER, 1889, p. 313]. Apparatus composed of two types of element; one long and slender, recurved to reclined, nongeniculate; the other reclined, more or less geniculate. Cusp acostate in most elements, slightly twisted in some. Basal cavity subtriangular in lateral view, moderately deep. [Apparatus reconstruction: LINDSTRÖM, 1971, 1973; VAN WAMEL, 1974. VAN WAMEL referred three different types of nongeniculate elements to the apparatus of *D. arcuatus,* but additional material is needed to confirm that reconstruction.] *Ord.,* Eu.-Asia-N. Am.-Australia.——FIG. 94,*2.* **D. arcuatus,* L.Ord. (*Didymograptus balticus* Subzone), Eu.(Sweden); *2a,b,* geniculate and nongeniculate elements, lat. views, ×50 (Bergström, n).

Family PROCONODONTIDAE Lindström, 1970

[Proconodontidae LINDSTRÖM, 1970, p. 429] [Materials for this family prepared by J. F. MILLER]

Nongeniculate coniform elements forming inferred unimembrate apparatuses; basal cavity extending to tip, white matter absent; basal cone prominent. *U.Cam., ?L.Ord.*

Proconodontus MILLER, 1969, p. 437 [**P. mülleri*; OD]. Large, proclined elements, essentially symmetrical but some slightly bent laterally; anterior keel absent in some; posterior keel variable in length (very short and present only near tip in some), serrate in some; cross section oval. [Based on stratigraphic occurrence and morphology, this genus appears to be the most primitive representative of the order Conodontophorida.] *U.Cam., ?L.Ord.,* N.Am.(N.Y., widespread in W. and SW. states of USA-Alberta-Dist.-Mackenzie, Arctic Islands)-Asia(Iran)-Australia(Queensl.).——FIG. 95,*2.* **P. muelleri,* holotype, U.Cam., USA (Utah); *2a-c,* ant. and lat. views, transv. sec., ×38 (Miller, 1969).

Family OISTODONTIDAE Lindström, 1970

[Oistodontidae LINDSTRÖM, 1970, p. 431]

Apparatus trimembrate, composed of a symmetry-transition series of coniform geniculate elements, two elements in some of modified tertiopedate and alate type; cusp long, slender, reclined to recurved; basal cavity shallow but extending along entire under side of element; modified tertiopedate and alate elements with prominent costae, developed as short lateral processes in some; other geniculate element acostate or multicostate; white matter abundant or missing. *L.Ord.-M.Ord.*

Oistodus PANDER, 1856, p. 27 [**O. lanceolatus*; SD ULRICH & BASSLER, 1926, p. 7]. Elements hyaline, lateral costae not developed as distinct processes, cusp with one or several lateral costae. Modified alate element nearly symmetrical. [Apparatus reconstruction: LINDSTRÖM, 1971.] *L.Ord.-M.Ord.,* Eu.-N.Am.-S.Am.——FIG. 96,*3.* **O. lanceolatus,* L.Ord., Eu.(Sweden), three types of elements

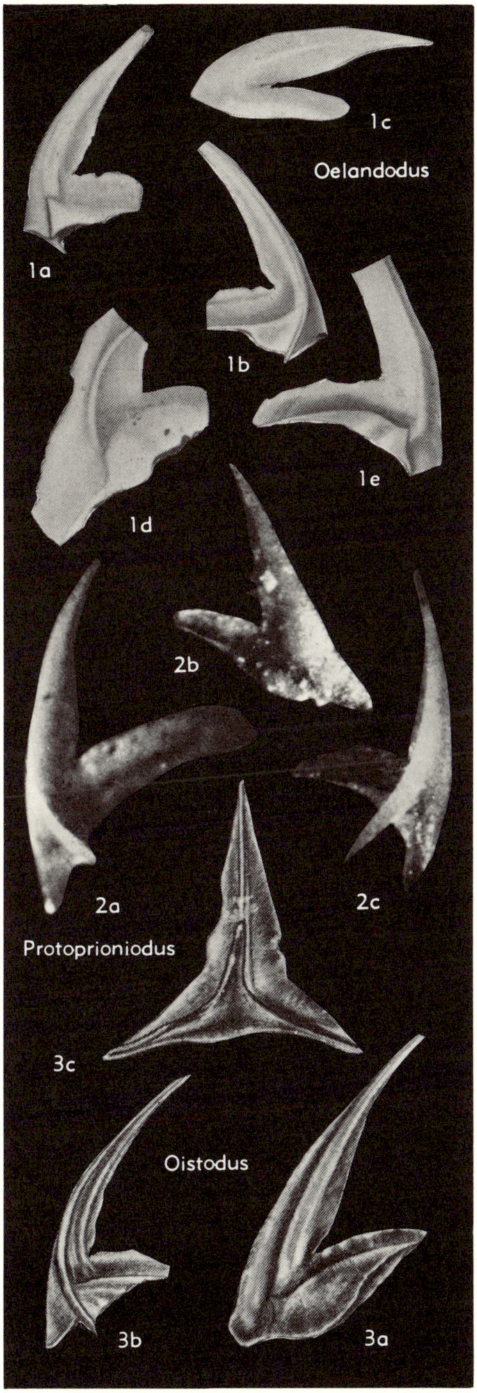

FIG. 96. Oistodontidae (p. *W*146-*W*147).

forming transition series; *3a,b,* lat. views, ×56, ×35; *3c,* post. view, ×110 (Van Wamel, 1974).

Oelandodus VAN WAMEL, 1974, p. 71 [**Oistodus elongatus* LINDSTRÖM, 1955, p. 574; OD]. Apparatus trimembrate, including three types of lamellar, geniculate, and coniform elements. Elements strongly compressed laterally, recurved to reclined, some with carinate cusp. Base long, basal cavity relatively shallow and subtriangular in lateral view. [Apparatus reconstruction: VAN WAMEL, 1974.] *L.Ord.,* Eu.-N.Am.——FIG. 96,*1.* **O. elongatus* (LINDSTRÖM), *Planilimbata* Ls., Eu.(Sweden); *1a-e,* lat. views of geniculate elements, all ×75 except *1c,* ×115 (Van Wamel, 1974).

Protoprioniodus MCTAVISH, 1973, p. 47 [**P. simplicissimus*; OD]. Elements with abundant white matter. Lateral costae forming short processes in some. Cusp in acostate element reclined. [*Protoprioniodus* and *Oelandodus* exhibit close similarity to *Oistodus* in important respects, and further study may show that these genera should be regarded as synonyms. Apparatus reconstruction: MCTAVISH, 1973.] *L.Ord.,* Australia-N.Am.-S.Am.——FIG. 96,*2.* **P. simplicissimus,* Emanuel F., Australia(W.Australia); *2a,* tertiopedate element holotype, lat. view; *2b,* geniculate element, lat. view; *2c,* alate element, lat. view; ×48 (McTavish, 1973).

Family STRACHANOGNATHIDAE Bergström, new

Apparatus basically unimembrate but with several types of closely similar, nongeniculate, lamellar, coniform elements characterized by slender, suberect to recurved cusp, short base, and presence of single denticle anterior to cusp, denticle may rival cusp in size; basal cavity of modest size, restricted to base, with apex beneath denticle in front of cusp. *Ord.*

Strachanognathus RHODES, 1955, p. 131 [**S. parvus*; OD]. Diagnosis as for family. [As noted by BERGSTRÖM (1962), there is a certain degree of intraspecific variation in morphology of elements of the type species of the genus; however, there is no evidence that its apparatus includes other types of elements.] *Ord.,* Eu.-N.Am.——FIG. 97,*2.* **S. parvus,* M.Ord.(Dalby Ls.), Eu. (Sweden); *2a,c,d,* lat. views; *2b,* ant. view; ×65 (Bergström, 1962).

Family TERIDONTIDAE Miller, new

[Materials for this family prepared by J. F. MILLER]

Nongeniculate coniform elements, proclined to reclined, forming unimembrate apparatuses or multimembrate apparatuses by symmetry transition. Sculpture consists

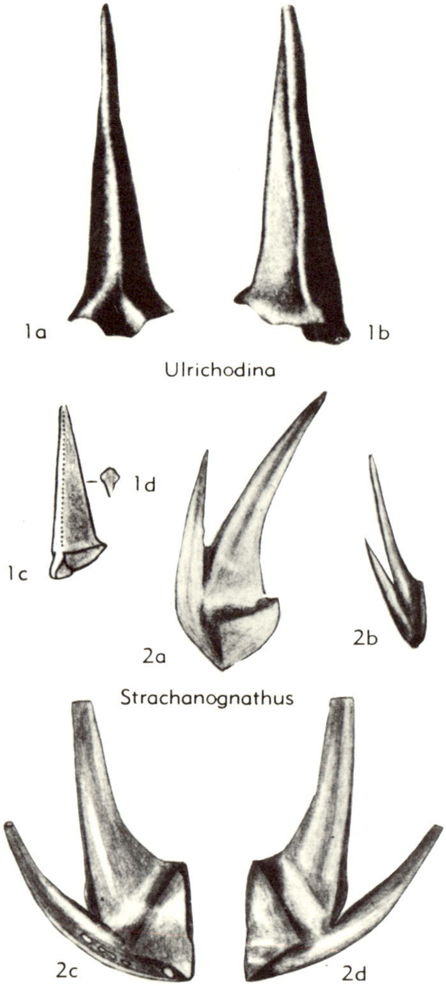

FIG. 97. Strachanognathidae, Ulrichodinidae (p. *W*147, *W*148).

of one or more costae on lateral or posterior faces, or costae absent; faint striae typically present on well-preserved specimens. Base round unless modified by costae; basal cavity shallow, with a white basal cone. Most of cusp composed of white matter. *U.Cam.-L.Ord.*

Teridontus MILLER, 1980, p. 33 [**Oneotodus nakamurai* NOGAMI, 1967, p. 216; OD]. Symmetrical coniform elements, usually erect to reclined, forming probable unimembrate apparatuses. Costae lacking, cross section circular to slightly oval; faint striae covering most of cusp in some. *U.Cam.-L.Ord.,* Asia-Australia-Eu.-N.Am.——FIG. 95,*1*. **T. nakamurai* (NOGAMI), L.Ord., USA (Texas); *1a-c,* lat. view, shape of basal cavity, transv. sec., ×154 (Miller, 1980).

Family ULRICHODINIDAE Bergström, new

Apparatus unknown, possibly unimembrate; elements nongeniculate, coniform, suberect, bilaterally symmetrical, with lateral carinae, rounded anterior margin, and sharp posterior margin; base low, not appreciably extended anteroposteriorly, with characteristic indentation anteriorly; basal cavity shallow, restricted to base. *L.Ord.,* N.Am.-Asia.

Ulrichodina FURNISH, 1938, p. 334 [**U. prima*; OD]. Diagnosis as for family. *L.Ord.,* N.Am.-?Asia(Sib.-Malaya).——FIG. 97,*1*. **U. prima,* Shakopee Dol., USA(Wisc.); *1a,b,* ant. and ant.-lat. views, ×63; *1c,d,* lat. view and cusp cross section, ×38 (Furnish, 1938).

Superfamily HIBBARDELLACEA Müller, 1956

[*nom. transl.* KLAPPER, herein, *ex* Hibbardellidae MÜLLER, 1956b, p. 824] [Superfamily diagnosis by GILBERT KLAPPER]

Apparatus basically seximembrate and composed of *P, M,* and *S* elements; *Pa* element unrecognized in some species. Platform development in pectiniform elements at most a lateral flange. Denticles discrete and not appreciably compressed in most species. Basal cavity large, at least under cusp. *M.Ord.-U.Penn.* [Prioniodinacea BASSLER, 1925 (*nom. transl.* LINDSTRÖM, 1970, *ex* Prioniodinidae BASSLER, 1925) does not compete as the name for this superfamily. According to the reviser principle, in a divided taxon the name must remain with a component that includes the type, and the apparatus of *Prioniodina subcurvata* BASSLER is unknown at present. Consequently, *Prioniodina* is treated here in Superfamily and Family Unknown and any higher taxonomic category based on it must remain bound to the genus.]

Family HIBBARDELLIDAE Müller, 1956

[Hibbardellidae MÜLLER, 1956b, p. 824] [Materials for this family prepared by GILBERT KLAPPER and S. M. BERGSTRÖM]

Characterized by a quinqui- or seximem-

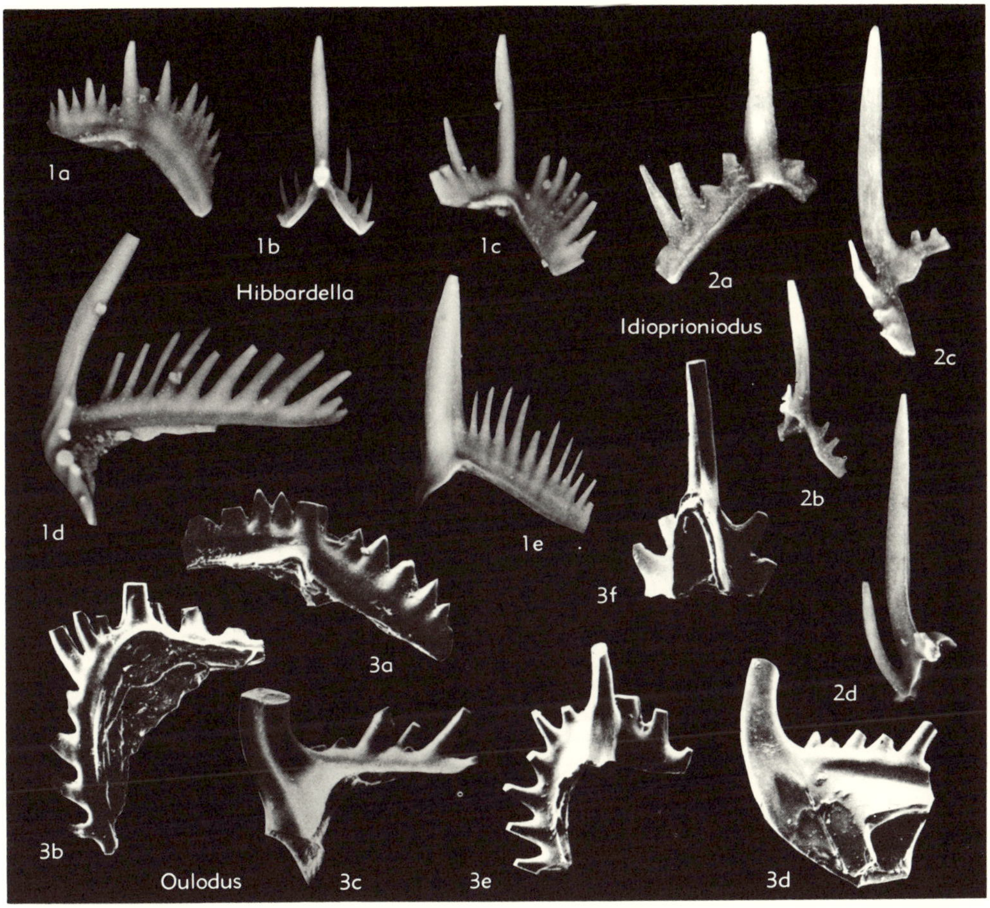

FIG. 98. Hibbardellidae (p. W149-W150).

brate apparatus; *Pa* element sigmoidal and carminate, angulate, or digyrate; *Pb* element angulate or digyrate; *M* element dolabrate or bipennate; *S* elements in a symmetry-transition series from alate through digyrate to bipennate units. *M.Ord.-U.Penn.*

Hibbardella BASSLER, 1925, p. 219 [**Prioniodus angulatus* HINDE, 1879, p. 360; OD]. Apparatus in some species quinquimembrate. *Sa* element with denticulate posterior process. [This process distinguishes the genus from *Oulodus*, the *Sa* element of which has an adenticulate posterior process. Reconstruction: KLAPPER & PHILIP, 1972, p. 101.] *M.Dev.-U.Dev.*, N.Am.-Australia.——FIG. 98,*1*. **H. angulata* (HINDE), U.Dev.(Frasn., Sadler Ls.), Australia(W. Australia); *1a, Pb* element, lat. view; *1b, Sa* element, post. view; *1c, Sb* element, lat. view; *1d, Sc* element, lat. view; *1e, M* element, lat. view; ×27 (Klapper & Philip, 1971).

Idioprioniodus GUNNELL, 1933, p. 265 [**I. typus*; OD] [=*Metalonchodina* BRANSON & MEHL, 1941c, p. 105; *Duboisella* RHODES, 1952, p. 895; *Neoprioniodus* RHODES & MÜLLER, 1956, p. 698]. Apparatus basically seximembrate. *Pb* element angulate, *M* element dolabrate, and *S* elements forming symmetry-transition series of 4 elements ranging from alate through digyrate to bipennate. Denticles massive, characteristically keeled, and somewhat compressed in *M* element. [Apparatus reconstruction: VON BITTER, 1972; BAESEMANN, 1973; MERRILL & MERRILL, 1974.] *Carb.*, Eu.; *U.Miss.-U.Penn.*, N.Am.——FIG. 98,*2*. **I. typus*, U.Penn.(Missour., Kansas City Gr.), USA(Kans.); *2a, Pb* element, lat. view; *2b, Sa* element, post. view; *2c, Sc* element, lat. view; *2d, Sb* element, lat. view; ×27 (Baesemann, 1973).

Oulodus BRANSON & MEHL, 1933b, p. 116 [**Cordylodus serratus* STAUFFER, 1930, p. 124; OD; =*O. mediocris* BRANSON & MEHL, 1933b] [=*Barbarodina* STAUFFER, 1935b; *Gyrognathus* STAUFFER,

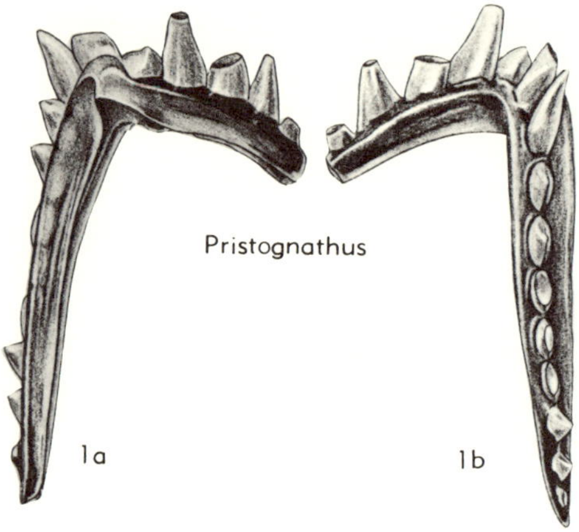

FIG. 99. Hibbardellidae (p. W150).

1935a; *Delotaxis* KLAPPER & PHILIP, 1971]. Apparatus in at least some species seximembrate. *Pa* elements angulate or digyrate (oulodontiform); latter with longer of two processes arched, straight or slightly bowed, forming angle of about 90° with downwardly directed, laterally strongly deflected, shorter process. *Pb* elements digyrate. *M* elements dolobrate or bipennate with wall of basal cavity expanded laterally toward one side. *S* elements forming transition series from alate through digyrate to dolabrate or bipennate. Denticles in all elements discrete, mostly peglike. Basal cavity large, extending along entire under side of processes. [Apparatus reconstruction: SWEET & SCHÖNLAUB, 1975.] *M.Ord.-L.Dev.*, N.Am.-Eu.-Asia(Sib.)-Australia.——FIG. 98,*3*. **O. serratus* (STAUFFER), M.Ord.(Decorah F.), USA(Iowa); *3a, ?Pa* element, lat. view, ×58; *3b, Pb* element, lat. view, ×44; *3c, M* element, lat. view, ×56; *3d, Sc* element, lat. view, ×42; *3e, Sb* element, post. view, ×58; *3f, Sa* element, post. view, ×60 (Sweet & Schönlaub, 1975).

Pristognathus STONE & FURNISH, 1959, p. 226 [**P. bighornensis*; OD]. Apparatus unknown; genus based on lamellar, twisted, digyrate elements with prominent lateral flange and central row of equal-sized denticles on upper side. Cusp about same size as process denticles. Basal cavity shallow, groovelike, extending along under side of most of unit. Denticles oval in cross section, confluent basally but free along most of element. *U.Ord.*, N.Am.——FIG. 99,*1*. **P. bighornensis*, Stony Mt. F., Can.; *1a,b,* lat. views, ×50 (Ethington & Furnish, 1960).

Superfamily GONDOLELLACEA Lindström, 1970

[Gondolellacea LINDSTRÖM, 1970, p. 438] [Diagnosis prepared by GILBERT KLAPPER]

Apparatus usually sexi- or septimembrate but reduced to *Pa* elements and thus unimembrate in some species. Platformed *Pa* elements segminiplanate in Gondolellidae and Xaniognathidae, pastiniplanate or anguliplanate in Ellisoniidae; bladelike *Pa* elements carminate, angulate, or segminate. *Pa* elements of this superfamily not closely related to *Pa* elements of Polygnathacea. Ramiform elements multidenticulate. *U. Carb.-U.Trias.*

Family GONDOLELLIDAE Lindström, 1970

[Gondolellidae LINDSTRÖM, 1970, p. 438] [Materials for this family prepared by GILBERT KLAPPER, R. L. AUSTIN, and F. H. T. RHODES]

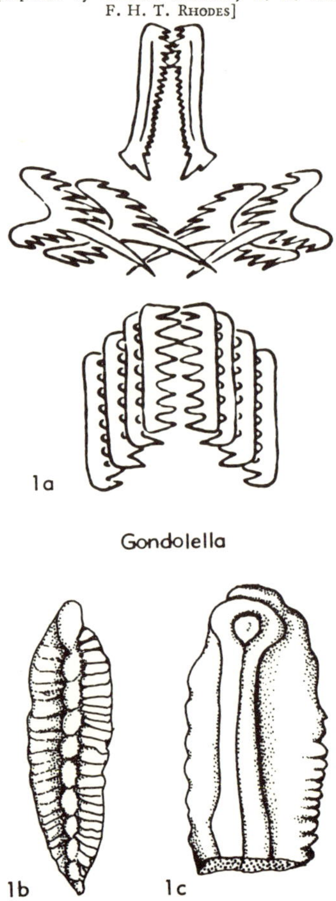

FIG. 100. Gondolellidae (p. W151, W152).

Apparatus apparently septimembrate, *Pa* element segminiplanate, *Pb* element angu-

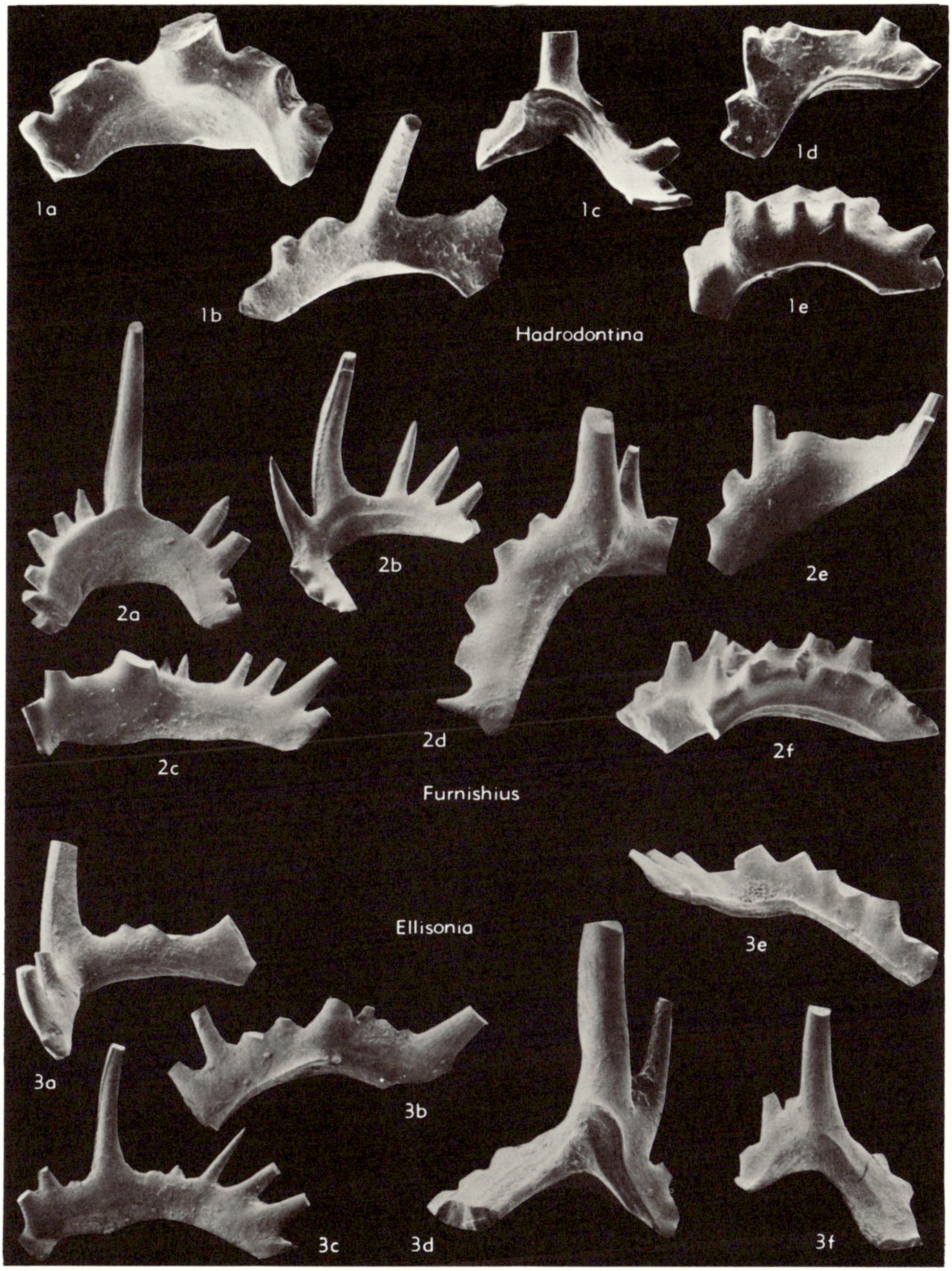

Fig. 101. Ellisoniidae (p. *W*152-*W*153).

late, *M* element ramiform, *S* elements form symmetry-transition series. *M.Penn.-L.Perm.*

Gondolella Stauffer & Plummer, 1932, p. 41 [**G. elegantula*; OD] [=*Illinella* Rhodes, 1952, p. 898]. Apparatus probably at least trimembrate; septimembrate according to von Bitter (1976a). *Pa* element segminiplanate, pectiniform. Free blade absent; long narrow gondola-shaped platform along entire axis; prominent reclined cusp terminal or near posterior termination; platform absent in some, when platform present, carina distinct with denticles at anterior; grooves and transverse ridges on platform; keel and basal pit prominent on lower side. *M.Penn.(Atokan)-L. Perm.*, N.Am.; *U.Carb.-L.Perm.*, W.Eu.-Afr.-Australia-Asia M.-N.Z.——FIG. 100,*1*. *G. curvata* STAUFFER & PLUMMER, M.Penn.(Labette Sh.), USA(Okla.); *1a,* apparatus based on natural assemblage, ×15; *1b,c,* upper and lower views of *Pa* elements, ×25, ×60 (Rhodes & Austin, n).

Family ELLISONIIDAE Clark, 1972

[*nom. transl.* SWEET, herein, *ex* superfamily Ellisoniacea CLARK, 1972a, p. 157] [Materials for this family prepared by W. C. SWEET]

Apparatus basically seximembrate, but reduced to *Pa* elements and thus unimembrate in some species. *Pa* elements angulate in species with seximembrate apparatus and pastiniplanate or anguliplanate with well-developed anterior and posterior processes in species with unimembrate apparatus. *U.Carb.-U.Trias.*

Ellisonia MÜLLER, 1956b, p. 822 [**E. triassica*; OD] [=*Neohindeodella* KOZUR, 1968; ?*Didymodella* MOSHER, 1969, *nom. subst. pro Dichodella* MOSHER, 1968, *non* SERPAGLI, 1967; ?*Oncadella* MOSHER, 1968; *Neoplectospathodus* KOZUR & MOSTLER, 1970; *Stepanovites* KOZUR, 1975a]. Apparatus seximembrate: *Pa* angulate; *Pb* digyrate, bowed out; *M* digyrate; *Sa* alate with long denticulate posterior process; *Sb* digyrate; *Sc* bipennate with long posterior process and shorter, laterally deflected anterior process. Denticles of elements representing late growth stages discrete; three distal denticles of posterior processes of *Sa* and *Sc* elements tending to be longer, stouter, and more widely spaced than more proximal denticles on same processes. [Type specimens of *Didymodella alternata* (MOSHER) and *Oncadella paucidentata* (MOSTLER) appear to be *Pb* and *Sc* elements, respectively, of the apparatus of a distinctive species that may, when better understood, be referable to *Ellisonia*.] *L.Penn.-U.Trias.*, N. Am.; *U.Carb.-U.Trias.*, Eu.-Asia-Australia.——FIG. 101,*3*. **E. triassica*, L.Trias., USA(Nev.); *3a, Sa* element, lat. view, ×63; *3b, Sb* element, post. view, ×46; *3c, Sc* element, lat. view, ×46; *3d, Pb* element, post. view, ×94; *3e, Pa* element, inner-lat. view, ×49; *3f, M* element, post. view, ×46 (Sweet, n).

Anastrophognathus BENDER, 1970, p. 500 [**A. sagittalis*; OD]. Apparatus unknown; genus based on apparently bowed pastinate element with posterior and outer-lateral processes of similar length and development joining straight anterior process to form near bilaterally symmetrical structure Y-shaped in upper or under view. [This genus may be related to *Furnishius* and *Pseudofurnishius*.] *M.Trias.*, Eu.-N.Am.-Asia——FIG. 102,*1*. **A. sagittalis*, Eu.(Greece); *1a,* upper side; *1b,* under side; about ×65 (Kozur & Mostler, 1972).

Furnishius CLARK, 1959, p. 310 [**F. triserratus*; OD] [=*Malaygnathus* IGO, KOIKE, & YIN, 1965]. Apparatus seximembrate: *Pa* digyrate with one long and one short lateral process twisting distally in opposite directions; *M* digyrate, bowed out; *Sa* alate with no posterior process; *Sb* digyrate with subequal lateral processes in essentially same plane; *Sc* bipennate with in-curved anterior process. Under side of all elements cuneiform with small basal pit and narrow grooves beneath processes, pit and grooves surrounded by prominent zone of recessive basal margin, upper edge of margin at midheight on process sides. [In some samples, the array of elements just described is accompanied by pastinate elements, which are structures on which *Furnishius* is based. Except for an irregularly denticulate outer lateral process, however, they are similar to the angulate *Pa* elements noted above. They may either have replaced angulate *Pa* elements in the apparatus of some *Furnishius* individuals or they may represent dimorphs in the skeletal apparatus of which they are the only mineralized elements.] *L.Trias.*, N.Am.-Asia.——FIG. 101,*2*. **F. triserratus*, USA (Nev.); *2a, Sa* element, post. view, ×46; *2b, Sb* element, oblique post. view, ×46; *2c, Sc* element, lat. view, ×46; *2d, Pb* element, post. view, ×63; *2e,* angulate *Pa* element, lat. view, ×46; *2f,* pastinate *Pa* element, lat. view, ×63 (Sweet, n).

Gladigondolella MÜLLER, 1962a, p. 116 [**Polygnathus tethydis* HUCKRIEDE, 1958; OD] [=*Cratognathodus* MOSHER, 1968]. Apparatus apparently unimembrate, composed of anguliplanate elements with thick, coarsely pitted, adenticulate platform and, on under side, longitudinally grooved keel extending full length of element and expanding to form small basal pit slightly posterior of midlength. [KOZUR & MOSTLER (1971) concluded that the apparatus of *G. tethydis* was septimembrate; however, the array of nonplatformed elements they listed as components forms a complete and typical seximembrate apparatus of *Ellisonia* type. Because anguliplanate elements assignable on form to *Gladigondolella* intergrade morphologically with closely similar angulate elements assumed to have occupied *Pa* positions in *Ellisonia* apparatuses, and because anguliplanate elements do not always occur with the array of ramiform elements included in *Gladigondolella* by KOZUR & MOSTLER, it is concluded that *Gladigondolella* had a unimembrate apparatus, as apparently did many other Permian and Triassic conodonts that developed from *Ellisonia* or *Cypridodella*.] *Trias.*, Eu.-Asia-N.Am.——FIG. 103,*1*. **G. tethydis* (HUCKRIEDE), M.Trias., Asia(Timor); *1a,* lat. view, ×34; *1b,* under side, ×36 (Sweet, n).

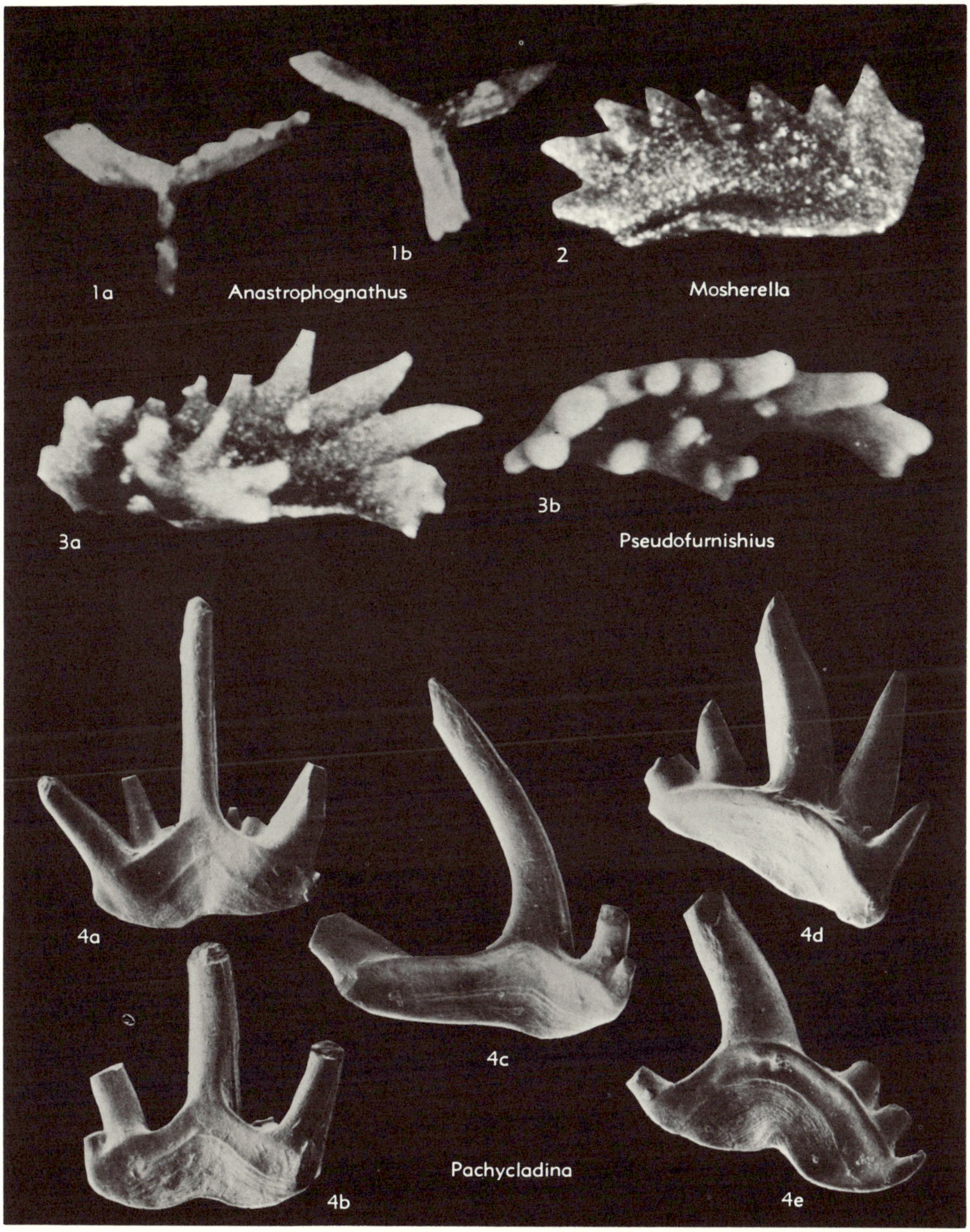

FIG. 102. Ellisoniidae (p. *W*152-*W*154).

Hadrodontina STAESCHE, 1964, p. 271 [**H. anceps*; OD]. Apparatus probably seximembrate, like that of *Furnishius,* but under sides of elements flattened rather than cuneiform, and dimorphic *Pa* element (=form-species *H. biserialis* STAESCHE, 1964) not pastinate but angulate, with secondary row of denticles on outer side parallel to main denticle series. *L.Trias.,* Eu.-Middle East.——FIG. 101,*1*. **H. anceps,* Eu.(Italy); *1a, Sa* element, post. view; *1b, Sb* element, post. view; *1c, M* element, post. view; *1d, Pa* element without secondary denticle row, inner-lat. view; *1e, Pa* element with secondary denticle row, outer-lat. view; ×47 (Sweet, n).

Mosherella KOZUR, 1972, p. 14 [**Neospathodus newpassensis* MOSHER, 1968, p. 931; OD]. Apparatus unimembrate, composed of angulate elements with long anterior and short posterior processes. [According to Kozur, *Mosherella* was derived from *Pseudofurnishius*, this being indicated by complete reduction in *Mosherella* elements of lateral platforms that characterize *Pseudofurnishius* elements.] *U.Trias.*, N.Am.-Eu.-Asia.——Fig. 102,*2*. **M. newpassensis* (Mosher), USA(Nev.); lat. view, ×94 (Sweet & others, 1971).

Pachycladina Staesche, 1964, p. 277 [**P. obliqua*; OD]. Apparatus seximembrate: *Pa* carminate to palmate, with smooth-surfaced, laterally expanded midlateral ribs; *Pb* digyrate, with processes of subequal length curving faintly in opposite directions distally and tending to develop laterally expanded, platformlike midlateral ribs; *M* digyrate, with processes of unequal length; *Sa* alate, with no posterior process; *Sb* digyrate, bowed, a slightly asymmetric version of *Sa; Sc* bipennate, with long posterior process and short, in-curved anterior process. Elements hyaline, with thick growth axes in all denticles. Lower surface of all elements cuneiform, with small basal pit and broadly expanded zones of recessive basal margin forming scarlike areas on inner and outer sides of *Pa* elements but only on inner sides of those in other positions. [The apparatus of *Pachycladina* is reminiscent of that of *Parachirognathus*, but appears to have been more differentiated, and the elements are more robust and have peglike rather than laterally compressed and fused denticles. These differences may prove to be of only specific significance and attributable to differences in geographic distribution. However, representatives of the two genera occur together in both North America and Europe, *Parachirognathus* dominating in the former and *Pachycladina* in the latter.] *L.Trias.*, Eu.-N.Am.——Fig. 102,*4*. **P. obliqua*, Eu.(Italy); *4a, Sa* element, post. view; *4b, Sb* element, post. view; *4c, Sc* element, lat. view; *4d, Pa* element, lat. view; *4e, M* element, post. view; ×42 (Sweet, n).

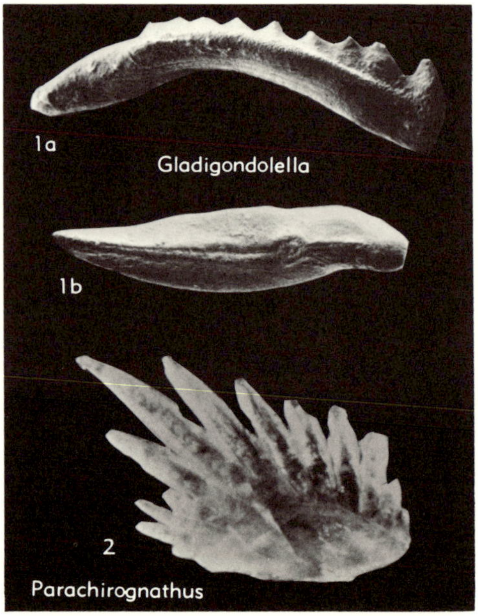

Fig. 103. Ellisoniidae (p. W152, W154).

Parachirognathus Clark, 1959, p. 311 [**P. ethingtoni*; OD]. Known apparatus bi- or trimembrate: *Sa* element alate with no posterior process, grading to closely similar, bowed bipennate *Sc* element. All elements hyaline, with small basal pits surrounded by narrow flattened areas grading on inner sides of elements into broad, semicircular zones of recessive basal margin, upper edges of margin marked by arched rib at base of denticle series. *L.Trias.*, N.Am.-Eu.——Fig. 103,*2*. **P. ethingtoni*, USA(Nev.); *?Sb* element, post. view, ×50 (Sweet & others, 1971).

Pseudofurnishius van den Boogaard, 1966, p. 5 [**P. murcianus*; OD]. Apparatus unimembrate, apparently composed entirely of pastiniplanate or stelliplanate elements with platform extensions only on lateral processes. *M.Trias.-U.Trias.*, Eu.-Middle East-Afr.——Fig. 102,*3*. **P. murcianus*, holotype, M.Trias., Eu.(Spain); *3a*, lat. view, ×102; *3b*, upper side, ×109 (van den Boogaard, 1966).

Family XANIOGNATHIDAE
Sweet, new

[Materials for this family prepared by W. C. Sweet]

Apparatus basically seximembrate, but reduced to *Pa* elements in some species and thus unimembrate. Carminate or angulate *Pa* elements of seximembrate species with short, fragile, unribbed posterior processes; segminate or segminiplanate *Pa* elements of unimembrate species lacking posterior processes. *Perm.-Trias.*

Xaniognathus Sweet, 1970, p. 261 [**X. curvatus*; OD]. Apparatus seximembrate: *Pa* carminate or angulate, with long, longitudinally ribbed anterior process and very short fragile, unribbed posterior process; *Pb* digyrate, bowed out, with one long and one very short lateral process; *M* digyrate; *Sa* alate, with long denticulate posterior process and cusp longer than any denticle of posterior series; *Sb* digyrate, arched; *Sc* bipennate, arched, with anterior process that may bifurcate distally. *U.Perm.-L.Trias.*, Asia-Eu.-N.Am.——Fig. 104,*2*. **X. curvatus*, L.Trias., Asia(Pak.); *2a, Sa* element, lat. view, ×86; *2b*, posteriorly incomplete *Sc* element, lat. view, ×78; *2c, Sb* element, post. view, ×64; *2d, Pb* element, oblique outer view, ×97; *2e, Pa* element, lat. view, ×119; *2f, M* ele-

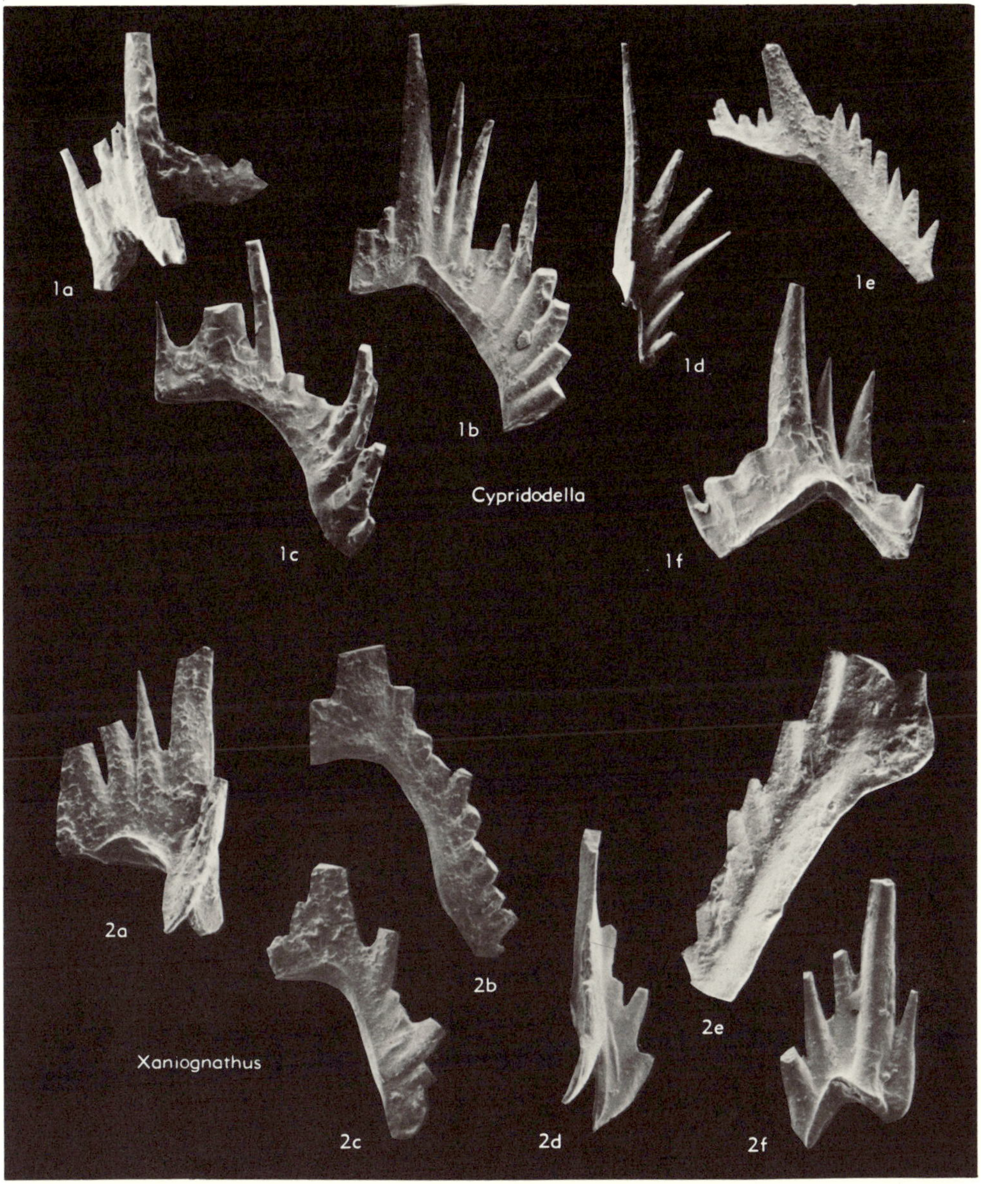

FIG. 104. Xaniognathidae (p. *W*154-*W*156).

ment, post. view, ×97 (Sweet, n).

Chirodella HIRSCHMANN, 1959, p. 71 [**Metalonchodina triquetra* TATGE, 1956, p. 137; OD] [?=*Cornudina* HIRSCHMANN, 1959]. Genus based on very small, outward bowed, digyrate elements with one long denticulate process and another evidenced only by an adenticulate costa or one bearing just one or two denticles; under side sharply keeled, with little indication of basal cavity or pit. [Elements of the type species are commonly associated with tiny, fragile digyrate and bipennate elements that bear numerous needlelike denticles, and with similarly small angulate or carminate elements with prominent cusp, very short processes, and flattened to broadly grooved under sides. The former have been referred by most authors to *Hindeodella* (or *Neohindeodella*), and the latter to *Cornudina*. It is probable that all these elements

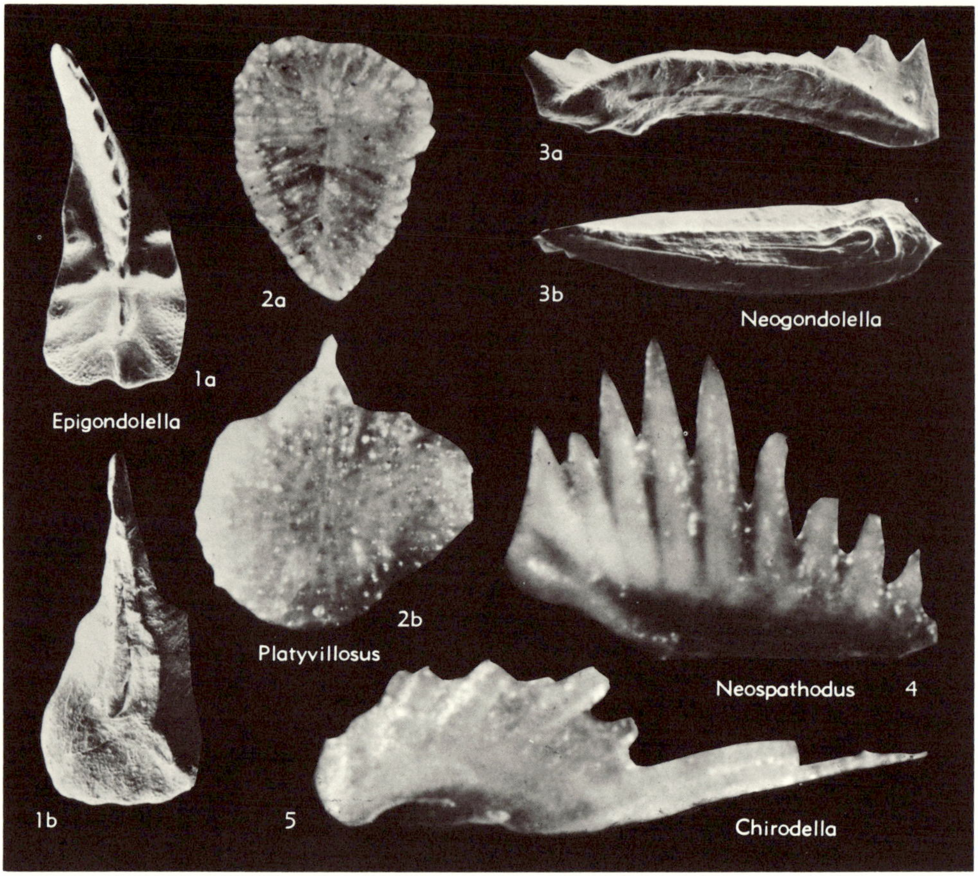

FIG. 105. Xaniognathidae (p. W155-W157).

represent the apparatus of a single species, which would be closely related to *Xaniognathus* and *Cypridodella*.] *M.Trias.-U.Trias.*, Eu.——FIG. 105, *5*. **C. triquetra* (TATGE), M.Trias., Eu.(W.Ger.); oblique outer-lat. view, ×200 (Sweet, n).

Cypridodella MOSHER, 1968, p. 920 [**C. conflexa*; OD] [=*Pollognathus* KOZUR, 1968; *Hibbardelloides* KOZUR & MOSTLER, 1970; *Grodella* KOZUR & MOSTLER, 1970; ?*Veghella* KOZUR & MOSTLER, 1970; *Pseudozarkodina* KOZUR, 1973]. Apparatus seximembrate: *Pa* angulate, with long, longitudinally ribbed anterior process and fragile, unribbed posterior process; *Pb* digyrate, bowed out, with one long and one very short lateral process; *M* digyrate; *Sa* alate, with long denticulate posterior process and cusp appreciably shorter than first or second denticle of posterior series; *Sb* digyrate, arched; *Sc* bipennate, arched; anterior process of either *Sa* or *Sb* element bifurcating distally in some. [This apparatus differs from that of *Xaniognathus* primarily in conformation of the *Sa* element.] *Trias.*, Asia-Eu.-N.Am.——FIG. 104, *1*. *C. magnidentata* (TATGE), M.Trias., Eu.(W. Ger.); *1a*, *Sa* element, lat. view; *1b*, *Sb* element, post. view; *1c*, *Sc* element, lat. view; *1d*, *Pb* element, oblique outer view; *1e*, *Pa* element, lat. view; *1f*, *M* element, post. view; *1a-d,f*, ×97; *1e*, ×62 (Sweet, n).

Epigondolella MOSHER, 1968, p. 935 [**Polygnathus abneptis* HUCKRIEDE, 1958, p. 156; OD] [=*Tardogondolella* BENDER, 1970; *Ancyrogondolella* BUDUROV, 1972; *Parvigondolella* KOZUR & MOCK, 1972; *Carinella* BUDUROV, 1973]. Apparatus unimembrate, apparently composed entirely of segminiplanate elements with more or less free anterior blade and platform margins marked by node- or spinelike projections or denticles (developed from pair of denticles projecting laterally and directed posteriorly). [Prominent zones of recessive basal margin apparently developed along a straight or bifurcate axis posterior to basal pit in intermediate to late stages of growth. Elements representing such growth stages were referred to *Ancyrogondolella* and *Carinella* by BUDUROV and have been included in *Gladigondolella* by some other authors. KOZUR & MOCK based *Parvigondolella* on platformless segminate elements interpreted as representatives of a species derived from *E. bidentata* MOSHER. Platform reduction, however, was evidently a phylogenetic tendency in *Epigondolella,* hence *Parvigondolella* is here regarded as merely the ultimate stage in the evolution of *Epigondolella,* not as a separate genus.] *M.Trias.-U.Trias.,* Eu.-Asia-N.Am.——FIG. 105,*1*. **E. abneptis* (HUCKRIEDE), M.Trias., Eu.(Aus.); *1a,* upper side; *1b,* under side; ×61 (Sweet, n).

Neogondolella BENDER & STOPPEL, 1965, p. 343 [**Gondolella mombergensis* TATGE, 1956; M] [=*Metapolygnathus* HAYASHI, 1968; *Paragondolella* MOSHER, 1968; *Celsigondolella* KOZUR, 1968]. Apparatus unimembrate, apparently composed entirely of segminiplanate elements supposed to have occupied *Pa* positions. [Elements typical of *Neogondolella* may have a relationship with *Cypridodella* and *Xaniognathus* like that between *Gladigondolella* and *Ellisonia,* or elements of *Neogondolella* and *Neospathodus* may represent dimorphs of species of *Cypridodella* or *Xaniognathus.* If the latter is true, taxonomy in the entire plexus will need great revision.] *Perm.–Trias.,* Eu.-Asia-N.Am.-Australia.——FIG. 105,*3*. **N. mombergensis* (TATGE), M.Trias., Eu.(W. Ger.); *3a,* lat. view, ×64; *3b,* under side, ×61 (Sweet, n).

Neospathodus MOSHER, 1968, p. 929 [**Spathognathodus cristagalli* HUCKRIEDE, 1958, p. 161; OD] [=*Neospathognathodus* BUDUROV, 1968, *nom. neg.; ?Misikella* KOZUR & MOCK, 1974; *Merrillina* KOZUR & MOCK, 1974]. Apparatus unimembrate, apparently composed entirely of segminate elements with distinct mid-lateral ribs; elements supposed to have occupied *Pa* position. [See comments under *Neogondolella.*] *Perm.-Trias.,* Asia-Eu.-N. Am.-Australia.——FIG. 105,*4*. **N. cristagalli* (HUCKRIEDE), L.Trias., Asia(Pak.); lat. view, ×100 (Sweet, 1970).

Platyvillosus CLARK, SINCAVAGE, & STONE, 1964, p. 376 [**P. asperatus*; OD] [=*Eurygnathodus* STAESCHE, 1964; *?Foliella* BUDUROV & PANTIC, 1973]. Apparatus unimembrate, apparently composed entirely of arched segminiplanate elements with irregularly nodose, radially or transversely ribbed upper surfaces, these surfaces subcircular, quadrate, or larchrymiform in upper view. Elements supposedly occupying *Pa* position. *L.Trias.,* N.Am.-Eu.——FIG. 105,*2a*. *P. costatus* (STAESCHE), USA(Nev.); upper side, ×50 (Sweet & others, 1971).——FIG. 105,*2b*. **P. asperatus,* USA (Nev.); upper side, ×50 (Sweet & others, 1971).

Sweetocristatus SZANIAWSKI in SZANIAWSKI & MATKOWSKI, 1979, *L.Perm.,* see addendum.

Superfamily POLYGNATHACEA Bassler, 1925

[*nom. transl.* LINDSTRÖM, 1970, p. 438, *ex* Polygnathidae BASSLER, 1925, p. 219] [Diagnosis by GILBERT KLAPPER]

Apparatus basically seximembrate, but may be reduced to *Pa* elements and thus unimembrate in specialized species. Ramiform elements multidenticulate and albid. [The apparatus is closely similar to that of Prioniodontacea, differing in that the *Pa* element is or can be derived, directly or indirectly, from the *Pa* element of multielement *Ozarkodina.*] *U.Ord.-L.Trias.*

Family KOCKELELLIDAE Klapper, new

[Materials for this family prepared by GILBERT KLAPPER]

Apparatus seximembrate: *Pa* element scaphate or planate; *Pb* angulate; *M* dolabrate; *Sa* symmetrical; *S* symmetry-transition series bearing discrete denticles. *Sil. (up.Llandov.-Ludlov.).*

Kockelella WALLISER, 1957, p. 34 [**K. variabilis*; OD]. *Pa* element carminiscaphate or stelliscaphate with posterior basal cavity; lateral processes characteristically developed. [Reconstruction: WALLISER, 1964, p. 14; BARRICK & KLAPPER, 1976.] *Sil.(up. Llandov.-Ludlov.),* Eu.-N. Afr.-Asia-N. Am.-Australia.——FIG. 106,*2*. *K. patula* WALLISER, Sil. (Wenlock., *K. amsdeni* Z., Clarita Fm.), USA (Okla.); *2a,b, Pa* element, upper and lower views; *2c, Pb* element, lat. view; *2d, M* element, lat. views; *2e, Sc* element, lat. view; *2f, Sb* element, lat. view; *2g, Sa* element, post. view; ×27 (Barrick & Klapper, 1976).

Ancoradella WALLISER, 1964, p. 28 [**A. ploeckensis*; OD]. *Pa* element modified stelliplanate, resembling that of late species of *Kockelella* with well-developed lateral processes, but unit not scaphate. [Only the *Pa* element is recognized.] *Sil.(Ludlov.),* Eu.-N.Am.-Australia.——FIG. 106, *1a,b*. **A. ploeckensis,* holotype, *A. ploeckensis* Z., Eu.(Aus., Carnic Alps); *1a,b, Pa* element, upper and lower views, ×27 (Walliser, 1964).

Family CRYPTOTAXIDAE Klapper & Philip, 1972

[Cryptotaxidae KLAPPER & PHILIP, 1972, p. 100] [Materials for this family prepared by GILBERT KLAPPER]

Apparatus seximembrate: *Pa* element carminate or carminiplanate; *Pb* carminate and sigmoidal; *M* dolabrate; *Sa* asymmetrical with inverted processes; *S* symmetry-transition series bearing discrete denticles. *M.Dev.-U.Dev.(Famenn.).*

Cryptotaxis KLAPPER & PHILIP, 1971, p. 444 [**Spathognathodus culminidirectus* SCOTT, 1961,

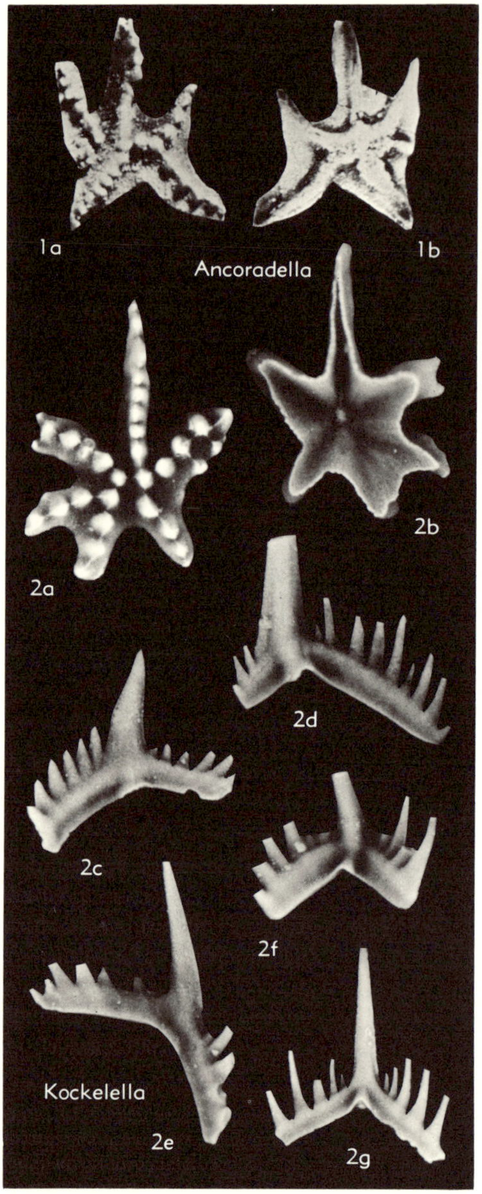

Fig. 106. Kockelellidae (p. W157).

p. 1226; OD]. *Pa* element carminate; *Pb* element carminate and slightly arched. [Reconstruction: Klapper & Philip, 1971.] *U.Dev.(Famenn.),* N.Am.——Fig. 107,*1*. **C. culminidirecta* (Scott), *Protognathodus* Fauna (Louisiana Ls.), USA (Mo.); *1a, Pa* element, lat. view; *1b, M* element, lat. view; *1c, Pb* element, lat. view; *1d, Sc* element, lat. view; *1e, Sb* element, lat. view; *1f, Sa* element, lat. view; ×27 (Klapper & Philip, 1971).

Parapolygnathus Klapper & Philip, 1971, p. 445 [**Polygnathus angusticostatus* Wittekindt, 1966, p. 631; OD]. Like *Cryptotaxis* but *Pa* element carminiplanate. [Reconstruction: Klapper & Philip, 1971; but see Sparling, 1981.] *M.Dev.-U.Dev.,* Eu.-N.Am.-Australia.——Fig. 107,*2*. **P. angusticostatus* (Wittekindt), M. Dev. (Eifel., Dundee Ls.), Can.(Ont.); *2a, Pa* element, upper view; *2b, Pb* element, lat. view; *2c, M* element, lat. view; *2d, Sc* element, lat. view; *2e, Sb* element, lat. view; *2f, Sa* element, lat. view; ×27 (Klapper & Philip, 1971).

Family CAVUSGNATHIDAE Austin & Rhodes, new

[Materials for this family prepared by R. L. Austin and F. H. T. Rhodes]

Apparatus seximembrate: pectiniform element scaphate; *Pb* element angulate; *M* element dolabrate or digyrate; *Sa* element alate and *Sc* bipennate, of two types. [Of the genera in this family, only *Adetognathus* and *Cavusgnathus* are known from multielement assemblages. The remaining genera are tentatively included on the basis of structural similarity.] *U.Dev.-L.Perm.*

Cavusgnathus Harris & Hollingsworth, 1933, p. 200 [**C. alta*; OD] [=*Lewistownella* Scott, 1934; *Windsorgnathus* Austin & Mitchell, 1975]. Known apparatus at least quinquimembrate: *Pa* scaphate, *Pb* angulate, *M* digyrate, *Sa* alate, *Sb* bipennate. Apparatus distinguished by *Pa* element having short free blade, conspicuous central trough, and transverse ridges; short fixed blade usually joining outer side of platform. [The *Pa* element resembles scaphate *Pa* element of *Adetognathus.*] *U.Miss.,* N.Am.; *U.Carb.,* S.Am.; *Carb. (Visean-Westphal.),* Eu.-Afr.-Asia M.-Australia-N.Z.——Fig. 108,*6*. *C. naviculus* Hinde, U.Miss. (Chester., Goddard Fm.), USA(Okla.); *6a-c, Pa* element, lat., upper, and lower views, ×31 (Lane and Straka, 1974).

Adetognathus Lane, 1967, p. 931 [**Cavusgnathus latus* Gunnell, 1933, p. 286; OD]. Apparatus seximembrate: *Pa* element scaphate; *Pb* angulate; *M* dolabrate; *Sa* alate, with posterior process; *Sc* bipennate, of two types. [Presence of denticulate posterior process in *Sa* elements of Missourian age distinguishes them from homologous elements in *Ozarkodina*. The *Pb* element can be distinguished from other homologous elements of Missourian age because it is less arched and has fewer but relatively larger denticles. *M* elements are distinguished from other homologous elements of Missourian age by the outward curvature of the anterolateral process, which differs from those of specimens of both *Ozarkodina* and *Idiognathodus*. One of the *Sc* elements is distinguished from homologous elements of *Ozarkodina* and *Idiognathodus* by strong downward deflection of the larger anterolateral process. In homologous elements of *Adetognathus,* elements may or may not have slight downward deflection of the anterolateral process. The second *Sc* element is distinguished by strong inward curvature of the posterior process, in contrast to presence or absence of slight inward curvature in other forms. The anterolateral process of the first *Sc* element may have a slight downward deflection, whereas the corresponding process in the second *Sc* element does not. The *Sa* element is distinguished from homologous elements in species of *Ozarkodina* of Missourian age by the presence of a denticulate posterior process. Apparatus reconstruction: BAESEMANN, 1973.] *U.Miss.-L.Perm.,* N.Am.-Eu. ——FIG. 108,*4a*. **A. lautus* (GUNNELL), L.Perm., USA(Okla.); *Pa* element, upper view, ×40 (Dunn, 1970).——FIG. 108,*4b*. *A. gigantus* (GUNNELL), L.Penn., USA(Texas); *Pa* element, lat. view, ×40 (Dunn, 1970).

Capricornognathus AUSTIN & MITCHELL, 1975, p. 47 [**Taphrognathus capricornis* DRUCE, 1970, p. 102; OD]. Apparatus unknown. Pectiniform element scaphate with free medial blade half length of specimen or less; free blade increases in elevation to posterior with conspicuous posterior denticle; platform with 1, 2, or 3 low parapets or transverse ridges and shallow medial trough; anterior carina and short posterior free blade present in some; basal cavity asymmetrical and flexed. [Pectiniform elements of *Capricornognathus* resemble those of *Cavusgnathus*.] *L.Carb.(low.Visean),* Eu.-Australia.——FIG. 108,*5*. **C. capricornis* (DRUCE), Eu.(Ire.); *5a,b, Pa* element, upper and lat. views, ×30 (Austin & Mitchell, 1975).

Cloghergnathus AUSTIN & MITCHELL, 1975, p. 48 [**C. globenskii*; OD]. Apparatus unknown. Pectiniform element scaphate; short free blade of uniform elevation, lateral in position, joining either right or left side of platform; lanceolate curved platform with medial trough and transverse ridges; flared asymmetrical basal cavity. [This element resembles the pectiniform *Pa* element of *Adetognathus,* which is probably a homeomorph.] *U.Miss.,* N.Am.; *L.Carb.(low.Visean),* Eu.——FIG. 108,*3*. **C. globenskii,* holotype, Eu.(Ire.); *3a,b, Pa* element, upper and lower views, ×40 (Austin & Mitchell, 1975).

Clydagnathus RHODES, AUSTIN, & DRUCE, 1969, p. 84 [**C. cavusformis*; OD]. Apparatus unknown. Pectiniform element scaphate with denticles of free blade highest at posterior end; short fixed blade joining platform on outer right margin; carina restricted to posterior, in some continuing as short free blade; medial trough; basal cavity asymmetrical. [This element resembles pectiniform elements of *Cavusgnathus* and *Adetognathus*.] *U.Dev.-L.Miss.,* N.Am.; *U.Dev.-L.Carb. (low.Tournais.),* Eu.——FIG. 108,*7*. *C. gilwernensis* RHODES, AUSTIN, & DRUCE, low.Tournais., Eu.(Wales); *7a,b, Pa* element, lat. and upper views, ×25 (Austin & Hill, 1973).

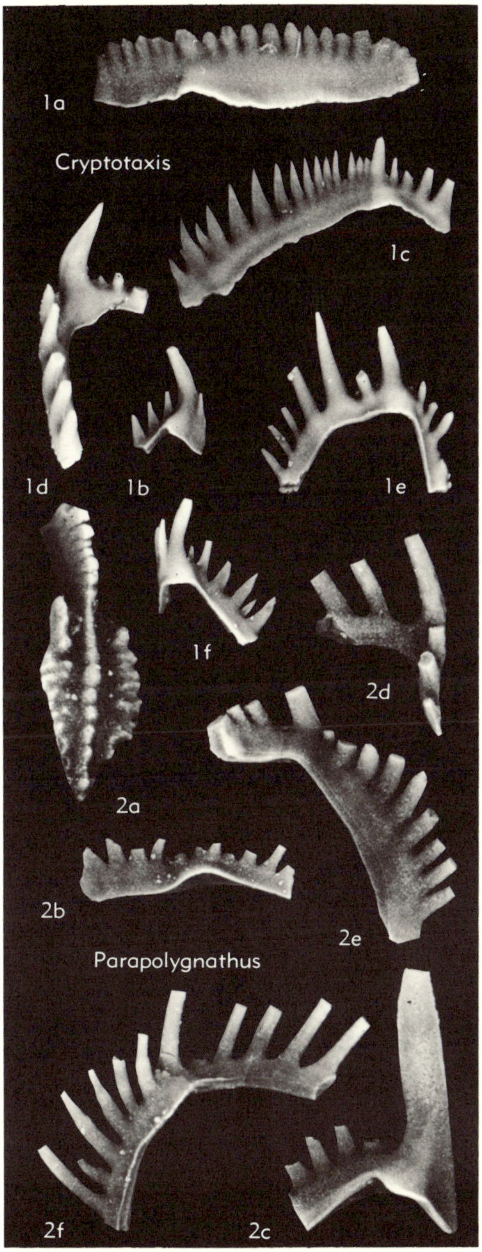

FIG. 107. Cryptotaxidae (p. *W*157-*W*158).

Patrognathus RHODES, AUSTIN, & DRUCE, 1969, p. 178 [**P. variabilis*; OD]. Apparatus unknown. Pectiniform element scaphate with free medial blade from half to one-third length of specimen;

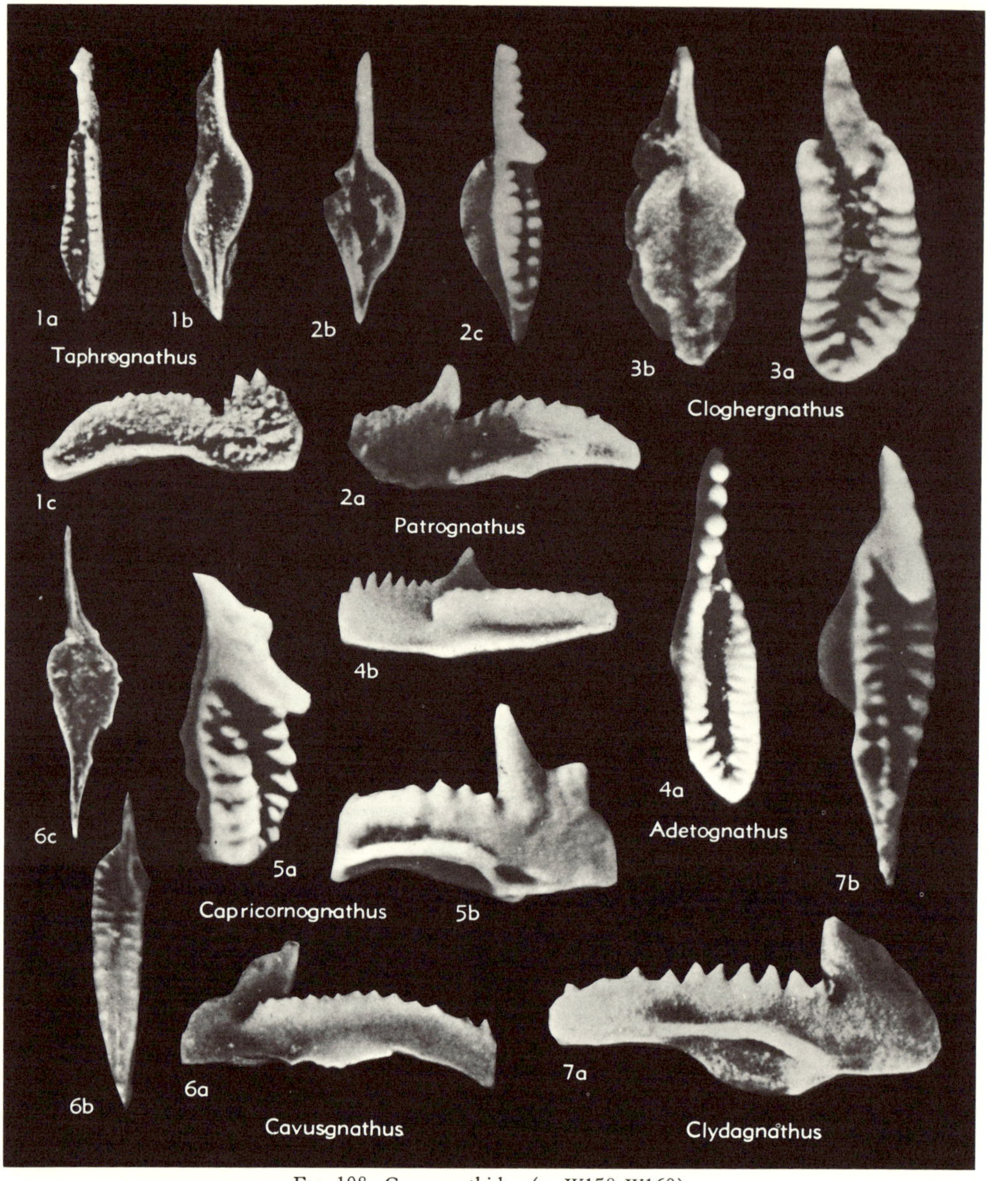

Fig. 108. Cavusgnathidae (p. *W*158-*W*160).

free blade having high posterior denticle; other blade denticles of uniform elevation; elongate symmetrical lanceolate platform having parapets; basal cavity narrow to flared. *U.Dev.-L.Miss.,* N.Am.; *U.Dev.-L.Carb.(low.Tournais.),* Eu.-Asia M.——Fig. 108,*2*. **P. variabilis,* low.Tournais., Eu.(Wales); *2a-c, Pa* element, lat., lower, and upper views, ×25 (Austin & Hill, 1973).

Taphrognathus Branson & Mehl, 1941a, p. 181, *non* Welles, 1947, an amphibian [**T. varians;* OD]. Apparatus unknown. Pectiniform element scaphate with free blade from half to one-third total length of specimen; right side of platform continuing as short carina; trough along midline of platform; ridges transverse. *Miss.(Keokuk F.-mid.St.Louis F.),* N.Am.; *L.Carb.,* Eu.——Fig. 108,*1*. **T. varians,* low.Visean, Eu.(Scot.); *1a-c, Pa* element, upper, lower, and lat. views, ×37 (Rhodes, Austin, & Druce, 1969).

Family IDIOGNATHODONTIDAE Harris & Hollingsworth, 1933

[*nom. transl. et correct.* HASS, 1959, p. 379, *pro* Idiognathinae HARRIS & HOLLINGSWORTH, 1933, p. 200] [Materials for this family prepared by R. L. AUSTIN and F. H. T. RHODES]

Apparatus either sexi- or septimembrate. Pectiniform elements scaphate *(Pa)* and angulate *(Pb); M* dolabrate; ramiform elements either alate *(Sa)* or bipennate *(Sa* and *Sc*). *U.Dev.-L.Perm.*

Idiognathodus GUNNELL, 1931, p. 249 [**I. claviformis*; OD] [=*Scottognathus* RHODES, 1953a *(partim), nom. subst. pro Scottella* RHODES, 1952, *non* ENDERLEIN, 1910, a dipteran]. Apparatus sexi- or septimembrate; *Pa* element scaphate, *Pb* angulate, *M* dolabrate, *Sa* alate, *Sb* bipennate, *Sc* bipennate. *Pa* element diagnostic; long free blade at least half length of element, carina partly or completely suppressed and transverse ridges strong, especially in posterior portion of upper surface. *L.Penn.-L.Perm.,* N.Am.; *U.Carb.(Namur.-Westphal.),* Eu.-Afr.-Asia M.-Australia-N.Z.-S.Am. ——FIG. 109,*7*. *I. delicatus* GUNNELL, L.Penn., USA(Ill.); *Pa* element, upper view, ×28 (Merrill & King, 1971); for a reconstruction of assemblage, see Fig. 53,*3*.

Gnathodus PANDER, 1856, p. 33, *non* FIEBER, 1866, an hemipteran [**G. mosquensis*; OD] [=*Dryphenotus* COOPER, 1939; *Westfalicus* MOORE & SYLVESTER-BRADLEY, 1957b]. Apparatus probably seximembrate; *Pa* element scaphate, *Pb* angulate, *M* dolabrate, *Sa* alate, *Sb* bipennate, *Sc* bipennate. *Pa* pectiniform element diagnostic; free blade medial, straight or curved, usually at least half length of element; carina distinct; sculpture of upper surface variable, either isolated nodes, parapets, adcarinal grooves, transverse ridges, or combination of these. *L.Miss.(Kinderhook.)-L.Penn.,* N.Am.; *L.Carb.(Tournais.-Namur.),* Eu.-Asia M.-Afr.-Australia-N.Z.-S.Am.——FIG. 109,*3*. *G. bilineatus* (ROUNDY), Visean, Eu.(Belg.); *Pa* element, upper view, ×63 (Austin & others, 1974); for a reconstruction of assemblage, see Fig. 53,*2*.

Idiognathoides HARRIS & HOLLINGSWORTH, 1933, p. 201 [**I. sinuata*; OD] [=*Polygnathodella* HARLTON, 1933; *Declinognathodus* DUNN, 1966; *Oxignathus* ELLISON, 1972]. Apparatus unknown, probably sexi- or septimembrate. Pectiniform element scaphate with medial or lateral blade almost half length of element, in some continuing as deflected carina merging with a parapet or terminating against parapet; sculpture of upper surface as parapets or transverse ridges and a trough. *L.Penn.-M.Penn.,* N.Am.; *U.Carb.(Namur.-Westphal.),* Eu.-Australia-Asia M.——FIG. 109,*5*. *I. noduliferus* (ELLISON & GRAVES), Eu.(G.Brit.); *Pa* element, upper view, ×28 (Austin, 1972).

Neognathodus DUNN, 1970, p. 336 [**Polygnathus bassleri* HARRIS & HOLLINGSWORTH, 1933, p. 198; OD]. Apparatus unknown, probably either sexi- or septimembrate. Pectiniform element scaphate

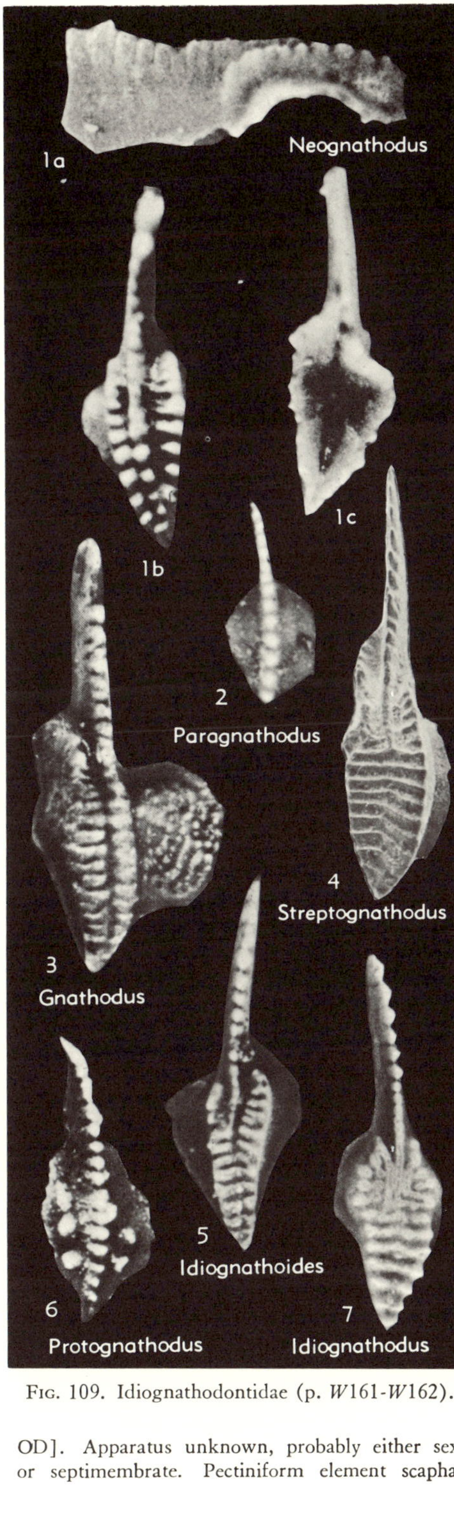

FIG. 109. Idiognathodontidae (p. W161-W162).

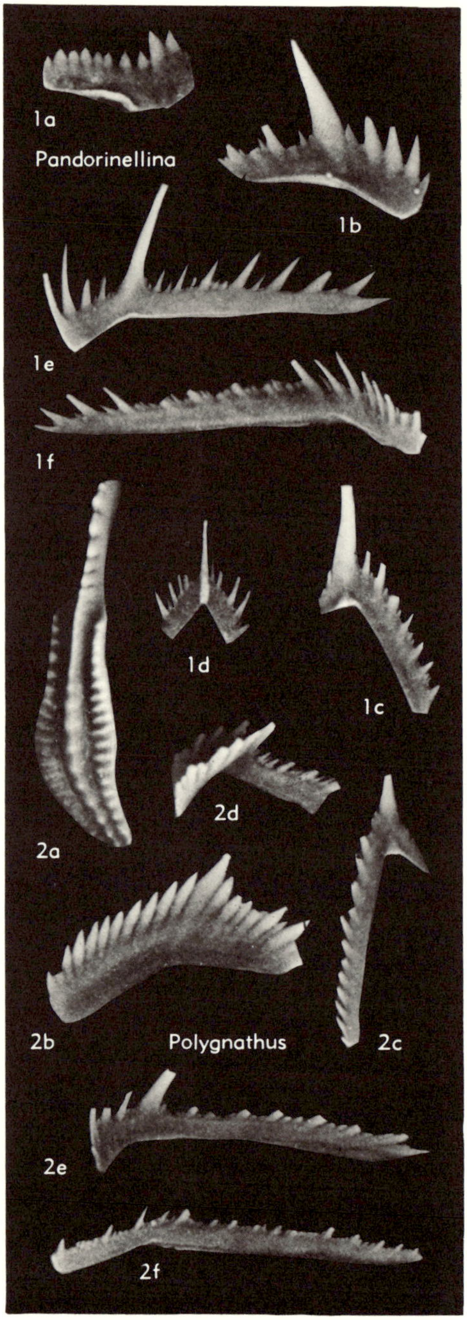

FIG. 110. Polygnathidae (p. W162-W164, W166).

with long free blade up to half length of element meeting platform centrally or subcentrally; platform in some reduced or absent on outer margin; parapets or transverse ridges flanking one or both sides of platform; carina extending to, or near, posterior tip of element; adcarinal grooves deep; basal cavity large, deep, asymmetrical. *L.Penn.-L.Perm.,* N.Am.; *U.Carb.(Namur.-Westphal.),* Eu. ——FIG. 109,*1*. **N. bassleri* (HARRIS & HOLLINGSWORTH), L.Penn., USA(Okla.); *1a-c, Pa* element, lat., upper, and lower views, ×40 (Dunn, 1970).

Paragnathodus HIGGINS, 1975, p. 70 [**Spathognathodus commutatus* BRANSON & MEHL, 1941c, p. 98; OD] [=*Paragnathodus* MEISCHNER, 1970, *nom. nud.*]. Apparatus unknown, possibly resembling that of *Gnathodus.* Pectiniform element scaphate with free medial blade equaling platform in length or longer; platform oval to circular in form, unornamented or ornamented with one or few nodes, nodes in some linear; carina commonly thickened. [This element resembles the pectiniform element of *Protognathodus.*] *U.Miss.-L.Penn. (Visean-low.Namur.),* N.Am.-Eu.——FIG. 109,*2*. **P. commutatus* (BRANSON & MEHL), Visean, Eu. (Eng.); *Pa* element, upper view, ×40 (Higgins, 1975).

Protognathodus ZIEGLER, 1969, p. 352 [**Gnathodus kockeli* BISCHOFF, 1957, p. 25; OD]. Apparatus unknown. Pectiniform element scaphate with free medial blade, usually half element length; round symmetrical platform smooth or ornamented with nodes; carina distinct; basal cavity large. *U.Dev.-L.Miss.(Tournais.),* N.Am.-Eu.-Afr.——FIG. 109, *6*. **P. kockeli* (BISCHOFF), U.Dev., Eu.(Ger.); *Pa* element, upper view, ×50 (Ziegler, 1969).

Streptognathodus STAUFFER & PLUMMER, 1932, p. 47 [**S. excellus*; OD] [=*Scottognathus* RHODES, 1953b *(partim), nom. subst. pro Scottella* RHODES, 1952, *non* ENDERLEIN, 1910, a dipteran]. Apparatus either sexi- or septimembrate; *Pa* element scaphate, *Pb* angulate, *M* dolabrate, *Sa* alate, *Sb* bipennate, *Sc* bipennate. *Pa* element with long free blade, platform with median trough and transverse ridges. [Median trough distinguishes *Streptognathodus* from *Idiognathodus.*] *L.Penn.-L. Perm.,* N. Am.-Eu.-Afr.-Australia-N. Z.——FIG. 109,*4*. *S. antiquus* (STAUFFER & PLUMMER), U. Penn., USA(Kans.); *Pa* element, upper view, ×52 (von Bitter, 1972).

Family POLYGNATHIDAE Bassler, 1925

[Polygnathidae BASSLER, 1925, p. 219] [=Spathognathodontidae HASS, *nom. transl.* LINDSTRÖM, 1970, p. 439, *ex* Spathognathodontinae HASS, 1959, p. 378] [Materials for this family prepared by GILBERT KLAPPER, R. L. AUSTIN, and F. H. T. RHODES]

Apparatus seximembrate; *Pa* element carminate, planate, or scaphate; *Pb* angulate or anguliplanate; *M* dolabrate; *S* symmetry-transition series bearing confluent denticles. *U.Ord.-U.Carb.*

Polygnathus HINDE, 1879, p. 361 [**P. dubius*; SD MILLER, 1889, p. 520; neotype selected by HUDDLE, 1970, p. 1037] [=*Hindeodella* BASSLER, 1925,

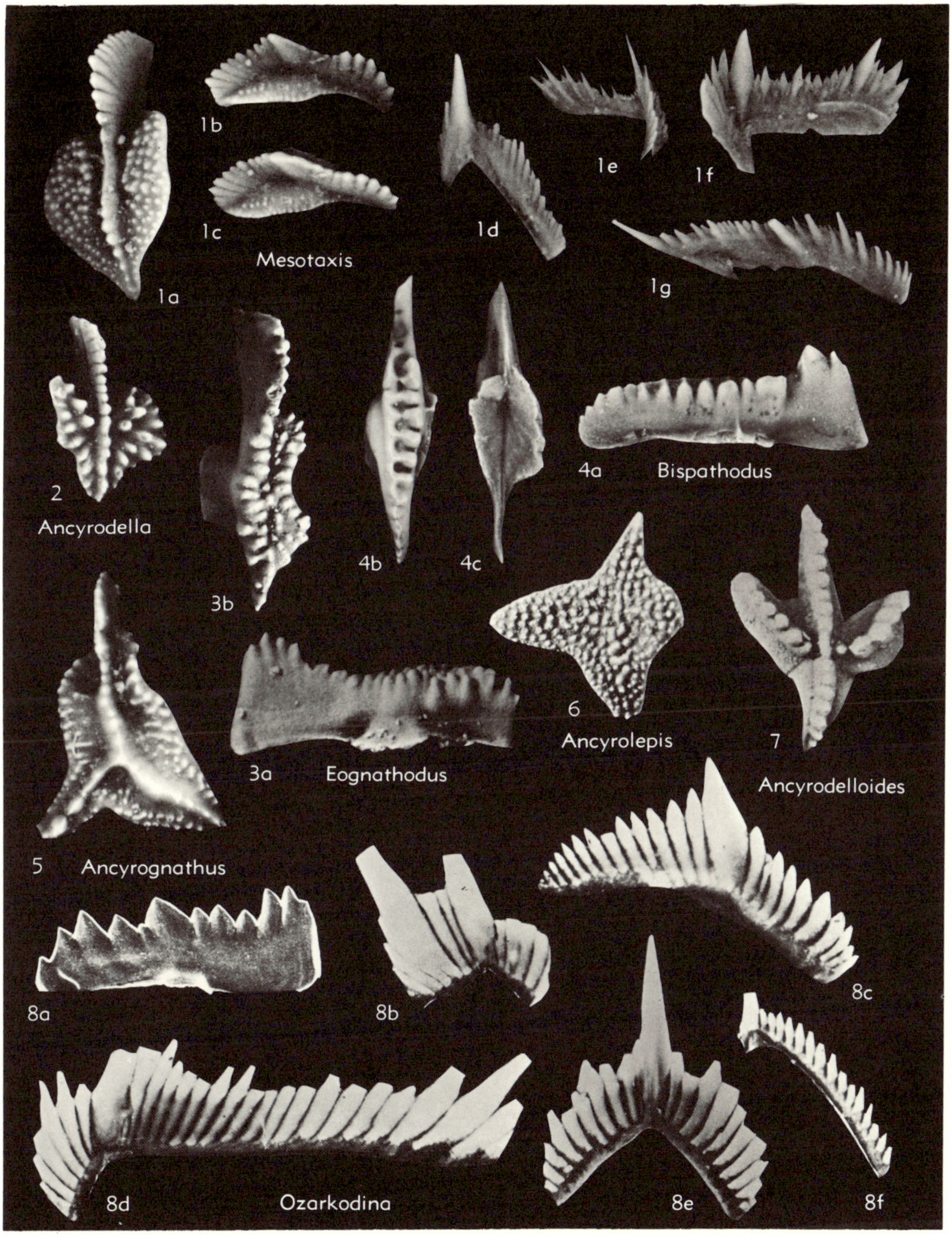

FIG. 111. Polygnathidae (p. *W*162-*W*163).

p. 219; *Ctenopolygnathus* MÜLLER & MÜLLER, 1957, p. 1084]. *Pa* element carminiplanate (carminiscaphate in earliest species); *Sa* element alate with denticulate posterior process. [Reconstruction: KLAPPER & PHILIP, 1971, 1972.] *L.Dev.* *(Ems.)-L. Miss.(low. Visean),* Eu.-N. Afr.-Asia-N. Am.-Australia.——FIG. 110,2. **P. dubius,* U.Dev. (L. *Mesotaxis asymmetrica* Z., Genundewa Ls.), USA(N.Y.); *2a, Pa* element, upper view; *2b, Pb* element, lat. view; *2c, M* element, lat. view; *2d,*

Sa element, lat. view; *2e, Sc* element, lat. view; *2f, Sb* element, lat. view; ×40 (Klapper & Philip, 1971).

Ancyrodella ULRICH & BASSLER, 1926, p. 48 [**A. nodosa*; OD] [=*Ancyropenta* MÜLLER & MÜLLER, 1957, p. 1092]. *Pa* element modified stelliplanate with posterior lobe and two anteriorly directed lobes. *Sa* element alate with denticulate posterior process. [Reconstruction: KLAPPER & PHILIP, 1972.] *U.Dev.(Frasn.),* Eu.-N.Afr.-Asia-N.Am.-Australia.——FIG. 111,*2*. *A. rotundiloba alata* GLENISTER & KLAPPER, U.Dev.(L. *Mesotaxis asymmetrica* Z., Snyder Creek Sh.), USA(Mo.); *Pa* element, upper view, ×27 (Klapper, n).

Ancyrodelloides BISCHOFF & SANNEMANN, 1958, p. 91 [**A. trigonicus*; OD]. Like *Ancyrodella* but upper platform surface of *Pa* element smooth adjacent to carina and secondary carinae. [Only *Pa* element is recognized.] *L.Dev.,* Eu.-Asia M.-N.Am.(Alaska).——FIG. 111,*7*. **A. trigonicus,* Tentaculitenkalk, Eu.(Ger.); *Pa* element, upper view, ×17 (Bischoff & Sannemann, 1958).

Ancyrognathus BRANSON & MEHL, 1934a, p. 240 [**A. symmetricus*; OD] [=*Ancyroides* MILLER & YOUNGQUIST, 1947, p. 504]. *Pa* element pastiniplanate; outer lateral lobe directed somewhat posteriorly, but absent in late species. [Only *Pa* element is recognized.] *U.Dev.(Frasn.-low.Famenn.),* Eu.-N.Afr.-N.Am.-Australia.——FIG. 111,*5*. *A. triangularis* YOUNGQUIST, U.Dev.(*A. triangularis* Z., Sweetland Creek Sh.), USA(Iowa); *Pa* element, upper view, ×27 (Klapper & Furnish, 1963).

Ancyrolepis ZIEGLER, 1959, p. 77 [**A. cruciformis*; OD]. *Pa* element modified pastiniplanate with weak inner lobe and strong outer lobe; secondary keel present at least on outer lobe. [Only *Pa* element is recognized.] *M.Dev.(Givet.)-U.Dev.(low.Famenn.),* Eu.-N.Am.-Australia.——FIG. 111, *6*. **A. cruciformis,* holotype, U.Dev.(L. *Palmatolepis crepida* Z.), Eu.(Ger., Dill syncline); *Pa* element, upper view, ×23 (Ziegler, 1959).

Bispathodus MÜLLER, 1962a, p. 114 [**Spathodus spinulicostatus* BRANSON, 1934, p. 305; OD]. *Pa* element carminate or carminiscaphate, commonly with accessory denticles on right side of blade; accessory denticles discrete or connected to main denticle row by transverse ridges. [Only *Pa* element is recognized.] *U.Dev.(Famenn.)-L.Miss.(Tournais.),* Eu.-Asia-N. Am.-Australia.——FIG. 111,*4*. *B. costatus* (BRANSON), holotype, L.Miss. (Hannibal F.), USA(Mo.); *4a-c, Pa* element in lat., upper, and lower views, ×23 (Ziegler, Sandberg, & Austin, 1974).

Eognathodus PHILIP, 1965, p. 99 [**E. sulcatus*; OD]. *Pa* element carminiscaphate (but cavity reduced in late species) with double denticle row; *Sa* element alate, lacking distinct posterior process. [Reconstruction: KLAPPER & PHILIP, 1971.] *L. Dev.-M. Dev.,* Eu.-Asia-N. Am.-Australia.——FIG. 111,*3*. **E. sulcatus,* holotype, L.Dev.(Coopers Creek F.), Australia(Vict.); *3a,b, Pa* element, lat. and upper views, ×23 (Philip, 1965).

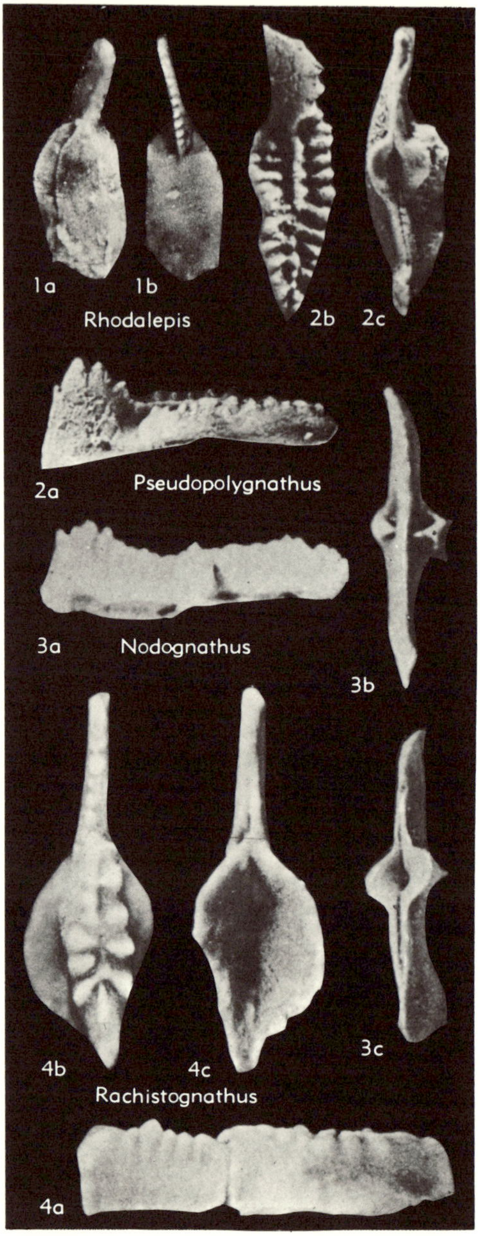

FIG. 112. Polygnathidae (p. W165, W166).

Hemilistrona CHAUFF & DOMBROWSKI, 1977, *U.Dev.,* see addendum.

Kimognathus MASHKOVA, 1978, *L.Dev.,* see addendum.

Mesotaxis KLAPPER & PHILIP, 1972, p. 100 [**Polygnathus asymmetricus* BISCHOFF & ZIEGLER, 1957, p. 88: OD]. Apparatus like *Polygnathus* but *Pb* element with well-developed platform ledge on inner side, *M* element with lower margin of both processes much straighter, and *Sc* element with posterior process higher. [Reconstruction: Klapper & Philip, 1972.] *U.Dev.(Frasn.),* Eu.-N.Am.-Australia.——Fig. 111,*1*. **M. asymmetrica,* U. Dev.(L. *Mesotaxis asymmetrica* Z., Gogo F.), Australia(W.Australia); *1a, Pa* element, upper view; *1b,c, Pb* element, oblique-lat. and upper views; *1d, M* element, lat. view; *1e, Sa* element, lat. view; *1f, Sc* element, lat. view; *1g, Sb* element, lat. view; ×27 (Klapper & Philip, 1971).

Nodognathus Cooper, 1939, p. 397 [**N. spicata;* OD]. Apparatus unknown. Pectiniform element carminiscaphate with free anterior and posterior blades; lateral expansions adjacent to prominent basal cavity commonly supporting long node or short transverse ridge. [*Nodognathus* may grade into *Pseudopolygnathus.*] *L.Miss.(Kinderhook.),* N.Am.——Fig. 112,*3*. **N. spicatus,* holotype, USA (Okla.); *3a-c,* lat., upper, and lower views, ×27 (Cooper, 1939).

Ozarkodina Branson & Mehl, 1933a, p. 51 [**O. typica;* OD; =*Hindeodella confluens* Branson & Mehl, 1933a, p. 45] [=*Plectospathodus* Branson & Mehl, 1933a, p. 47; *Spathognathodus* Branson & Mehl, 1941c, p. 98, *nom. subst. pro Spathodus* Branson & Mehl, 1933a, p. 46, *non* Boulenger, 1900, a fish; *Ctenognathodus* Fay, 1959, p. 195, *nom. subst. pro Ctenognathus* Pander, 1856, p. 32, *non* Fairmaire, 1843, a beetle]. *Pa* element carminate or carminiscaphate; *Sa* element alate and lacking distinct posterior process. [Reconstruction: Jeppsson, 1969; Lindström, 1970, p. 439-440.] *U.Ord.-L.Dev.,* Eu.-N.Afr.-Asia-N.Am.-Australia.——Fig. 111,*8*. **O. confluens* (Branson & Mehl), Sil.(Ludlov.), Eu.(Sweden, Scania); *8a, Pa* element, lat. view, ×45; *8b, Sb* element, lat. view; *8c, Pb* element, lat. view; *8d, Sc* element, lat. view; *8e, Sa* element, post. view; *8f, M* element, lat. view; all ×27 except *8a* (Jeppsson, 1975).

Palmatolepis Ulrich & Bassler, 1926, p. 49 [**P. perlobata;* OD] [=*P. (Manticolepis)* Müller, 1956a, p. 16; *P. (Deflectolepis)* Müller, 1956a, p. 16; *P. (Panderolepis)* Helms, 1963, p. 467; *P. (Conditolepis)* Boogaard & Kuhry, 1979, p. 50; *Klapperina* Lane, Müller, & Ziegler, 1979, p. 217]. Apparatus like that of *Mesotaxis* but carminiplanate *Pa* element with distinct central (azygous) node; *Sc* element with anticusp joining posterior process at much lower angle than that formed by two processes in *Sc* element of *Mesotaxis.* [Reconstruction: Klapper & Philip, 1972.] *U.Dev.,* Eu.-N.Afr.-Asia-N.Am.-Australia. ——Fig. 113,*2a*. **P. perlobata perlobata* Ulrich & Bassler, U.Dev.(L. *Palmatolepis marginifera* Z., Bugle Gap Ls.), Australia(W.Australia); *Pa* element, upper view, ×30 (Glenister & Klapper, 1966).——Fig. 113,*2b*. *Palmatolepis* sp.; *M* element, lat. view, ×30 (Glenister & Klapper, 1966).

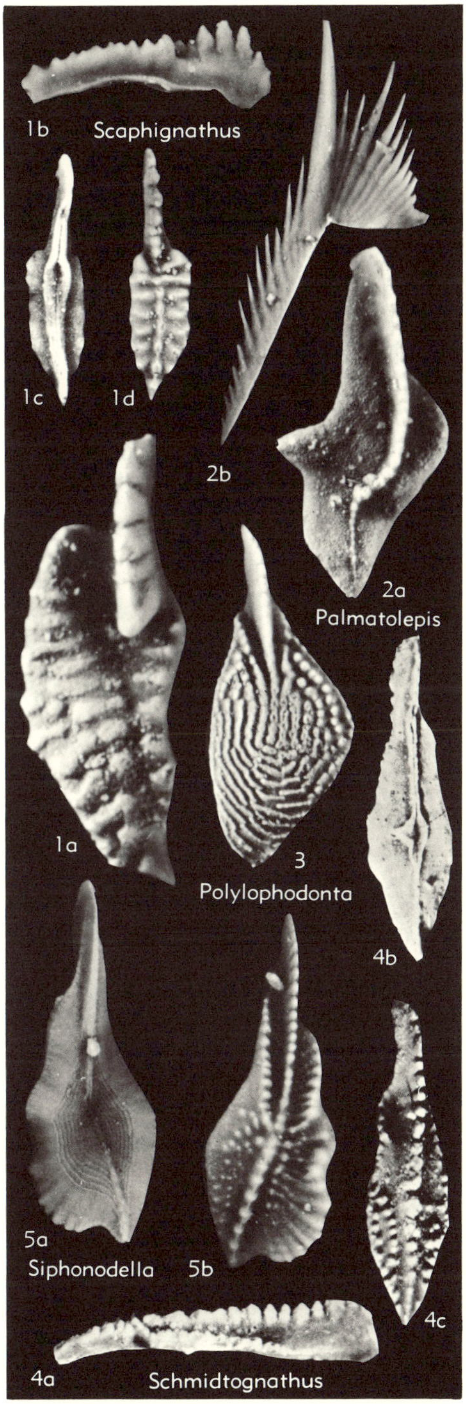

Fig. 113. Polygnathidae (p. W156, W166).

Pandorinellina MÜLLER & MÜLLER, 1957, p. 1082, *nom. subst. pro Pandorina* STAUFFER, 1940, p. 428, *non* BORY DE ST. VINCENT, 1827, a protozoan, *nec* SCACCHI, 1833, a molluscan [**Pandorina insita* STAUFFER, 1940, p. 429; OD] [=*Criteriognathus* WALLISER, 1972, p. 78]. *Pa* element carminate; *Sa* element alate with denticulate posterior process. [Reconstruction: KLAPPER & PHILIP, 1972.] *Dev.*, Eu.-N.Afr.-Asia-N.Am.-Australia.——FIG. 110,*1*. **P. insita* (STAUFFER), U.Dev.(*P. insita* Fauna, State Quarry Ls.), USA(Iowa); *1a, Pa* element, lat. view; *1b, Pb* element, lat. view; *1c, M* element, lat. view; *1d, Sa* element, post. view; *1e, Sc* element, lat. view; *1f, Sb* element, lat. view; all ×40 (Klapper & Philip, 1971).

Polylophodonta BRANSON & MEHL, 1934a, p. 242 [**Polygnathus gyratilineatus* HOLMES, 1928, p. 31; OD; =*Polygnathus pergyratus* HOLMES in BUTTS, 1926, p. 160; for discussion of type species, see GLENISTER & KLAPPER, 1966, p. 831]. *Pa* element like that of *Polygnathus* (especially *P. nodocostatus* group) but platform upper surface with strong concentric arrangement of ridges or rows of nodes. [Only *Pa* element is recognized.] *U.Dev.(Famenn.)*, Eu.-N.Am.-Australia.——FIG. 113,*3*. *P. confluens* (ULRICH & BASSLER), L. *Palmatolepis marginifera* Z.(Virgin Hills F.), Australia(W.Australia); *Pa* element, upper view, ×30 (Glenister & Klapper, 1966).

Pseudopolygnathus BRANSON & MEHL, 1934b, p. 297 [**P. prima*; OD] [=*Macropolygnathus* COOPER, 1939]. Apparatus unknown. Pectiniform element scaphate with free medial blade half to one-third element length; platform symmetrical or asymmetrical with nodes or sturdy ridges, carina distinct; basal cavity usually prominent with longer dimension generally transverse to element axis, small basal pit rare. *U.Dev.-L.Miss.(Kinderhook.* or *Tournais.)*, N.Am.-Eu.-Australia-S.Am.-Africa-Asia M.——FIG. 112,*2*. **P. primus*, low.Tournais., Eu.(Wales); *2a-c*, lat., upper, and lower views, ×32 (Rhodes, Austin, & Druce, 1969).

Rhachistognathus DUNN, 1966, p. 1301 [**R. primus*; OD]. Apparatus unknown but probably either sexi- or septimembrate. Pectiniform element scaphate with laterally compressed, long, free blade of uniform elevation, joining platform at midlength of element; platform lanceolate with parapets or discontinuous carina, parapet or nodes often radiating outward. *U.Carb.(Namur.-Westphal.)*, N.Am.-Eu.——FIG. 112,*4*. **R. primus*, L.Penn. (Bird Spring F.), USA(Nev.); *4a-c*, lat., upper, and lower views, ×40 (Dunn, 1966).

Rhodalepis DRUCE, 1969, p. 116 [**R. inornata*; OD]. *Pa* element like that of *Polygnathus* but upper surface lacking carina and nodes or ridges, and lower surface with broad, flat inverted area (similar to pseudokeel in early species of *Siphonodella*). [Only *Pa* element is recognized.] *U.Dev. (Famenn.)*, Australia.——FIG. 112,*1*. **R. inornata*, low. Ningbing Ls., Australia(W. Australia, Bonaparte Gulf basin); *1a,b*, lower and upper views, magnification not stated (Druce, 1969).

Scaphignathus HELMS, 1959, p. 655 [**S. velifer*; M; for discussion see BEINERT & others, 1971, p. 82-83]. *Pa* element carminiplanate with blade commonly offset from carina. [Only *Pa* element is recognized.] *U.Dev.(Famenn.)*, Eu.-N.Am.-Australia.——FIG. 113,*1a*. **S. velifer, S. velifer* Z.(Virgin Hills F.), Australia(W.Australia); *Pa* element, upper view, ×45 (Glenister & Klapper, 1966).——FIG. 113,*1b-d*. *S. subserratus* (BRANSON & MEHL), *Scaphignathus subserratus-Pelekysgnathus inclinatus* Fauna (Trident Mbr., Three Forks F.), USA(Mont.); *Pa* element, lat., lower, and upper views, ×28 (Beinert & others, 1971).

Schmidtognathus ZIEGLER, 1966, p. 664 [**S. hermanni*; OD]. *Pa* element carminiplanate like that of *Polygnathus* but pit larger, asymmetrical, with distinct constriction in outer margin. [Only *Pa* element is recognized with certainty.] *?M.Dev., U. Dev.(Frasn.)*, Eu.-N.Am.——FIG. 113,*4*. **S. hermanni*, holotype, disputed M.Dev.-U.Dev. boundary interval (Flinzkalk, *Schmidtognathus hermanni–Polygnathus cristatus* Z.), Eu.(Ger., Rhenish Slate Mts.); *4a-c, Pa* element, lat., lower, and upper views, ×23 (Ziegler, 1966).

Siphonodella BRANSON & MEHL, 1944, p. 245, *nom. subst. pro Siphonognathus* BRANSON & MEHL, 1934b, p. 295, *non* RICHARDSON, 1858, a fish [**Siphonognathus duplicata* BRANSON & MEHL, 1934b, p. 296; OD]. *Pa* element carminiplanate like that of *Polygnathus* but keel generally absent just posterior of basal pit. [Only *Pa* element is recognized.] *U.Dev.(up.Famenn.)-L.Miss.(Tournais.)*, Eu.-N.Am.-Australia.——FIG. 113,*5*. *S. crenulata* (COOPER), L.Miss.(basal Lodgepole Ls.), USA(Mont.); *5a,b, Pa* element, lower and upper views, ×27 (Klapper, 1971).

Tortodus WEDDIGE, 1977, *M.Dev.*, see appendix.

Family ANCHIGNATHODONTIDAE Clark, 1972

[*nom. transl.* herein, SWEET & CLARK, *ex* superfamily Anchignathodontacea CLARK, 1972a, p. 157] [Materials for this family prepared by W. C. SWEET and D. L. CLARK]

Apparatus of generalized, long-ranging species seximembrate, but reduced to quadrimembrate or *Pa* elements and thus unimembrate in specialized species. Scaphate *Pa* element with broadly expanded base and short anterior free blade, smooth on upper side in all species with seximembrate apparatuses, bearing denticles, nodes, or transverse ridges in majority of species with unimembrate apparatuses. *Pb* element typically bowed and angulate, with anterior process shorter than posterior process; *M* elements digyrate; *Sa* elements alate, without posterior process in most species; *Sb* digyrate, with one lateral process deflected strongly upward and posteriorly; *Sc* bipennate. *L.Carb.-L.Trias.*

Hindeodus Rexroad & Furnish, 1964, p. 671 [**Trichonodella imperfecta* Rexroad, 1957; OD; =*Spathognathodus cristulus* Youngquist & Miller, 1949, p. 621] [=*Anchignathodus* Sweet, 1970]. Apparatus seximembrate: *Pa* element scaphate, *Pb* angulate, *M* dolabrate or digyrate with one lateral process adenticulate, *Sa* alate with no posterior process, *Sb* digyrate, and *Sc* bipennate. [The originally designated type species was based on the *Sa* element of a seximembrate apparatus including elements of *Spathognathodus cristulus* Youngquist & Miller, 1949 *(Pa)*, *Ozarkodina curvata* Rexroad, 1958b *(Pb)*, *Neoprioniodus camurus* Rexroad, 1957 *(M)*, *Falcodus? alatoides* Rexroad & Burton, 1961 *(Sb)*, and *Hindeodella* sp. of Rexroad & Furnish, 1964 *(Sc)*; hence, *H. cristulus* (Youngquist & Miller) has priority as the type species of multielement *Hindeodus*.] *L.Miss.-L.Trias.*, N. Am.-Asia-Eu.-S. Am.——Fig. 114,*1*. **H. cristulus* (Youngquist & Miller), Miss., USA(Iowa); *1a, Sa* element, post. view; *1b, Sb* element, post. view; *1c, Sc* element, lat. view; *1d, M* element, post.-lat. view; *1e, Pb* element, outer lat. view; *1f, Pa* element, lat. view; ×50 (Rexroad & Furnish, 1964).

Aethotaxis Baesemann, 1973, p. 697 [**A. advena*; OD]. Apparatus quadrimembrate: *Sa* symmetrical; *Sb* nearly symmetrical, with short posterior process; *Sc* bipennate, of two forms distinguished by 90° flexure of anterior process and position of posterior process just behind cusp. [The X element of Baesemann was not considered by him to be closely comparable to previously described elements; however, it bears a striking resemblance to some forms of *Apatognathus* and therefore occupies an *Sa* position.] *U.Penn.*, N.Am.(Kans.).——Fig. 115,*1*. **A. advena; 1a, Sa* element, holotype, lat. view; *1b,c, Sc* elements, lat. views; *1d, Sb* element, lat. view; *1e, Sa* element, post. view; ×34 (Baesemann, 1973).

Diplognathodus Kozur & Merrill in Kozur, 1975a, p. 9 [**Spathognathodus coloradoensis* Murray & Chronic, 1965, p. 606; OD]. Apparatus seximembrate, *Pa* scaphate with free blade equal in length to fused or partly fused carina of platform; basal cavity subelliptical in outline, deepest portion located behind anterior blade; *Pb* angulate, *M* dolobrate, *Sa* alate with well-developed posterior process, *Sb* and *Sc* bipennate. *U.Penn.-L.Perm.*, N.Am.; *U.Carb.-U.Perm.*, Eu.-Asia.——Fig. 115,*2*. *D.* sp., L.Perm., USA(Nev.); *2a, Pa* element, lat. view, ×110; *2b, Pb* element, lat. view, ×110; *2c, M* element, lat. view, ×80; *2d, Sa* element, post. view, ×175; *2e, Sc* element, lat. view, ×70 (Clark, n).

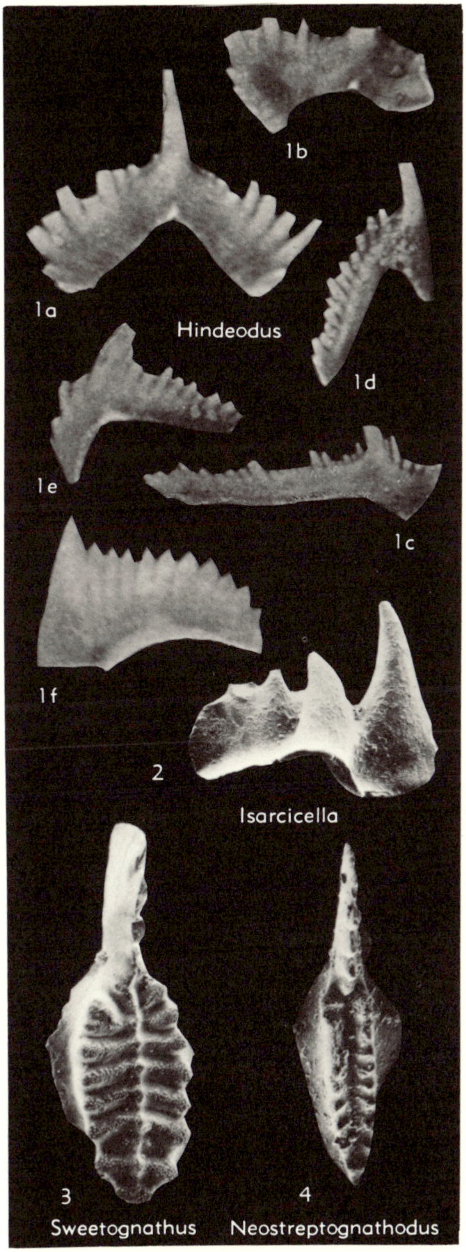

Fig. 114. Anchignathodontidae (p. W167-W169).

Iranognathus Kozur, Mostler, & Rahimi-Yazd, 1976, *U.Perm.*, see addendum.

Isarcicella Kozur, 1975a, p. 11 [**Spathognathodus isarcicus* Huckriede, 1958; OD]. Apparatus unknown, but probably unimembrate with variable single element. Genus based on tiny, subquadrate, posteriorly truncate, scaphate elements with bases

FIG. 115. Anchignathodontidae (p. W167).

broadly expanded laterally; may be smooth on upper surfaces or bear 1 or 2 long denticles on either or both sides. [Laterally adenticulate elements, which are numerous in known collections, have been referred by several authors to *Anchignathodus* SWEET, 1970 a junior subjective synonym of *Hindeodus* REXROAD & FURNISH, 1964. Such elements resemble laterally denticulate forms closely, are common in samples that lack any ramiform components of *Hindeodus,* and are thus included here in *Isarcicella.*] *L.Trias.,* Eu.-Asia-Middle East.——FIG. 114,*2*. **I. isarcica* (HUCKRIEDE), Eu.(Italy); lat. view, ×65 (Sweet, n).

Neostreptognathodus CLARK, 1972a, p. 155 [**Streptognathodus sulcoplicatus* YOUNGQUIST, HAWLEY, & MILLER, 1951, p. 363; OD] [?=*Vjalovites* KOZUR in KOZUR & MOSTLER, 1976]. Apparatus unknown. *Pa* element scaphate, pectiniform with slender free anterior blade from one-third to one-half total length of element; trough along midline; edges smooth to parapet-form. [This element is a homeomorph of Carboniferous *Streptognathodus.*] *Perm.,* N.Am.-S.Am.-Eu.——FIG. 114,*4*. **N. sulcoplicatus* (YOUNGQUIST, HAWLEY, & MILLER), L.Perm., USA(Wyo.); upper view, ×80 (Clark, n).

Rabeignathus KOZUR, 1978, *L.Perm.,* see addendum.

Sweetognathus CLARK, 1972a, p. 155 [**Spatho-

gnathodus whitei RHODES, 1963, p. 104; OD]. Apparatus probably unimembrate; pectiniform element scaphate with short free anterior blade in young forms; blade approaching length of total unit in older forms. Faint rostrum in juveniles, developing to heavy rostrum and carina at maturity. *L.Perm., ?U.Perm.,* N.Am.-S.Am.-Asia. ——FIG. 114,*3*. **S. whitei* (RHODES), L.Perm., USA(Nev.); upper view, ×65 (Clark, n).

Superfamily UNKNOWN

Family BACTROGNATHIDAE Lindström, 1970

[Bactrognathidae LINDSTRÖM, 1970, p. 441] [Materials for this family prepared by R. L. AUSTIN and F. H. T. RHODES]

Apparatus unknown. Pectiniform element carminate, pastiniplanate, pastiniscaphate, or scaphate. [Possibly a member of the superfamily Polygnathacea.] *U.Dev.-L.Miss.*

Bactrognathus BRANSON & MEHL, 1941c, p. 98 [**B. hamata*; OD]. Apparatus unknown. Pectiniform element carminate with straight anterior process and laterally deflected posterior process that is only one-third as long as anterior process. Under side of both processes longitudinally grooved; basal cavity cup-shaped. *L.Miss.(Osag. or Tournais.),* N.Am.-Eu.——FIG. 116,*1*. **B. hamatus,* holotype, Pierson Ls., USA(Mo.); *1a-c,* upper, lower, and lat. views, ×25 (Branson & Mehl, 1941c).

Apatella CHAUFF & KLAPPER, 1978, *U.Dev.,* see addendum.

Doliognathus BRANSON & MEHL, 1941c, p. 100 [**D. lata*; OD]. Apparatus unknown. Pectiniform element pastiniplanate with long and straight anterior process; posterior and lateral processes shorter, diverging at angle of approximately 120°; platforms decreasing uniformly in width to pointed extremities but extending to end of processes; transverse ridges or parapets at margin; carina of anterior process distinct, low, nodes fused, becoming prominent anteriorly to form blade; generally fewer nodes forming carina of posterior and lateral processes; basal pit small and situated at junction of processes; keel under each process. *L.Miss.(low.Osag. or up.Tournais.),* N.Am.-Afr.-Eu.——FIG. 116,*3*. **D. latus,* Pierson Ls., USA (Mo.); *3a,* paratype, upper view; *3b,c,* holotype, upper and lower views; ×17 (Branson & Mehl, 1941c).

Dollymae HASS, 1959, p. 394 [**D. sagittula*; OD]. Apparatus unknown. Element pastiniscaphate with anterior free blade bearing either 1 or 2 median rows of denticles; terminal cusp at posterior extremity; 2 denticulate lateral processes, commonly developed at lateral side of free blade; basal cavity large, apex located near posterior of element. *L.Miss.(Osag. or up.Tournais.),* N.Am.-Eu.-Asia

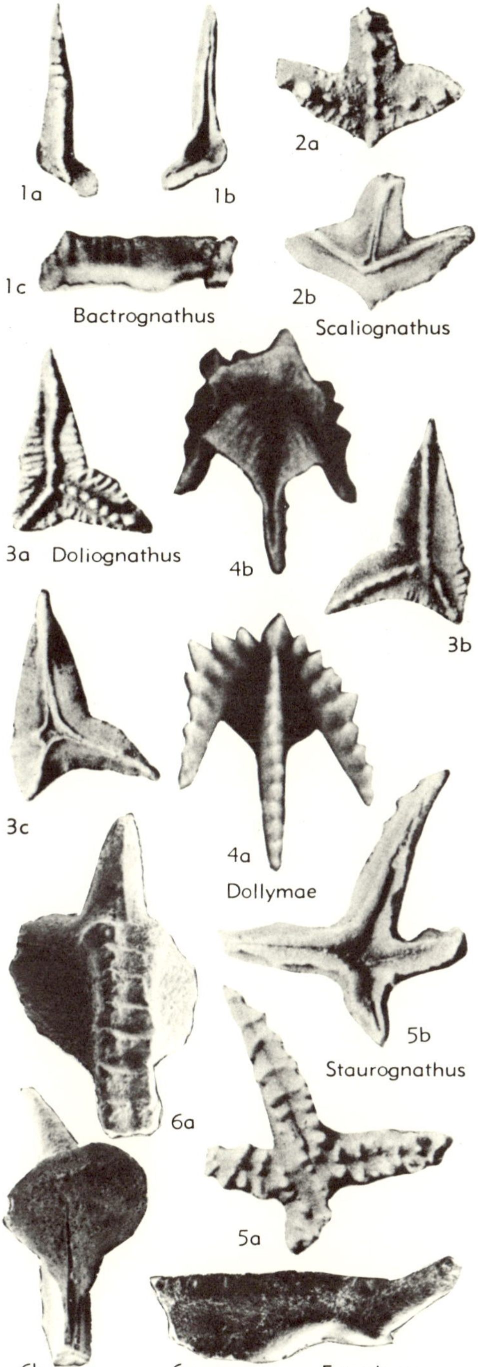

FIG. 116. Bactrognathidae (p. *W*169-*W*170).

Minor.——FIG. 116,*4*. **D. sagittula,* Chappel Ls., USA(Texas); *4a,b,* upper and lower views, ×30

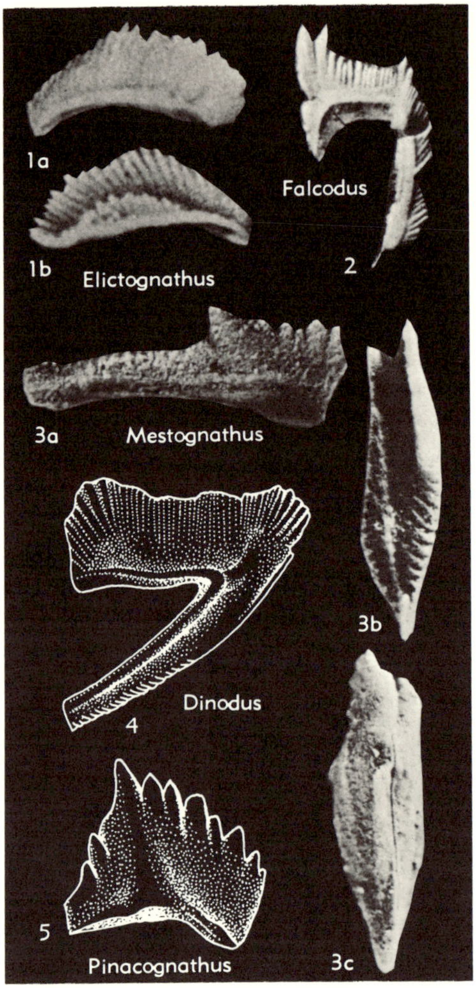

FIG. 117. Elictognathidae, Mestognathidae (p. W170-W172).

(Hass, 1959).

Eotaphrus PIERCE & LANGENHEIM, 1974, p. 155 [*E. *burlingtonensis*; OD]. Apparatus unknown. Pectiniform element segminiscaphate with distinct, low-crowned denticles; posterior cusp bladelike; basal cavity open, elongate, gradually tapering toward anterior. *L.Miss.(up.Tournais.),* N.Am.-Eu.——FIG. 116,*6*. **E. burlingtonensis,* Bullion Ls., USA(Nev.); *6a-c,* upper, lower, and lat. views, ×40 (Pierce & Langenheim, 1974).

Scaliognathus BRANSON & MEHL, 1941c, p. 101 [*S. *anchoralis*; OD]. Apparatus unknown. Pectiniform element planate and anchor-shaped with anterior and two lateral processes, each bearing median row of nodes and faint marginal transverse ridges; basal cavity near posterior end. *L.Miss.(low.Osag. or up.Tournais.),* N.Am.-Eu.-S.Am.-Australia-Afr.——FIG. 116,*2*. **S. anchoralis,* holotype, Pierson Ls., USA(Mo.); *2a,b,* upper and lower views, ×25 (Branson & Mehl, 1941c).

Staurognathus BRANSON & MEHL, 1941c, p. 102 [*S. *cruciformis*; OD]. Apparatus unknown. Pectiniform element scaphate with 4 processes; anterior process longest, straight or gently curved; posterior process shortest and deflected laterally relative to anterior process; 2 lateral processes of unequal length; all processes tapering uniformly to bluntly pointed tips; upper surface of processes bearing median grooves and transverse ridges. *L.Miss. (low.Osag. or up.Tournais.),* N.Am.-Eu.-Afr.——FIG. 116,*5*. **S. cruciformis,* holotype, Sycamore Ls., USA(Okla.); *5a,b,* upper and lower views, ×25 (Branson & Mehl, 1941c).

Family ELICTOGNATHIDAE
Austin & Rhodes, new

[Materials for this family prepared by R. L. AUSTIN & F. H. T. RHODES]

Total apparatus unknown. Ramiform element digyrate; pectiniform element anguliplanate. [Possibly a member of the superfamily Polygnathacea.] *U.Dev.-L.Miss.*

Elictognathus COOPER, 1939, p. 386 [**Solenognathus bialata* BRANSON & MEHL, 1934b, p. 273; OD] [=*Solenodella* BRANSON & MEHL, 1944, *nom. subst. pro Solenognathus* BRANSON & MEHL, 1934b, *non* AGASSIZ, 1846, *nec* BLEEKER, 1856-57, *nec* PICTET & HUMBERT, 1866, all fishes]. Apparatus unknown. Pectiniform element anguliplanate, compressed, anterior process directed downward relative to posterior process. Element slightly arched; basal part of posterior extremity flexed inward in some; inner side near lower margin in some with narrow platform and denticulate parapet; cusp prominent, or 2 or 3 prominent denticles; basal cavity elongate and small, keel distinct. *L.Miss.(Kinderhook. or Tournais.),* N.Am.-Eu.-Australia.——FIG. 117,*1*. **E. bialata* (BRANSON & MEHL), USA(Okla.); *1a,b,* outer and inner lat. views, ×36 (Cooper, 1939).

Dinodus COOPER, 1939, p. 386 [**D. leptus*; OD]. Apparatus unknown. Ramiform element digyrate with laterally compressed denticles; thin anterior process turning down and curving beneath remainder of element; unit broadest adjacent to lower margin; denticles needlelike, closely set, laterally confluent; cusp indistinct; basal cavity small. *L. Miss.,* N.Am.-Eu.——FIG. 117,*4*. *D. fragosus* (BRANSON), Houy F., USA(Texas); lat. view, ×25 (Hass, 1962).

Falcodus HUDDLE, 1934, p. 87 [**F. angulus*; OD]. Apparatus unknown. Ramiform element digyrate; thin denticulate posterior process continuing to near downward deflection, where 1 or 2 large denticles common; denticles laterally confluent along length; lower margin of posterior process

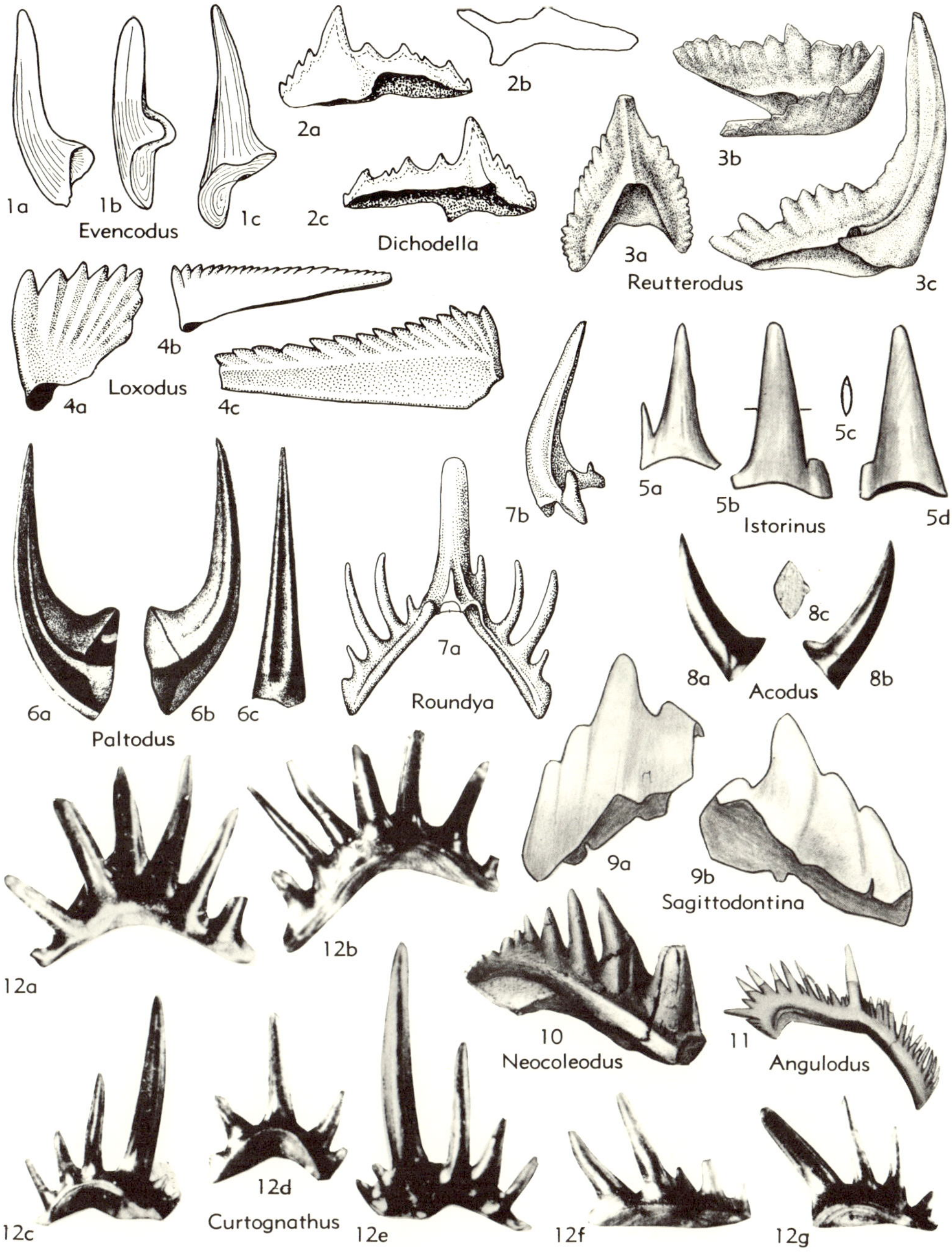

FIG. 118. Family Unknown (p. W172-W178).

angled downward from basal cavity to posterior end; thin anterior process angled downward about 90°; denticles laterally compressed and closely set; basal pit small. *U.Dev.-L.Miss.(Kinderhook. or low.Tournais.),* N.Am.-Eu.——FIG. 117,2. **F. angulus,* holotype, up. New Albany Sh., USA (Ind.); outer lat. view, ×25 (Huddle, 1934).

Pinacognathus BRANSON & MEHL, 1944, p. 244, *nom. subst. pro Pinacodus* BRANSON & MEHL, 1934b, *non* DAVIS, 1883, a fish [**Pinacodus pro-*

fundus Branson & Mehl, 1934b; OD]. Apparatus unknown. Pectiniform element angulate with prominent main cusp; anterior process high, short, compressed, bearing fused denticles; posterior process shorter, denticulate upper edge falling steeply in elevation toward posterior; basal cavity small and expanded laterally. *L.Miss.(Kinderhook. or low.Dinant.),* N.Am.-Eu.——Fig. 117,*5*. **P. profunda* (Branson & Mehl), Bachelor F., USA (Mo.); lat. view, ×30 (Austin, n).

Family MESTOGNATHIDAE Austin & Rhodes, new

[Materials for this family prepared by R. L. Austin & F. H. T. Rhodes]

Apparatus unknown. Pectiniform element planate with short, free, lateral blade, distinctive carina on platform, small basal cavity. *U.Miss.(Visean)-L.Penn.(Namur.).*

Mestognathus Bischoff, 1957, p. 36 [**M. beckmanni*; OD]. Apparatus unknown. Pectiniform element planate; short free blade on lateral margin; highest at posterior; carina present and distinct in most specimens; transverse ridges on posterior platform; small basal pit and groove on lower surface. [The element resembles the pectiniform *Pa* element of *Cavusgnathus,* but the small basal cavity is distinctive.] *U.Miss.(Visean)-L.Penn. (Namur.),* N.Am.(Can.)-Eu.——Fig. 117,*3*. **M. beckmanni,* (Dinant.), Eu.(Wales); *3a-c,* lat., upper, and lower views, ×245 (Rhodes, Austin, & Druce, 1969).

Family UNKNOWN

[Materials for this section prepared by R. L. Austin, S. M. Bergström, D. L. Clark, Gilbert Klapper, and F. H. T. Rhodes]

Brief entries for several genera proposed since completion of the text and not assignable to family are included in the addendum.

Acodus Pander, 1856, p. 21 [**A. erectus*; SD Ulrich & Bassler, 1926, p. 7] [?=*Acontiodus* Pander, 1856]. Apparatus unknown; genus based on nongeniculate, coniform elements with slightly recurved cusp, short base, and shallow basal cavity. Cusp somewhat compressed laterally with anterior and posterior edges; one lateral face with prominent central carina, other face evenly convex. [Estonian specimens identified as *A. erectus* have recently been illustrated by Viira (1975, fig. 17a, 17b). They appear to be the first new specimens of this species figured since Pander's time. As noted by Sweet & Bergström (1972), data from Estonia (Viira, 1966) suggest that elements of *A. erectus* are associated with, and have the same range as, those of *Acontiodus latus* Pander, 1856, type species of *Acontiodus.* Accordingly, it is possible that all these elements were components of the same apparatus and *Acodus* and *Acontiodus* are synonyms; see also Lindström (1973).] *L.Ord.-Sil.(Llandov.), ?Sil.(Wenlock.),* Eu.-Asia (Sib.)-?N.Am.——Fig. 118,*8*. *A. acutus* Pander, L.Ord.(Pakerort Stage, ?Varangu Mbr.), Eu.(Baltic); *8a,b,* lat. views; *8c,* cross section of cusp; magnification unknown (Pander, 1856).

Angulodus Huddle, 1934, p. 76 [**A. demissus*; OD] [=*Cervicornoides* Stauffer, 1938, p. 424]. *Sb* element bipennate with downward projecting anterior process in vertical plane of posterior process. Denticles confluent and commonly alternating in size. *Dev.-Penn.,* Eu.-N.Am.-Australia.——Fig. 118,*11*. **A. demissus,* holotype, U.Dev.(low.New Albany Sh.), USA(Ind.); *Sb* element, lat. view, ×17 (Huddle, 1934).

Apatognathus Branson & Mehl, 1934a, p. 201 [**A. varians*; OD]. *Sa* element with anterior and posterior processes forming symmetrical arch, but cusp twisted posteriorly; both processes flexed inward, commonly with cyclic alternation of diagonally offset denticles. *U.Dev.-U.Miss.,* Eu.-N. Am.-Australia.——Fig. 119,*3*. **A. varians,* U.Dev. (*Bispathodus costatus* Z.), Australia(W.Australia); *Sa* element, inner view, ×45 (Glenister & Klapper, 1966).

Branmehla Hass, 1959, p. 381 [**Spathodus inornatus* Branson & Mehl, 1934a, p. 185]. *Pa* element carminate with basal cavity near posterior end; other elements unknown. *U.Dev.(up. Famenn.)-L.Miss.(Tournais.),* Eu.-N.Am.——Fig. 120,*5*. **B. inornata* (Branson & Mehl), holotype, U.Dev.(Saverton Sh.), USA(Mo.); *Pa* element, lat. view, ×27 (Klapper, n).

Bryantodus Bassler, 1925, p. 219 [**B. typicus*; OD]. Anguliplanate *Pb* element with massive cusp, narrow platform ledges, and downward projecting apical lip. *U.Dev.(Frasn.),* N.Am.(N.Y.).——Fig. 120,*12*. **B. typicus,* lectotype, Rhinestreet Sh., USA(N.Y.); *Pb* element, lat. view, ×20 (Huddle, 1968).

Caenodontus Behnken, 1975, p. 298 [**C. serrulatus*; OD]. Apparatus unimembrate; erect, nongeniculate, coniform element with denticulate posterior edge; anterior edge rounded; basal cavity occupying more than half of element. [The genus is homeomorphic with Ordovician *Belodina.*] *U. Perm.,* N.Am.——Fig. 119,*1*. **C. serrulatus,* USA (Texas); lat. view, ×100 (Behnken, 1975).

Cornuodus Fåhraeus, 1966, p. 20 [**C. erectus*; OD]. Apparatus unknown but apparently unimembrate; coniform elements lamellar, nongeniculate, recurved to reclined, with more or less oval cross section and relatively shallow basal cavity. No costae or carinae. *M.Ord.,* Eu.——Fig. 121,*1*. **C. erectus,* Eu.(Sweden); *1a,* topotype, lat. view, ×75; *1b,c,* long. and transv. cross sections, ×75 (Bergström, n).

Coryssognathus Link & Druce, 1972, p. 31 [**C.*

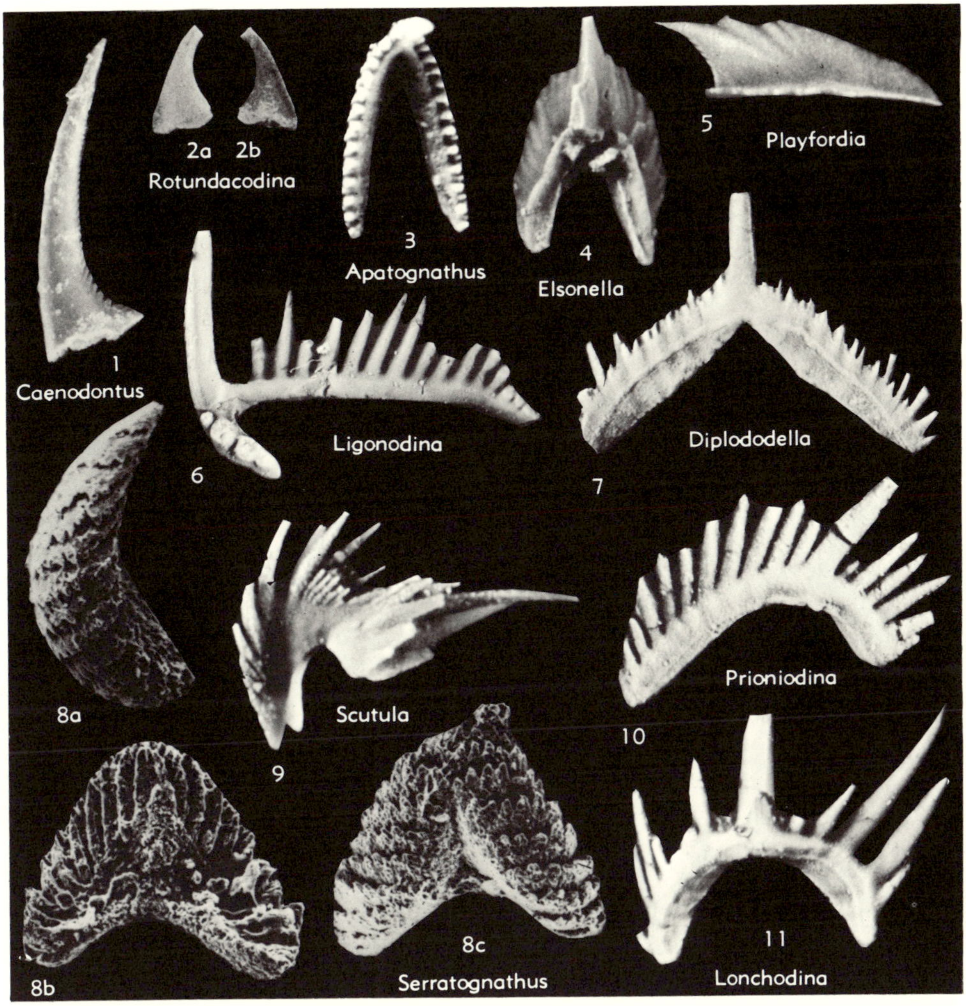

FIG. 119. Family Unknown (p. *W*172-*W*179).

dentatus; OD; =*Pelekysgnathus dubius* JEPPSSON, 1972, p. 62]. *Pa* element resembling that of *Pelekysgnathus*, but cusp at anterior end. [If apparatus includes elements identified by JEPPSSON (1972) as *Distomodus dubius* (RHODES), as suggested by COOPER (1974b), *Coryssognathus* should then be referred to the Distomodontidae.] *Sil. (Ludlov.)*, Eu.-Australia.——FIG. 120,*3*. **C. dubius* (JEPPSSON), holotype, Eu.(Sweden, Scania); *Pa* element, lat. view, ×27 (Jeppsson, 1972).

Curtognathus BRANSON & MEHL, 1933b, p. 87 [**Curtognathus typa*; OD] [=*Polycaulodus* BRANSON & MEHL, 1933b; *Trucherognathus* BRANSON & MEHL, 1933b; *Cardiodella* BRANSON & MEHL, 1944, *nom. subst. pro Cardiodus* BRANSON & MEHL, 1933b, *non* TROUESSART, 1881, a mammal]. Apparatus unknown, of elements similar to the following form-genera of BRANSON & MEHL: *Cardiodella, Curtognathus, Polycaulodus, Microcoelodus, Trucherognathus*. Elements forming morphologically intergradational series, apparently consistently associated. Elements fibrous (hyaline), pectiniform, under side flat or weakly excavated, denticles discrete, rounded in cross section. Cardiodelliform elements broadly triangular to V-shaped in upper view, slightly arched, with central cusp and a few denticles on each process. Curtognathiform elements strongly arched with several subequal denticles on upper side and no prominent cusp. Polycaulodontiform elements almost straight with few denticles on upper side. Trucherognathiform elements essentially straight with somewhat irregularly oriented denticles on upper side. *M. Ord.*, N.Am.——FIG. 118,*12*. *C. tumidus* BRANSON & MEHL, Glenburnie Sh., Can.(Ont.); *12a,b*, cardiodelliform element, ant. and post. views; *12c,e*,

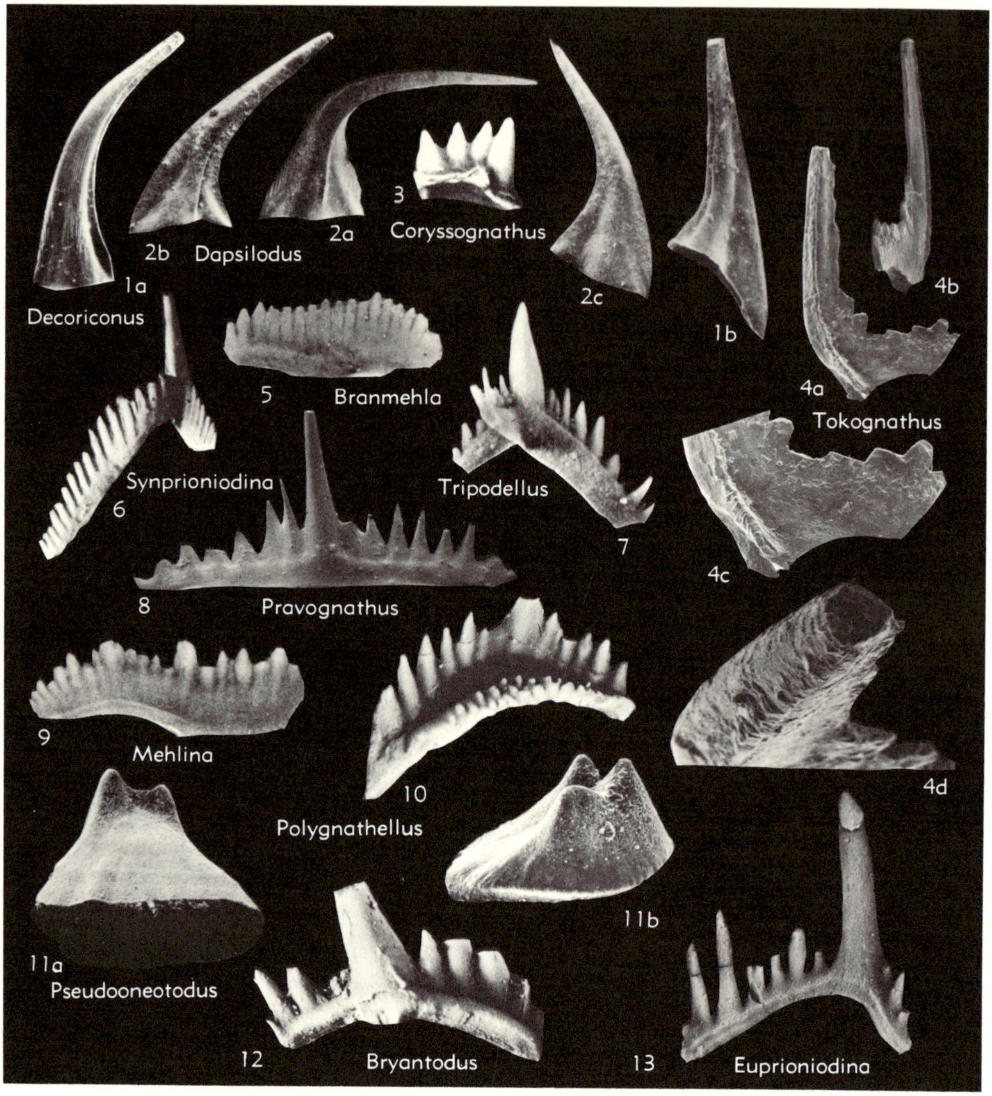

FIG. 120. Family Unknown (p. W172-W179).

microcoelodontiform element, post. and ant. views; *12d,* curtognathiform element, post. view; *12f,* trucherognathiform element, lat. view; *12g,* polycaulodontiform element, lat. view; ×40 (Votaw, 1971).

Dapsilodus Cooper, 1976, p. 211 [**Distacodus obliquicostatus* Branson & Mehl, 1933a, p. 41; OD]. Apparatus multimembrate, composed of lamellar coniform elements, consisting of acodontiform *M* element and *S* symmetry-transition series. *Sa* element modified distaconodontiform element with lateral costae even with or extending behind posterior keel. *Sb* and *Sc* elements slightly and strongly twisted distaconodontiform elements respectively. [Apparatus reconstruction: Serpagli, 1971; Cooper, 1976; Barrick, 1977.] *Sil.,* N. Am.-Eu.-N.Afr.-Australia.——Fig. 120,*2*. **D. obliquicostatus* (Branson & Mehl), Sil.(Wenlock., St. Clair Ls.), USA(Ill.); *2a, Sb-Sc* element, lat. view, ×43; *2b,c, M* elements, inner lat. and outer lat. views, ×37, ×33 (Cooper, 1976).

Decoriconus Cooper, 1975, p. 992 [**Paltodus costulatus* Rexroad, 1967, p. 40; OD]. Multimembrate apparatus comprising symmetry-transition series of small, twisted, generally striate, lamellar coniform elements. Each element bearing narrow longitudinal groove near posterior keel on both sides. Basal cavity shallow, extending one-third of element height. [Apparatus reconstruction: COOPER, 1975; BARRICK, 1977.] *U.Ord.(Ashgill.)-Sil.,* N.Am.-Eu.——FIG. 120,*1*. *D. fragilis* (BRANSON & MEHL), Sil.(Wenlock., St. Clair Ls.), USA (Ill.); *1a, Sa-Sb* element, lat. view; *1b, Sc* element, lat. view; ×93 (Cooper, 1976).

Dichodella SERPAGLI, 1967, p. 62 [**D. exilis*; OD]. Apparatus unknown; genus based on lamellar, pectiniform, pastinate elements with distinct cusp and anterior and posterior processes. Anterior process bladelike with central row of subequal denticles laterally partly confluent. Posterior process somewhat wider, slightly flaring laterally, with denticles of irregular size. Anterior margin of cusp extending downward into adenticulate offset or short lateral process. Basal cavity large and wide, extending over entire under side of element. [Representatives of *D. exilis* show some similarity to *Pa* elements of *Amorphognathus,* but are distinguished by lack of well-developed platforms and lateral processes. *Dichodella* has denticulation similar to that of pastinate elements of *Phragmodus* and *Plectodina* and is distinguished from *Prioniodus (Baltoniodus)* by the weak development of the third process.] *U.Ord.,* Eu.(Italy, Eng.).——FIG. 118,*2*. **D. exilis,* U.Ord.(Ashgill.), Eu.(Italy, Carnic Alps); *2a,c,* lat. views; *2b,* lower view; ×66 (Serpagli, 1967).

Diplododella BASSLER, 1925, p. 219 [**D. bilateralis*; OD]. *Sa* element alate with denticulate posterior process; denticulation like that of *Angulodus. U.Dev.(Famenn.),* N.Am.(Ala.).——FIG. 119,*7*. **D. bilateralis,* holotype, Gassaway Mbr., Chattanooga Sh., USA(Ala.); *Sa* element, ant. view, ×30 (Huddle, 1968).

Elsonella YOUNGQUIST, 1945, p. 358 [**E. prima*; OD] [=*Neorhipidognathus* MOUND, 1968, p. 494]. *Sa* element with highly fused denticles; surface covered with minute pits (LINDSTRÖM, 1964, p. 157). *U.Dev.,* N.Am.——FIG. 119,*4*. **E. prima,* lectotype, Frasn.(Amana beds), USA (Iowa); *Sa* element, post. view, ×34 (Klapper, 1966).

Euprioniodina BASSLER, 1925, p. 219 [**E. deflecta*; OD]. *M* element dolabrate with both processes having discrete denticles. *U.Dev.(Frasn.),* N.Am. (N.Y.).——FIG. 120,*13*. **E. deflecta,* lectotype, Rhinestreet Sh., USA(N.Y.); *M* element, lat. view, ×20 (Huddle, 1968).

Evencodus MOSKALENKO, 1970, p. 42 [**E. sibiricus*; OD]. Apparatus not known with certainty; genus based on nongeniculate, hyaline coniform elements with suberect to proclined, in cross section rounded, laterally multicostate cusp, very shallow basal cavity; anterobasal portion of unit characteristically extended into tonguelike structure directed downward. [MOSKALENKO (1972, fig. 5) referred three types of coniform elements to the apparatus of *E. sibiricus.* Other forms assigned

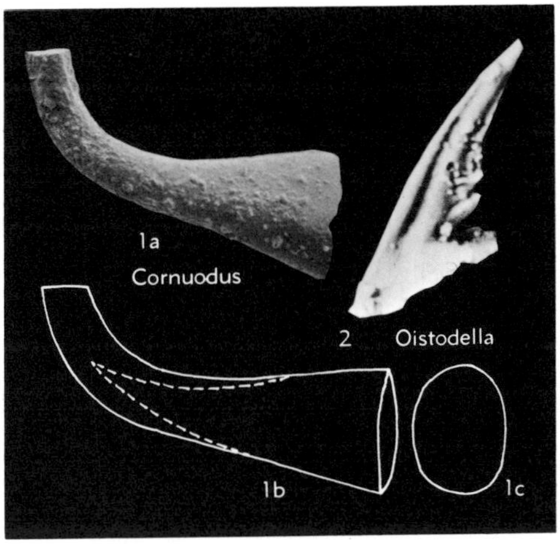

FIG. 121. Family Unknown (p. *W*172, *W*177).

to *Evencodus* have an appearance that does not exclude fish affinities.] *M.Ord.,* Asia(Sib.).——FIG. 118,*1*. **E. sibiricus,* Krivoluksky horiz., USSR (Sib.); *1a-c,* three types of coniform elements, lat. views, approx. ×20 (Moskalenko, 1972).

Geniculatus HASS, 1953, p. 77 [**Polygnathus? claviger* ROUNDY, 1926; OD]. Apparatus unknown. Pectiniform element anguliplanate; geniculate, asymmetrical, broadest at vertex where cusp and basal cavity located; with a few large, medial, distinct denticles; keel along midline of under side. *U.Miss.(up.Dinant.),* N.Am.-Asia M.-Eu. ——FIG. 122,*9*. **Geniculatus claviger* (ROUNDY), Barnett F., USA(Texas); upper view, ×21 (Hass, 1953).

Hindeodelloides HUDDLE, 1934, p. 48 [**H. bicristatus*; OD]. Apparatus unknown. Ramiform element bipennate; posterior process long, denticulate, anterior process shorter, both with closely set denticles, denticles alternating in length and size in some. *L.Miss.(low.Dinant.),* N.Am.-Eu.-Afr.——FIG. 122,*3*. **H. bicristatus,* New Albany Sh., USA (Ind.); outer lat. view, ×25 (Hass, 1962).

Hindeodina HASS, 1959, p. 382 [**H. simplaria*; OD]. Apparatus unknown. Ramiform element bipennate with elongate straight to slightly curved posterior process bearing discrete denticles; short denticulate anterior process deflected downward and occasionally inward. Cusp indistinct and distinguished only through position above basal cavity; lower margin sharp-edged; lips of basal pit extremely small or absent. *U.Dev.-U.Miss.,* N.Am.-Eu.——FIG. 122,*4a*. **H. simplaria,* holotype, L.Miss.(Chappel Ls.), USA(Texas); inner lat. view, ×30 (Hass, 1959).——FIG. 122,*4b*. *H. uncata,* holotype, L.Miss.(Chappel Ls.), USA (Texas); upper view, ×30 (Hass, 1959).

Istorinus KNÜPFER, 1967, p. 31 [**I. erectus*; OD]. Apparatus unknown; elements with suberect, lat-

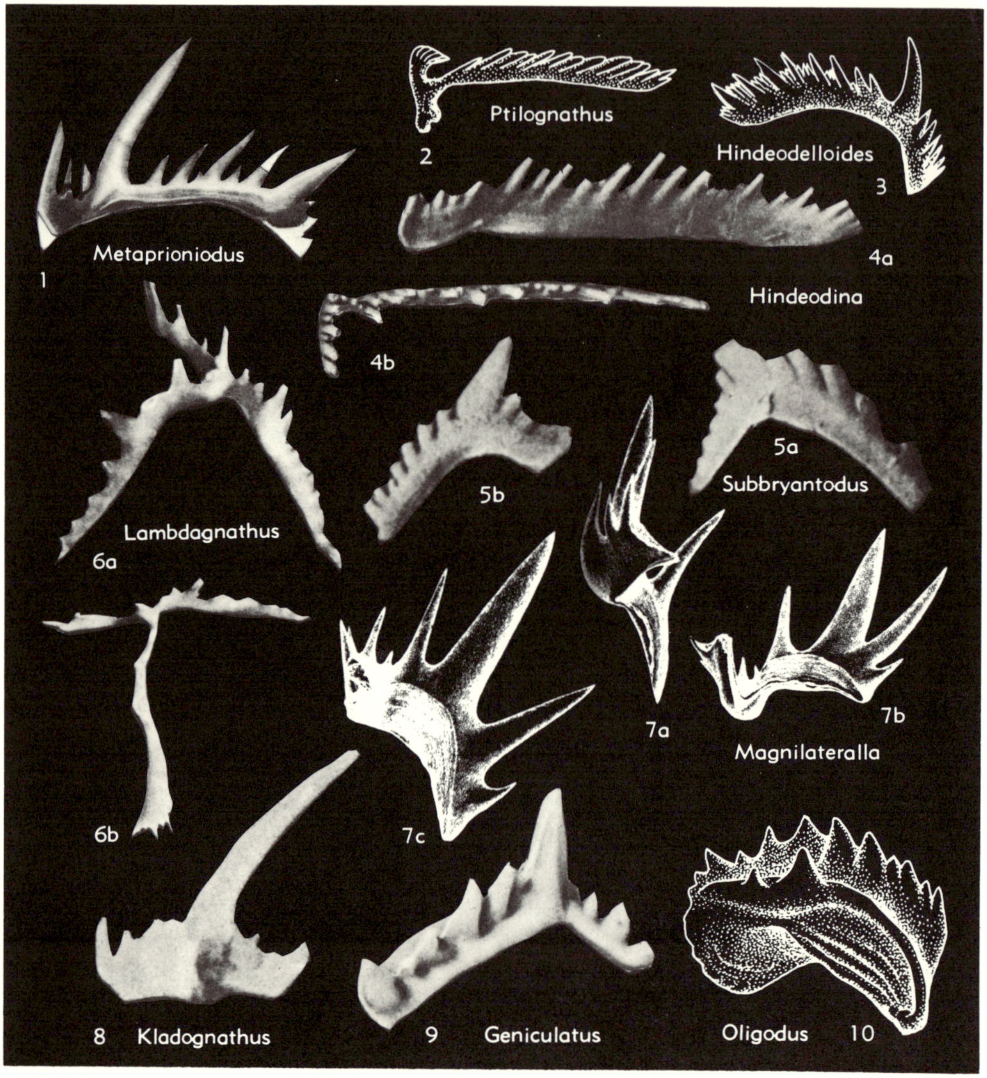

FIG. 122. Family Unknown (p. W175, W179).

erally compressed cusp with sharp anterior and posterior edges; small suberect denticle in front of cusp. Basal cavity deep, extending through entire unit. [Original illustrations (KNÜPFER, 1967, pl. 1) show the types of the genus to be essentially complete and not fragments of ramiform elements.] *U.Ord.*, Eu.(Thuringia).——FIG. 118,*5*. **I. erectus; 5a,b,d,* lat. views, small denticle anterior to main cusp; *5c,* cross-section shape; ×59 (Knüpfer, 1967).

Kladognathus REXROAD, 1958b, p. 19 [**Cladognathus prima* REXROAD, 1957, p. 28; OD] [=*Cladognathodus* REXROAD & COLLINSON, 1961 (obj.), *nom subst. pro Cladognathus* REXROAD, 1957, *non* BURMEISTER, 1847, a beetle]. Apparatus unknown, although unlikely to contain pectiniform element. Ramiform element bipennate with long, thin, posterior process surmounted by discrete denticles; long reclined cusp at anterior extremity of posterior process. Anterior process bifid, separated into subsidiary processes at point 1 or 2 denticles anterior of cusp. Under side of bars grooved. *U.Miss. (Chester. or up.Dinant.),* N.Am.-Eu.——FIG. 122, *8*. **K. primus* (REXROAD), USA(Ill.); inner lat. view, ×72 (Rexroad, 1958b).

Lambdagnathus REXROAD, 1958b, p. 19 [**L. fragilidens*; OD]. Apparatus unknown. Ramiform element tertiopedate. Arched denticulate posterior process longest and deepest of 3 processes; lateral and anterolateral processes relatively thin and deep, denticulate, of unequal length, developed at right angles to posterior process; reclined cusp same elevation as proximal denticles of the 3 processes; basal pit triangular; under side of specimen more or less sharp edged. *U.Miss.(Chester. or up.Dinant.),* N.Am.(Ill., Ind., Ky.)-Eu.——Fig. 122,*6*. **L. fragilidens,* holotype, Glen Dean Ls., USA(Ill.); *6a,b,* oblique and upper views, ×40 (Rexroad, 1958b).

Ligonodina Bassler, 1925, p. 218 [**L. pectinata*; OD]. *Sc* element bipennate, with anterolateral process and its denticles flexed posteriorly, denticles discrete. *U.Dev.(Frasn.),* N.Am.(N.Y.).——Fig. 119,*6*. **L. pectinata,* lectotype, Rhinestreet Sh., USA(N.Y.); *Sc* element, lat. view, ×20 (Huddle, 1968).

Lonchodina Bassler, 1925, p. 219 [**L. typicalis*; OD]. *Sb* element with anterior and posterior processes incurved distally; denticles discrete. *U. Dev.(Frasn.),* N.Am.(N.Y.).——Fig. 119,*11*. **L. typicalis,* lectotype, Rhinestreet Sh., USA(N.Y.); *Sb* element, lat. view, ×20 (Huddle, 1968).

Loxodus Furnish, 1938, p. 338 [**L. bransoni*; OD]. Apparatus unknown, may be unimembrate. Elements lamellar, segminate, strongly compressed laterally, highest anteriorly, denticulate, without distinct cusp. Denticles relatively short, confluent along most of element length, about equal in size. Basal cavity narrow but slightly expanded at anterior end. [*Loxodus* elements are not known to be associated with other ramiform or pectiniform elements that might be parts of the same apparatus.] *L.Ord.,* N.Am.-Asia(Sib.).——Fig. 118,*4*. **L. bransoni, Oneota Dol.,* USA(Iowa); *4a,* ant. part of element, lat. view, ×48; *4b,c,* lat. views, ×25, ×48 (Hass, 1962).

Magnilateralla Rexroad & Collinson, 1963, p. 11 [**M. robusta*; OD]. Apparatus unknown but unlikely to contain pectiniform element. Ramiform element bipennate; anterolateral process large and denticulate, directed down and to the rear, largest denticles part way back on process; posterior process gently curved and denticulate; basal pit generally present at junction of processes, grooves extending in some along under side of either process. *U.Miss.(up.Dinant.),* N.Am.-Eu.——Fig. 122,*7*. **M. robusta,* holotype, Glen Dean F., USA (Ill.); *7a-c,* outer lat., lower, and post. views, ×40 (Rexroad & Collinson, 1963).

Mehlina Youngquist, 1945, p. 363 [**M. irregularis*; OD; =*M. gradata* Youngquist, 1945, p. 363; first reviser, Müller & Müller, 1957, p. 1083]. *Pa* element carminate with basal pit at midlength and inverted basal cavity posteriorly. [Other elements are unknown.] *U.Dev.(Frasn.)-L.Miss. (Tournais.),* Eu.-Australia-N.Am.——Fig. 120,*9*. **M. gradata* (syntype of *M. irregularis*), U.Dev. (Frasn., Amana beds), USA(Iowa); lat. view, ×27 (Klapper, n).

Metaprioniodus Huddle, 1934, p. 57 [**M. biangulatus*; OD]. Apparatus unknown. Ramiform element bipennate; anterior process short, denticulate; posterior process elongate, curved, with discrete denticles, largest denticle near posterior deflection; cusp long and reclined; basal pit small. *U.Dev.-L.Miss.(Kinderhook. or low.Dinant.),* N.Am.-Eu.——Fig. 122,*1*. **M. biangulatus,* holotype, L.Miss. (New Albany Sh.), USA(Ind.); lat. view, ×18 (Huddle, 1934).

Neocoleodus Branson & Mehl, 1933a, p. 24 [**N. spicatus*; OD]. Apparatus unknown; genus based on fibrous, hyaline, angulate, pectiniform elements with narrow basal cavity. Denticles discrete, rounded in cross section, decreasing in size posteriorly. *M.Ord.,* N.Am.——Fig. 118,*10*. **N. spicatus,* Harding Ss., USA(Colo.); lat. view, ×17 (Branson & Mehl, 1933a).

Oistodella Bradshaw, 1969, p. 1155 [**O. pulchra*; OD]. Apparatus unknown. Elements lamellar, basically coniform, geniculate, with denticles along posterior cusp margin; laterally compressed and costate on one side with reclined, relatively robust cusp and anteroposteriorly extended base that flares to one side. Basal cavity shallow but occupying entire length of base. Denticles distinct, although short and confluent along part of element; size decreasing upward on cusp. [Apart from denticulation, representatives of *Oistodella pulchra* show considerable similarity to *"Oistodus" angulatus* Bradshaw, 1969, with which they are associated in the type strata, and it is possible that these forms belong to the same apparatus.] *L.Ord., ?M.Ord.,* N.Am.——Fig. 121,*2*. **O. pulchra,* L. Ord.(Fort Peña F.), USA(Texas); lat. view, ×63 (Bradshaw, 1969).

Oligodus Cooper, 1939, p. 398 [**O. curtus*; OD]. Apparatus unknown. Pectiniform element angulate with fused denticles. Process curved inward, especially posterior to basal cavity; flange on under side prominent, nodose near posterior extremity; under side wide and deep, especially in posterior half. [Affinity is doubtful; the genus is like *Pinacognathus* according to Cooper, 1939; possibly synonymous with *Bryantodus* according to Rhodes, 1953a.] *L.Miss.(Kinderhook.),* N.Am.——Fig. 122,*10*. **O. curtus,* USA(Okla.); inner lat. view, ×40 (Hass, 1962).

Paltodus Pander, 1856, p. 24 [**P. subaequalis*; SD Ulrich & Bassler, 1926, p. 7]. Apparatus apparently bimembrate, including geniculate and nongeniculate lamellar coniform elements. Geniculate elements with anteroposteriorly extended base and reclined cusp. Nongeniculate elements with triangular base and suberect to recurved cusp, cusp costate in some. Basal cavity relatively wide, flaring toward one side, restricted to base. [This interpretation follows Lindström (1971, 1973). Apparatus reconstruction: Lindström, 1971.] *L. Ord.,* Eu.-N.Am.-Australia.——Fig. 118,*6*. **P. subaequalis,* L.Ord.(?Pakerort Stage), Eu.(Baltic);

6a,b, lat. views; *6c,* anterior view, magnification unknown (Pander, 1856).

Playfordia GLENISTER & KLAPPER, 1966, p. 827 [**Pelekysgnathus? primitivus* BISCHOFF & ZIEGLER, 1957, p. 83; OD]. *Pa* element with single row of denticles and large basal cavity. [This element suggests reference to Icriodontidae; other elements are unknown.] *U.Dev.(Frasn.),* Eu.-N.Am.-Australia.——FIG. 119,*5.* **P. primitiva* (BISCHOFF & ZIEGLER), L. *Mesotaxis asymmetrica* Z. (Gogo F.), Australia(W.Australia); lat. view, ×45 (Glenister & Klapper, 1966).

Polygnathellus BASSLER, 1925, p. 220 [**P. typicalis*; OD] [=*Nothognathella* BRANSON & MEHL, 1934a, p. 226]. *Pb* element anguliplanate with platform ledges on both sides; inner ledge commonly broader. [Possibly this element is part of the apparatus of a *Palmatolepis* or *Mesotaxis* species.] *U.Dev.,* Eu.-N.Am.-Australia.——FIG. 120,*10.* **P. typicalis,* lectotype, Frasn.(Rhinestreet Sh.), USA (N.Y.); lat. view, ×20 (Huddle, 1968).

Pravognathus STAUFFER, 1936, p. 79, *nom. subst. pro Heterognathus* STAUFFER, 1935b, *non* GIRARD, 1854, a fish, *nec* SCHMARDA, 1859, a rotifer, *nec* KING, 1864, a beetle, *nec* REY, 1888, a beetle] [**Heterognathus idoneus* STAUFFER, 1935b; OD]. Apparatus not known with certainty. According to WEBERS (1966, p. 45), apparatus of type species includes two kinds of rather long, laterally compressed, carminate elements, each essentially straight with slightly twisted and arched central region; cusp longer than denticles, slender, subcentral, erect; numerous short, discrete denticles of somewhat variable size anterior and posterior to cusp; basal cavity extending along entire length of element, largest beneath cusp. *M.Ord.,* N.Am. ——FIG. 120,*8.* **P. idoneus* (STAUFFER), Platteville Fm., USA(Minn.); lat. view, ×74 Bergström, n).

Prioniodina BASSLER, 1925, p. 219 [**P. subcurvata*; OD] [=*Prioniodella* BASSLER, 1925, p. 219]. Angulate *Pb* element like that of *Lonchodina* but unit not so strongly bowed inwardly and denticles more uniform. *U.Dev.(Frasn.),* N.Am.(N.Y.). ——FIG. 119,*10.* **P. subcurvata,* lectotype, Rhinestreet Sh., USA(N.Y.); *Pb* element, lat. view, ×30 (Huddle, 1968).

Pseudooneotodus DRYGANT, 1974, p. 66 [**Oneotodus? beckmanni* BISCHOFF & SANNEMANN, 1958, p. 98; OD]. Characterized by short, stout cones with deep basal cavity, walls that thicken apically, and relatively smooth surface. One to three apical denticles present. [Diagnosis: BARRICK & KLAPPER, 1976.] *U.Ord.-L.Dev.,* Eu.-N.Am.——FIG. 120,*11a.* *P. bicornis* DRYGANT, Sil.(Wenlock., *Kockelella amsdeni* Z., Clarita F.), USA(Okla.); lat. view, ×64 (Barrick, 1977).——FIG. 120,*11b.* *P. tricornis* DRYGANT, Sil.(up.Llandov.-low.Wenlock., *Pterospathodus amorphognathoides* Z., Clarita F.), USA(Okla.); lat. view, ×100 (Barrick, 1977).

Ptilognathus ELIAS, 1956, p. 114 [**P. fayi*; OD]. Apparatus unknown. Ramiform element alate; posterior process long with laterally confluent, posteriorly directed denticles; lateral processes denticulate and short. *U.Miss.,* N.Am.——FIG. 122,*2.* **P. fayi,* Goddard Sh., USA(Okla.); lat. view, ×20 (Hass, 1962).

Reutterodus SERPAGLI, 1974, p. 79 [**R. andinus*; OD]. Apparatus apparently trimembrate, including nongeniculate coniform element and two types of ramiform elements, all lamellar and forming symmetry-transition series. Coniform elements asymmetrical, robust, suberect, laterally compressed, with conspicuous longitudinal lateral costa near anterior margin. Ramiform elements asymmetrical, digyrate to bipennate, with 1 or 2 multidenticulate, lateral processes and no posterior process; denticles of irregular size and confluent at least basally; basal cavity developed as subapical pit and shallow groove along under side of processes. [Apparatus reconstruction: SERPAGLI, 1974.] *L.Ord.,* S.Am. (Arg.)-?N.Am.(Texas).——FIG. 118,*3.* **R. andinus,* San Juan Ls., S.Am.(Arg.); *3a,b,* ramiform elements, post. and lat. views, ×67, ×49; *3c,* coniform element, oblique post. view, ×49 (Serpagli, 1974).

Rotundacodina CARLS & GANDL, 1969, p. 206 [**R. noguerensis*; OD]. Simple coniform element. [This element may be part of an *Icriodus* or *Pelekysgnathus* apparatus, or both.] *L.Dev.,* Eu.-N.Am.——FIG. 119,*2.* **R. noguerensis,* Nogueras beds, Eu.(Sp., Aragón); *2a,b, Pb* element, outer and inner lat. views, ×30 (Carls & Gandl, 1969).

Roundya HASS, 1953, p. 88 [**R. barnettana*; OD]. Apparatus unknown, but probably multimembrate. Ramiform element alate; lateral processes bearing discrete robust denticles and forming anterior arch; posterior process short and denticulate; basal cavity large. *?L.Miss., U.Miss.-L.Penn.,* N.Am.-Eu.-Afr.-Asia M.-Australia.——FIG. 118,*7.* **R. barnettana,* U.Miss.(Barnett F.), USA(Texas); *7a,b,* post. and lat. views, ×25 (Hass, 1962).

Sagittodontina KNÜPFER, 1967, p. 37 [**S. robusta*; OD]. Apparatus unknown; genus based on fragmentary angulate, pectiniform elements with stout, subcentral cusp and short, denticulate anterior and posterior processes. Basal cavity deep, extending over entire under side of element. *U.Ord.,* Eu. (Thuringia).——FIG. 118,*9.* **S. robusta; 9a,b,* lat. views, ×54 (Knüpfer, 1967).

Scutula SANNEMANN, 1955, p. 154 [**S. venusta*; OD] [?=*Avignathus* LYS & SERRE, 1957, p. 797; ?*Gnamptognathus* ZIEGLER, 1958, p. 53]. Modified tertiopedate, quadriramate, and multiramate symmetrical and asymmetrical elements (of a probable symmetry-transition series) characterized by thin, confluent denticles. [These elements are possibly part of the apparatus of a *Palmatolepis* species.] *U.Dev.,* Eu.-N.Afr.-N.Am.-Australia.——FIG. 119,*9.* *S. bipennate* SANNEMANN, L. *Palmatolepis marginifera* Z. (Bugle Gap Ls.), Australia

(W.Australia); quadriramate element, lat. view, ×45 (Glenister & Klapper, 1966).

Serratognathus LEE, 1970, p. 335 [**S. bilobatus*; OD]. Apparatus unknown; may be unimembrate. Elements bilaterally symmetrical, alate, anteriorly convex and posteriorly concave, lacking basal cavity; convex side with about 10 horizontal rows of short, densely spaced denticles; concave side with numerous radially disposed ribs; posterior process missing; midportion of anterior and posterior face smooth, without denticles and ribs. *L.Ord.*, Asia (Korea-China).——FIG. 119,*8*. **S. bilobatus*; lat., post., and ant. views, magnification unknown (Lee, 1970).

Subbryantodus BRANSON & MEHL, 1934b, p. 285 [**S. arcuatus*; OD]. Apparatus unknown. Pectiniform element angulate with short, straight posterior process; longer anterior process directed downward; denticles laterally confluent; basal cavity small, widest beneath center of distinct main cusp. *U.Dev.-L.Miss.*, cosmop.——FIG. 122,*5*. **S. arcuatus*, L.Miss.(Bachelor F.), USA(Mo.); *5a,b*, lat. views, ×40 (Branson & Mehl, 1934b).

Synprioniodina BASSLER, 1925, p. 219 [**S. alternata*; OD]. *M* element dolabrate; both processes with confluent denticles, characteristically alternating in size. *U.Dev.(Famenn.)*, N.Am.(Ala.).——FIG. 120,*6*. **S. alternata*, holotype, Gassaway Mbr., Chattanooga Sh., USA(Ala.); lat. view, ×20 (Huddle, 1968).

Tokognathus NIEPER, 1969, p. O12 [**T. proclinatus*; OD]. Apparatus unknown; genus based on symmetrical lamellar elements with moderately deep, unexpanded, elongated base with triangular cross section. Cusp slightly proclined to erect, with smooth anterior margin and anterolateral costa on each lateral face. Posterior process long with numerous subequal, laterally compressed, basally fused denticles. [The type of *T. proclinatus* is similar to elements referred to *Oepikodus* cf. *O. quadratus* by MCHARGUE (1974, pl. 2, fig. 6), and also to elements of *Belodella* ETHINGTON (1959a). Thus, *Tokognathus* may be a synonym of either *Oepikodus* or *Belodella*.] *L.Ord.*, Australia (Queensl.).——FIG. 120,*4*. **T. proclinatus*, Nora F.; *4a,b*, lat. and post.-lat. views, ×171; *4c*, base in lat. view, ×278; *4d*, cusp, upper-lat. view, approx. ×970 (Bergström, n).

Tripodellus SANNEMANN, 1955, p. 155 [**T. flexuosus*; OD]. *S* element modified tertiopedate with anterior, lateral, and posterior processes. [This element is possibly part of the apparatus of a *Palmatolepis* species.] *U.Dev.(Famenn.)*, Eu.-Australia.——FIG. 120,*7*. *T. robusta* BISCHOFF, L. *Palmatolepis marginifera* Z., Virgin Hills F., Australia(W.Australia); lat. view, ×30 (Glenister & Klapper, 1966).

PROBLEMATIC NAMES

A number of generic names published in articles on conodonts are no longer used by conodont workers. Some of these names are based on nonexistent types, some are for specimens that definitely are not conodonts, and others are for specimens of uncertain or unknown taxonomic status. Among these are the following:

Archeognathus CULLISON, 1938
Arcugnathus COOPER in COOPER and SLOSS, 1943
Astacoderma HARLEY, 1861
Bransonella HARLTON, 1933
Centrognathodus BRANSON and MEHL, 1944
Centrognathus BRANSON and MEHL, 1934a
Coleodus BRANSON and MEHL, 1933a
Cornuramia SMITH, 1907
Dermatolithis EHRENBERG, 1854
Fortscottella GUNNELL, 1931
Gnathodella MATERN, 1933
Hamulosodina COOPER, 1931
Holmesella GUNNELL, 1931
Icriodina BRANSON and BRANSON, 1947
Icthyodus HARRIS and HOLLINGSWORTH, 1933
Lepodus BRANSON and MEHL, 1933a
Lepognathodus FAY, 1958
Ligonodinoides STAUFFER, 1938
Lonchodus PANDER, 1856
Multidentodus HARLTON, 1933
Nurrella POMESANO CHERCHI, 1967
Pachysomia SMITH, 1907
Palmatodella BASSLER, 1925 [see BOOGAARD & KUHRY, 1979, p. 26]
Panderodella BASSLER, 1925
Prionognathodus FAY, 1958
Prionognathus PANDER, 1856
Ptiloncodus HARRIS, 1962
Rhombocorniculum WALLISER, 1958
Scolopodella STAUFFER and PLUMMER, 1932
Scotlandia COSSMANN, 1909
Stephanodella MATERN, 1933
Subprioniodus SMITH, 1907
Telumodina COOPER, 1931
Valentia SMITH, 1907

ADDENDUM: Conodont Genera Proposed Since Text Completion

Antognathus LIPNYAGOV in KOZITSKAYA *et al.*, 1978, p. 17 [**A. volnovachensis*; OD]. *U.Dev.(up. Famenn.)*. Family Icriodontidae.

Apatella CHAUFF & KLAPPER, 1978, p. 153 [**A. ziegleri*; OD]. *U.Dev.(up.Famenn.)*. Family Bactrognathidae.

Complexodus DZIK, 1976, p. 423 [**Balognathus pugionifer* DRYGANT, 1974; OD]. *M.Ord.* Family unknown.

Erraticodon DZIK, 1978, p. 64 [**E. balticus*; OD]. *M.Ord.* Family unknown.

Gapparodus ABAIMOVA, 1978, p. 79 [**Hertzina?*

bisulcata MÜLLER, 1959; OD]. *M.Cam.-U.Cam.* Family Furnishinidae.

Hemilistrona CHAUFF & DOMBROWSKI, 1977, p. 111 [**H. depkei*; OD]. *U.Dev.(up.Famenn.).* Family Polygnathidae.

Iranognathus KOZUR, MOSTLER, & RAHIMI-YAZD, 1976, p. 7 [**I. unicostatus*; OD]. *U.Perm.* Family Anchignathodontidae.

Johnognathus MASHKOVA, 1977, p. 127 [**J. huddlei*; OD]. *Sil.(up.Llandov.-low.Wenlock.).* Family Pterospathodontidae.

Kimognathus MASHKOVA, 1978, p. 93 [**K. alexeii*; OD]. *L.Dev.* Family Polygnathidae.

Laterignathus ARISTOV & ALEKSEEV, 1976, p. 192 [**L. barskovi*; OD]. *L.Carb.(Tournais.).* Family unknown.

Macerodus FÅHRAEUS & NOWLAN, 1978, p. 461 [**M. dianae*; OD]. *L.Ord.* Family unknown.

Parabelodina SWEET, 1979, p. 64 [**P. denticulata*; OD]. *U.Ord.* Family unknown.

Pavlovites KOZUR in KOZUR & MOSTLER, 1976, p. 21 [**P. artinskiensis*; OD]. *L.Perm.(up.Artinsk.).* Family unknown. [Taxonomic status uncertain.]

Polonodus DZIK, 1976, p. 423 [**Ambalodus clivosus* VIIRA, 1975; OD]. *L.Ord.-M.Ord.* Family unknown.

Pseudobelodina SWEET, 1979, p. 68 [**Belodina kirki* STONE & FURNISH, 1959; OD]. *M.Ord.-U.Ord.* Family unknown.

Pseudopanderodus LANDING, 1979, p. 1025 [**P. fisheri*; OD]. *U.Cam.* Family ?Oneotodontidae.

Rabeignathus KOZUR, 1978, p. 144 [**Gnathodus bucaramangus* RABE, 1977; OD]. *L.Perm.* Family Anchignathodontidae.

Sannemannia AL-RAWI, 1977, p. 58 [**S. pesanseris*; OD]. *L.Dev.-M.Dev.(Eifel.).* Family Icriodontidae.

Scalpellodus DZIK, 1976, p. 421 [**Protopanderodus larus* VAN WAMEL, 1974; OD]. *M.Ord.* Family unknown.

Spinodus DZIK, 1976, p. 424 [**Cordylodus spinatus* HADDING, 1913; OD]. *M.Ord.* Family unknown.

Sweetocristatus SZANIAWSKI in SZANIAWSKI & MATKOWSKI, 1979, p. 253 [**S. arcticus*; OD]. *L.Perm. (up.Artinsk.).* Family Xaniognathidae.

Tasmanognathus BURRETT, 1979, p. 32 [**T. careyi*; OD]. *M.Ord.* Family unknown.

Tortodus WEDDIGE, 1977, p. 326 [**Polygnathus kockelianus* BISCHOFF & ZIEGLER, 1957; OD]. *M.Dev.(Eifel.-Givet.).* Family Polygnathidae.

REFERENCES

Abaimova, G. P., 1972, *Raschlenenie i korrelyatsiya Ordovikskikh otlozheniy yugo-vostoka Sibirskoy platformy po konodontam:* Tr. Sib. Nauchno-Issled. Inst. Geol. Geofiz. Miner. Syrya, v. 146, p. 65-67. [*Examination and correlation of near-East Ordovician deposits of the Siberian platform by conodonts.*]

1975, *Ranneordovikskie konodontiy srednego techeniya r. leny:* Tr. Sib. Nauchno-Issled. Inst. Geol. Geofiz. Miner. Syrya, v. 207, p. 1-129, pl. 1-10. [*Early Ordovician conodonts from the River Lena.*]

1978, *Pozdnekembriyskiye konodonty Tsentralnogo Kazakhastana:* Paleontol. Zh., no. 4, p. 77-87, 3 text-fig., pl. 7, 8. [*Late Cambrian conodonts from Central Kazakhstan.*]

Aldridge, R. J., 1972, *Llandovery conodonts from the Welsh Borderland:* Bull. Br. Mus. Nat. Hist. Geol., v. 22, no. 2, p. 127-226, pl. 1-9.

1974, *An amorphognathoides Zone conodont fauna from the Silurian of the Ringerike area, South Norway:* Nor. Geol. Tidsskr., v. 54, p. 295-303, 1 text-fig.

1975, *The stratigraphic distribution of conodonts in the British Silurian:* J. Geol. Soc. London, v. 131, p. 607-618, 2 text-fig., pl. 1-3.

Al-Rawi, Dhya, 1977, *Biostratigraphische Gliederung der Tentaculiten-Schichten des Frankenwaldes mit Conodonten und Tentaculiten (Unter- und Mittel-Devon; Bayern, Deutschland):* Senckenbergiana Lethaea, v. 58, p. 25-[80], 7 text-fig., 8 pl.

Aristov, V. A., & **Alekseev, A. S.**, 1976 (1978), *Late Tournaisian conodonts from the Scaliognathus anchoralis Zone of Eastern Kazakhstan:* Dokl. Acad. Sci. USSR, Earth Sci. Sect., v. 229, no. 1-6, p. 229-232, 1978 transl. of Dokl. Akad. Nauk SSSR, v. 229, no. 1, p. 189-192.

Armstrong, W. G., & **Tarlo, L. B. H.**, 1966, *Amino-acid components in fossil calcified tissues:* Nature, v. 210, no. 5035, p. 481-482.

Austin, R. L., 1972, *Problems of conodont taxonomy:* Geol. Palaeontol., Spec. Vol. 1, p. 115-126, 2 text-fig., 2 pl.

1973a, *Phylogeny and homeomorphy of conodonts in the Lower Carboniferous:* in F. H. T. Rhodes (ed.), Conodont paleozoology, Geol. Soc. Am., Spec. Pap. 141, p. 105-116, 2 text-fig.

1973b, *Modification of the British Avonian conodont zonation and a reappraisal of European Dinantian conodont zonation and correlation:* Ann. Soc. Geol. Belg., Ann., v. 96, p. 523-532.

1975, *Dinantian conodont biostratigraphy in Europe:* Geol. Soc. Am. Abstr. Programs, v. 7, no. 6, p. 713-714.

———, **Conil, Raphael, Groessens, Eric**, & **Pirlet, H.**, 1974, *Étude biostratigraphique de l'encrinite de Tramaka:* Bull. Soc. Belge Géol. Paléontol. Hydrol., v. 83, p. 113-129, pl. 1-3.

———, & **Hill, P. J.**, 1973, *A lower Avonian (K zone) conodont fauna from near Tintern, Monmouthshire, Wales:* Geol. Palaeontol., v. 7, p. [23]-34, 2 text-fig., pl. 1, 2.

———, & **Mitchell, M.**, 1975, *Middle Dinantian platform conodonts from Co. Fermanagh and Co. Tyrone, Northern Ireland:* Bull. Geol. Surv. G. B., no. 55, p. 43-54, 2 text-fig., pl. 1-2.

———, & **Rhodes, F. H. T.**, 1969, *A conodont assemblage from the Carboniferous of the Avon Gorge, Bristol:* Palaeontology, v. 12, p. 400-405, 1 text-fig.

Avcin, M. J., & **Norby, R. D.**, 1973, *Scottognathus (Pennsylvanian conodont), a generic concept:* Geol. Soc. Am. Abstr. Programs, v. 5, no. 4, p. 296.

Babcock, L. C., 1976, *Conodont paleoecology of the Lamar Limestone (Permian), Delaware basin, west Texas:* in C. R. Barnes (ed.), Conodont paleoecology, Geol. Soc. Can., Spec. Pap. 15, p. 279-294.

Baesemann, J. F., 1973, *Missourian (Upper Pennsylvanian) conodonts of northeastern Kansas:* J. Paleontol., v. 47, p. 689-710, 2 text-fig., pl. 1-3.

Barnes, C. R., 1967, *A questionable natural conodont assemblage from Middle Ordovician limestone, Ottawa, Canada:* J. Paleontol., v. 41, p. 1557-1560, 2 text-fig.

———, & **Fåhraeus, L. E.**, 1975, *Provinces, communities, and the proposed nektobenthic habit of Ordovician conodontophorids:* Lethaia, v. 8, p. 133-149, 5 text-fig.

———, & **Nowlan, G. S.**, 1975, *Conodonts: A thing of the past—and of the future:* Geosci. Can., v. 2, p. 85-89, 6 text-fig.

———, **Rexroad, C. B.**, & **Miller, J. F.**, 1973, *Lower Paleozoic conodont provincialism:* in F. H. T. Rhodes (ed.), Conodont paleozoology, Geol. Soc. Am. Spec. Pap. 141, p. 157-190, 16 text-fig.

———, **Sass, D. B.**, & **Monroe, E. A.**, 1970, *Preliminary studies of the ultrastructure of selected Ordovician conodonts:* R. Ont. Mus. Life Sci. Contrib., no. 76, p. 1-24, pl. 1-10.

1973, *Ultrastructure of some Ordovician conodonts:* in F. H. T. Rhodes (ed.), Conodont paleozoology, Geol. Soc. Am. Spec. Pap. 141, p. 1-30, 8 text-fig.

———, ———, & **Poplawski, M. L. S.**, 1973, *Conodont ultrastructure: The family Panderodontidae:* R. Ont. Mus. Life Sci. Contrib., no. 90, p. 1-36, 11 text-fig.

———, & **Slack, D. J.**, 1975, *Conodont ultrastructure: The subfamily Acanthodontinae:* R. Ont. Mus. Life Sci. Contrib., no. 106, p. 1-21, 6 text-fig.

Barnett, S. G., 1971, *Biometric determination of the evolution of Spathognathodus remscheidensis: A method for precise intrabasinal correlations in the northern Appalachians:* J. Paleontol., v. 45, p. 274-300, 17 text-fig., pl. 35-37.

Barrick, J. E., 1977, *Multielement simple-cone conodonts from the Clarita Formation (Silurian), Arbuckle Mountains, Oklahoma:* Geol. Palaeontol., v. 11, p. 47-68, 1 text-fig., 3 pl.

———, **& Klapper, Gilbert,** 1976, *Multielement Silurian (late Llandoverian-Wenlockian) conodonts of the Clarita Formation, Arbuckle Mountains, Oklahoma, and phylogeny of Kockelella:* Geol. Palaeontol., v. 10, p. 59-98, 5 text-fig., 4 pl.

Barskoy, I. S., Alekseev, A. S., & Goreva, N. V., 1977, *The taxonomic and nomenclatural status of conodonts of the genus Gnathodus Pander, 1856:* Acad. Sci. USSR, Paleontol. J., no. 4, p. 131-134, transl. of Akad. Nauk SSSR, Paleontol. Zh., v. 11, no. 4, p. 518-521, pl. 8.

Bassler, R. S., 1925, *Classification and stratigraphic use of the conodonts:* Geol. Soc. Am. Bull., v. 36, p. 218-220.

Beckmann, Heinz, 1949, *Conodonten aus dem Iberger Kalk (Ober-Devon) des Bergischen Landes und ihr Feinbau:* Senckenbergiana, v. 30, no. 1-3, p. 153-168, pl. 1-4.

1952, *Zur Anwendung von Essigsäure in der Mikropaläontologie:* Palaeontol. Z., v. 26, p. 138-139.

Behnken, F. H., 1975, *Leonardian and Guadalupian (Permian) conodont biostratigraphy in western and southwestern United States:* J. Paleontol., v. 49, p. 284-315, 11 text-fig., 2 pl.

Beinert, R. J., Klapper, Gilbert, Sandberg, C. A., & Ziegler, Willi, 1971, *Revision of Scaphignathus and description of Clydagnathus? ormistoni n. sp. (Conodonta, Upper Devonian):* Geol. Palaeontol., v. 5, p. 81-91, 1 text-fig., 2 pl.

Bender, Hans, 1970, *Zur Gliederung der Mediterranen Trias II. Die Conodontenchronologie der Mediterranen Trias:* Ann. Geol. Pays Hell., v. 19, p. 465-540.

———, **& Stoppel, Dieter,** 1965, *Perm-conodonten:* Geol. Jahrb., v. 82, p. 331-364, pl. 14-16.

Bengtson, Stefan, 1976, *The structure of some Middle Cambrian conodonts, and the early evolution of conodont structure and function:* Lethaia, v. 9, p. 185-206, 12 text-fig.

Bergström, S. M., 1962, *Conodonts from the Ludibundus Limestone (Middle Ordovician) of the Tvären area (S. E. Sweden):* Ark. Mineral. Geol., v. 3, no. 1, p. 1-61, 5 text-fig., pl. 1-5.

1964, *Remarks on some Ordovician conodont faunas from Wales:* Lunds Univ. Arsskr. Avd. 2, no. 3, p. 1-66, 22 text-fig.

1968, *Biostratigraphy of the Lower Ordovician sequence at Skattungbyn, Dalarna:* Geol. Foeren. Stockholm Foerh., v. 90, p. 454.

1971, *Conodont biostratigraphy of the Middle and Upper Ordovician of Europe and eastern North America:* in W. C. Sweet & S. M. Bergström (eds.), Symposium on conodont biostratigraphy, Geol. Soc. Am. Mem. 127, p. 83-164, 14 text-fig., 2 pl.

1973, *Biostratigraphy and facies relations in the lower Middle Ordovician of easternmost Tennessee:* Am. J. Sci., v. 273-A, p. 261-293.

———, **Carnes, J. B., Ethington, R. L., Votaw, R. B., & Wigley, P. B.,** 1974, *Appalachignathus, a new multielement conodont genus from the Middle Ordovician of North America:* J. Paleontol., v. 48, p. 227-235, 5 text-fig., 1 pl.

———, **Epstein, A. G., & Epstein, J. B.,** 1972, *Early Ordovician North Atlantic Province conodonts in eastern Pennsylvania:* U.S. Geol. Surv. Prof. Pap. 800D, p. D37-D44, 3 text-fig.

———, **& Sweet, W. C.,** 1966, *Conodonts from the Lexington Limestone (Middle Ordovician) of Kentucky and its lateral equivalents in Ohio and Indiana:* Bull. Am. Paleontol., v. 50, no. 229, p. 271-441, 13 text-fig., pl. 28-35.

Bischoff, Günther, 1957, *Die Conodonten-Stratigraphie des rheno-herzynischen Unterkarbons mit Berücksichtigung der Wocklumeria-Stufe und der Devon/Karbon-Grenze:* Abh. Hess. Landesamtes Bodenforsch., no. 19, p. 1-64, pl. 1-6.

1973, *On the nature of the conodont animal:* Geol. Palaeontol., v. 7, p. 147-174, 1 text-fig., pl. 1-6.

———, **& Sannemann, Dietrich,** 1958, *Unterdevonische Conodonten aus dem Frankenwald:* Notizbl. Hess. Landesamtes Bordenforsch. Wiesbaden, v. 86, p. 87-110, pl. 12-15.

———, **& Ziegler, Willi,** 1957, *Die Conodontenchronologie des Mitteldevons und des tiefsten Oberdevons:* Abh. Hess. Landesamtes Bodenforsch., v. 22, 136 p., 16 text-fig., 21 pl.

Bitter, P. H. von, 1972, *Environmental control of conodont distribution in the Shawnee Group (Upper Pennsylvanian) of eastern Kansas:* Univ. Kans. Paleontol. Contrib. Artic. 59, 105 p., 16 pl.

1974, *Element composition and micromorphology of species of Gondolella from the Middle and Upper Pennsylvanian of the central U.S.A.:* Geol. Soc. Am. Abstr. Programs, v. 6, no. 6, p. 551.

1976a, *The apparatus of Gondolella sublanceolata Gunnell (Conodontophorida, Upper Pennsylvanian) and its relationship to Illinella typica Rhodes:* R. Ont. Mus. Life Sci. Contrib., no. 109, p. 1-44, 18 text-fig.

1976b, *Paleoecology and distribution of Windsor Group (Visean-?early Namurian) conodonts, Port Hood Island, Nova Scotia, Canada:* in C. R. Barnes (ed.), Conodont paleoecology, Geol. Soc. Can., Spec. Pap. 15, p. 225-241.

Boersma, K. Th., 1973, *On the basal structure of Siphonodella cooperi Hass, 1959, and Siphonodella lobata (Branson & Mehl), 1934:* Leidse Geol. Meded., v. 49, p. 39-57, pl. 1-4.

Boogaard, M. van den, 1966, *Post-Carboniferous conodonts from south-eastern Spain:* Proc. K. Ned. Akad. Wet. Ser. B, v. 69, p. 1-8, pl. 1-2.

———, & **Kuhry, B.**, 1979, *Statistical reconstruction of the Palmatolepis apparatus (Late Devonian conodontophorids) at the generic, subgeneric, and specific level:* Scr. Geol., v. 49, 57 p., 28 text-fig.

Bradshaw, L. E., 1969, *Conodonts from the Fort Peña Formation (Middle Ordovician), Marathon Basin, Texas:* J. Paleontol., v. 43, p. 1137-1168, 4 text-fig., pl. 131-137.

———, **Noel, J. A.**, & **Larson, R. J.**, 1972, *Neutron activation analysis of selected conodonts:* in F. H. T. Rhodes (ed.), Conodont paleozoology, Geol. Soc. Am. Spec. Pap. 141, p. 67-84.

Branson, C. C., 1957, *Comment on the Moore/Sylvester-Bradley "Parataxa Plan":* Bull. Zool. Nomencl., v. 15, Doc. 1/25, p. 169.

———, & **Mankin, D. J.**, 1964, *Composition of conodonts:* Okla. Geol. Notes, v. 24, p. 296-302.

Branson, E. B., & **Branson, C. C.**, 1947, *Lower Silurian conodonts from Kentucky:* J. Paleontol., v. 21, p. 549-556, 1 text-fig., pl. 81, 82.

———, & **Mehl, M. G.**, 1933a, *Conodont studies no. 1: Conodonts from the Harding sandstone of Colorado; Bainbridge (Silurian) of Missouri; Jefferson City (Lower Ordovician) of Missouri:* Univ. Mo. Stud., v. 8, no. 1, p. 5-72, pl. 1-4.

1933b, *Conodont studies no. 2: Conodonts from the Joachim (Middle Ordovician) of Missouri; Conodonts from the Plattin (Middle Ordovician) of Missouri; Conodonts from the Maquoketa-Thebes (Upper Ordovician) of Missouri; A study of Hinde's types of conodonts preserved in the British Museum:* Univ. Mo. Stud., v. 8, no. 2, p. 77-167, 3 text-fig., pl. 5-12.

1934a, *Conodont studies no. 3: Conodonts from the Grassy Creek shale of Missouri:* Univ. Mo. Stud., v. 8, 1933, no. 3, p. 171-259, 3 text-fig., pl. 13-21.

1934b, *Conodont studies no. 4: Conodonts from the Bushberg sandstone and equivalent formations of Missouri:* Univ. Mo. Stud., v. 8, 1933, no. 4, p. 265-300, pl. 22-24.

1936, *Geological affinities and taxonomy of conodonts:* Geol. Soc. Am. Proc. Abstr., 1935, p. 436 (also Pan. Am. Geol., v. 65, p. 233).

1938, *The conodont genus Icriodus and its stratigraphic distribution:* J. Paleontol., v. 12, p. 156-166, 1 text-fig., pl. 26.

1940, *Caney conodonts of Upper Mississippian age:* Denison Univ. Sci. Lab. J., v. 35, p. 167-178, pl. 5.

1941a, *Conodonts from the Keokuk formation:* Denison Univ. Sci. Lab. J., v. 35, 1940, p. 179-188, pl. 1-6.

1941b, *The recognition and interpretation of mixed conodont faunas:* Denison Univ. Sci. Lab. J., v. 35, 1940, p. 195-209.

1941c, *New and little known Carboniferous conodont genera:* J. Paleontol., v. 15, p. 97-106, pl. 19.

1943, *Ordovician conodont faunas from Oklahoma:* J. Paleontol., v. 17, p. 374-387, pl. 63, 64.

1944, *Conodonts:* in H. W. Shimer & R. R. Shrock, Index fossils of North America, p. 235-246, pl. 93, 94, Wiley & Sons (New York).

1948, *Conodont homonyms and names to replace them:* J. Paleontol., v. 22, p. 527-528.

———, ———, & **Branson, C. C.**, 1951, *Richmond conodonts of Kentucky and Indiana:* J. Paleontol., v. 25, p. 1-17, 1 text-fig., pl. 1-4.

Branson, E. R., 1934, *Conodonts from the Hannibal formation of Missouri:* Univ. Mo. Stud., v. 8, Oct., 1933, p. 301-334, pl. 25-28.

Bryant, W. L., 1921, *The Genesee conodonts:* Bull. Buffalo Soc. Nat. Sci., v. 13, p. 1-59, 7 text-fig., pl. 1-16.

Budurov, K. J., 1968, *Conodonten-fauna der nicht Metamorphen Trias Bulgariens:* Univ. Sofia Geol. Geog. diss., p. 1-15.

1972, *Ancyrogondolella triangularis gen. et sp. n. (Conodonta):* Mitt. Ges. Geol. Bergbaustud. Wien, v. 21, p. 853-860.

1973, *Carinella n. gen. und revision der Gattung Gladigondolella (Conodonta):* Dokl. Bolg. Akad. Nauk, v. 26, p. 799-802.

———, & **Pantic, S.**, 1973, *Conodonten aus den Campiller Schichten von Brassina (Westserbien):* Izv. Geol. Inst. Bulg. Akad. Nauk. Ser. Paleontol., v. 22, p. 49-64.

Bultynck, P., 1976, *Le Silurien supérieur et le Devonien inférieur de la Sierra de Guadarrama (Espagne centrale), Troisième partie: Éléments icriodiformes, pelekysgnathiformes et polygnathiformes:* Bull. Inst. R. Sci. Nat. Belg., v. 49 (Sci. Terre no. 5), 74 p., 10 text-fig., 11 pl.

Burnley, G. I., 1938, *The conodonts of the shale overlying the Lexington Coal Bed of Lafayette County and Jackson County, Missouri:* Unpubl. MA thesis, Univ. Missouri (Columbia).

Burrett, C. F., 1979, *Tasmanognathus: A new Ordovician conodontophorid genus from Tasmania:* Geol. Palaeontol., v. 13, p. 31-38, 4 text-fig., 1 pl.

Butler, Malcolm, 1973, *Lower Carboniferous conodont faunas from the Eastern Mendips, England:* Palaeontology, v. 16, p. 477-517, pl. 56-59.

Butts, C., 1926, *The Paleozoic rocks,* in G. I. Adams et al., Geology of Alabama: Ala. Geol. Surv. Spec. Rep. 14, p. 41-230, pl. 3-76.

Carls, Peter, & **Gandl, Josef**, 1969, *Stratigraphie und Conodonten des Unter-Devons der Östlichen Iberischen Ketten (NE-Spanien):* Neues Jahrb. Geol. Palaeontol. Abh., v. 132, p. 155-218, 3 text-fig., pl. 15-20.

Carnes, J. B., 1975, *Conodont biostratigraphy in the lower Middle Ordovician of the western Appalachian thrust-belts in northeastern Tennessee:* Unpubl. Ph.D. diss., 291 p., 8 pl., The Ohio State University (Columbus).

Chamberlain, C. K., & **Clark, D. L.**, 1973, *Trace fossils and conodonts as evidence for deep-water deposits in the Oquirrh Basin of Central Utah:*

J. Paleontol., v. 47, p. 663-682, 7 text-fig., pl. 1-3.

Chauff, K. M., & Dombrowski, Anna, 1977, *Hemilistrona, a new conodont genus from the basal shale member of the Sulphur Springs Formation, east-central Missouri:* Geol. Palaeontol., v. 11, p. 109-120, 1 text-fig., 2 pl.

———, **& Klapper, Gilbert,** 1978, *New conodont genus Apatella (Late Devonian), possible homeomorph Bactrognathus (Early Carboniferous, Osagean Series), and homeomorphy in conodonts:* Geol. Palaeontol., v. 12, p. 151-164, 3 pl.

Clark, D. L., 1959, *Conodonts from the Triassic of Nevada and Utah:* J. Paleontol., v. 33, p. 305-312, 1 text-fig., pl. 44, 45.

1972a, *Early Permian crisis and its bearing on Permo-Triassic conodont taxonomy:* Geol. Palaeontol., Spec. Vol. 1, p. 147-158, 11 text-fig.

1972b, *Conodont paleoecology:* Geol. Soc. Am. Abstr. Programs, v. 4, p. 312.

1974, *Factors of Early Permian conodont paleoecology in Nevada:* J. Paleontol., v. 48, p. 710-720, 3 text-fig., pl. 1, 2.

1977, *Late Paleozoic and Triassic conodont biostratigraphy: Correlations around the expanding Atlantic Ocean:* in F. M. Swain (ed.), Stratigraphic micropaleontology of Atlantic Basin and borderlands, Elsevier Dev. Paleontol. Stratigr. no. 6, p. 111-136.

———, **& Behnken, F. H.,** 1971, *Conodonts and biostratigraphy of the Permian:* in W. C. Sweet & S. M. Bergström (eds.), Symposium on conodont biostratigraphy, Geol. Soc. Am. Mem. 127, p. 415-439, 4 text-fig., 2 pl.

———, **& Miller, J. F.,** 1969, *Early evolution of conodonts:* Geol. Soc. Am. Bull., v. 80, p. 125-134, 8 text-fig.

———, **& Müller, K. J.,** 1968, *The basal opening of conodonts:* J. Paleontol., v. 42, p. 561-570, 6 text-fig.

———, **& Rosser, S. V.,** 1976, *Analysis of paleoecologic factors associated with the Triassic Parachirognathus/Furnishius conodont fauna in Utah and Nevada:* in C. R. Barnes (ed.), Conodont paleoecology, Geol. Soc. Can., Spec. Pap. 15, p. 295-311.

———, **Sincavage, J. P., & Stone, D. D.,** 1964, *New conodont from the Lower Triassic of Nevada:* J. Paleontol., v. 38, p. 375-377, 1 text-fig., pl. 60.

Clarke, J. M., 1885, *On the higher Devonian faunas of Ontario County, New York:* U.S. Geol. Surv., Bull., v. 3, no. 16, p. 35-120, pl. 1-3.

Collinson, C. W., 1963, *Techniques for the collecting and processing of conodonts:* Ill. State Geol. Surv. Circ. 343, p. 1-16.

1965, *Conodonts:* in Bernhard Kummel & D. M. Raup (eds.), Handbook of paleontological techniques, p. 94-102, W. H. Freeman & Co. (San Francisco).

———, **Avcin, M. J., Norby, R. D., & Merrill, G. K.,** 1972, *Pennsylvanian conodont assemblages from LaSalle County, northern Illinois:* Ill. State Geol. Surv. Guideb. Ser., no. 10, p. 1-37, 18 text-fig.

———, **Rexroad, C. B., & Thompson, T. F.,** 1971, *Conodont zonation of the North American Mississippian:* in W. C. Sweet & S. M. Bergström (eds.), Symposium on conodont biostratigraphy, Geol. Soc. Am. Mem. 127, p. 353-394, 8 text-fig.

Conway Morris, Simon, 1976, *A new Cambrian lophophorate from the Burgess Shale of British Columbia:* Palaeontology, v. 19, p. 199-222, pl. 30-34.

Cooper, B. J., 1974a, *New forms of Belodella (Conodonta) from the Silurian of Australia:* J. Paleontol., v. 48, p. 1120-1125, 1 text-fig., 1 pl.

1974b, *Studies of multielement Silurian conodonts:* Unpubl. Ph.D. diss., 197 p., 10 text-fig., 7 pl., The Ohio State University (Columbus).

1975, *Multielement conodonts from the Brassfield Limestone (Silurian) of southern Ohio:* J. Paleontol., v. 49, p. 984-1008, 3 text-fig., 3 pl.

1976, *Multielement conodonts from the St. Clair Limestone (Silurian) of southern Illinois:* J. Paleontol., v. 50, p. 205-217, 2 text-fig., 2 pl.

1977, *Toward a familial classification of Silurian conodonts:* J. Paleontol., v. 51, p. 1057-1071, 2 pl.

Cooper, C. L., 1931, *New conodonts from the Woodford formation of Oklahoma:* J. Paleontol., v. 5, p. 230-243, pl. 28.

1939, *Conodonts from a Bushberg-Hannibal horizon in Oklahoma:* J. Paleontol., v. 13, p. 379-422, pl. 38-47.

———, **& Sloss, L. L.,** 1943, *Conodont fauna and distribution of a Lower Mississippian black shale in Montana and Alberta:* J. Paleontol., v. 17, p. 168-176, 1 text-fig., pl. 28, 29.

Croneis, Carey, 1938, *Utilitarian classification for fragmentary fossils:* J. Geol., v. 46, p. 975-984.

Cullison, J. S., 1938, *Dutchtown fauna of southeastern Missouri:* J. Paleontol., v. 12, p. 219-228, pl. 29.

Demanet, Felix, 1939, *Filtering appendices on the branchial arches of Coelacanthus lepturus Agassiz:* Geol. Mag., v. 76, no. 899, p. 215-219, 2 text-fig.

Denham, R. L., 1944, *Conodonts:* J. Paleontol., v. 18, p. 216-218, 1 text-fig.

Diebel, Kurt, 1956, *Conodonten in der Oberkreide von Kamerun:* Geologie, v. 5, p. 424-450, pl. 1-6.

Dow, V. E., 1965, *Magnetic separation of conodonts:* in Bernhard Kummel & D. M. Raup (eds.), Handbook of paleontologic techniques, p. 263-267, W. H. Freeman & Co. (San Francisco).

Druce, E. C., 1969, *Devonian and Carboniferous conodonts from the Bonapart Gulf basin, northern Australia:* Aust. Bur. Miner. Resour. Geol. Geophys. Bull. 98, 242 p., 33 text-fig., 43 pl.

1970, *Lower Carboniferous conodonts from the*

Northern Yarrol basin, Queensland: Aust. Bur. Miner. Resour. Geol. Geophys. Bull. 108 (1967), p. 91-113, pl. 15-18.

1973, *Upper Paleozoic and Triassic conodont distribution and the recognition of biofacies:* in F. H. T. Rhodes (ed.), Conodont paleozoology, Geol. Soc. Am. Spec. Pap. 141, p. 191-237, 1 text-fig.

———, & **Jones, P. J.**, 1971, *Cambro-Ordovician conodonts from the Burke River structural belt, Queensland:* Aust. Bur. Miner. Resour. Geol. Geophys. Bull. 110, p. 1-159, pl. 1-20.

———, **Rhodes, F. H. T.**, & **Austin, R. L.**, 1972, *Statistical analysis of British Carboniferous conodont faunas:* Q.J. Geol. Soc. London, v. 128, p. 53-70, 8 text-fig.

1974, *Recognition, evolution and taxonomy of Lower Carboniferous conodont assemblages:* J. Paleontol., v. 48, p. 387-402, 7 text-fig.

Drygant, D. M., 1974, *Prostye Konodonty Silura i nizov Devona Volyno-Podolya:* Paleontol. Sb. Lvov Univ., v. 10, p. 64-70, 2 pl. [*Simple conodonts of the Silurian and lowermost Devonian of the Volyn-Podolian area.*]

Dubois, E. P., 1943, *Evidence on the nature of conodonts:* J. Paleontol., v. 17, p. 155-159, pl. 25.

Dunn, D. L., 1966, *New Pennsylvanian platform conodonts from southwestern United States:* J. Paleontol., v. 40, p. 1294-1303, 2 text-fig., pl. 157, 158.

1970, *Middle Carboniferous conodonts from western United States and phylogeny of the platform group:* J. Paleontol., v. 44, p. 312-342, 11 text-fig., pl. 61-64.

Durden, C. J., 1969, *Gnathostomulida: Is there a fossil record?:* Science, v. 164, p. 855.

Dzik, Jerzy, 1976, *Remarks on the evolution of Ordovician conodonts:* Acta Palaeontol. Pol., v. 21, no. 4, p. 395-455, 36 text-fig., pl. XLI-XLIV.

1978, *Conodont biostratigraphy and paleogeographical relations of the Ordovician Mójcza Limestone (Holy Cross Mts., Poland):* Acta Palaeontol. Pol., v. 23, no. 1, p. 51-72, 6 text-fig., pl. 12-15.

Ehrenberg, C. G., 1854, *Über den Grünsand und seine Erläuterung des organischen Lebens:* Abh. K. Akad. Wiss. Berlin, 1855, p. 85-176, pl. 1-7.

Eichenberg, W., 1930, *Conodonten aus dem Culm des Harzes:* Palaeontol. Z., v. 12, p. 177-182, text-fig, 1-30.

Eisenack, Alfred, 1973, *Kleinorganismen als Zerstörer säurefester organischer Substanzen und von Biophosphaten:* Palaeontol. Z., v. 47, p. 8-16, pl. 1, 2.

Elias, M. K., 1956, *Upper Mississippian and Lower Pennsylvanian formations of south-central Oklahoma:* in I. C. Hicks *et al.* (eds.), Petroleum geology of southern Oklahoma, A symposium, v. 1, p. 56-134, 2 text-fig., 6 pl., Ardmore Geologica Society.

Ellison, S. P., Jr., 1941, *Revision of the Pennsylvanian conodonts:* J. Paleontol., v. 15, p. 107-143, pl. 20-23.

1944, *The composition of conodonts:* J. Paleontol., v. 18, p. 133-140.

1972, *Conodont taxonomy in the Pennsylvanian:* Geol. Palaeontol., Spec. Vol. 1, p. 127-146, 2 text-fig., pl. 1.

Epstein, A. G., Epstein, J. B., & **Harris, L. D.**, 1974, *Incipient metamorphism, structural anomalies, and oil and gas potential in the Appalachian basin determined from conodont color:* Geol. Soc. Am. Abstr. Programs, v. 6, no. 7, p. 723-724.

1975, *Conodont color alteration—An index to diagenesis of organic matter:* Am. Assoc. Pet. Geol. and Soc. Econ. Paleontol. Mineral., Ann. Mtg. Abstr., v. 2, p. 21-22 (Dallas, Texas).

1977, *Conodont color alteration—An index to organic metamorphism:* U.S. Geol. Surv. Prof. Pap. 995, 27 p., 20 text-fig.

Ethington, R. L., 1959a, *Conodonts of the Ordovician Galena formation:* J. Paleontol., v. 33, p. 257-292, 2 text-fig., pl. 39-41.

1959b, *Rhynchognathodus, new name for the Ordovician conodont Rhynchognathus, preoccupied:* J. Paleontol., v. 33, p. 1128.

———, & **Brand, Uwe**, 1981, *Oneotodus simplex (Furnish) and the genus Oneotodus (Conodonta):* J. Paleontol., v. 55, p. 239-247, 2 text-fig.

———, & **Clark, D. L.**, 1971, *Lower Ordovician conodonts in North America:* in W. C. Sweet & S. M. Bergström (eds.), Symposium on conodont biostratigraphy, Geol. Soc. Am. Mem. 127, p. 63-82, 2 text-fig., 2 pl.

———, & **Furnish, W. M.**, 1959, *Ordovician conodonts from northern Manitoba:* J. Paleontol., v. 33, p. 540-546, 1 text-fig., pl. 73.

1960, *Upper Ordovician conodonts from southern Manitoba:* J. Paleontol., v. 34, p. 265-274, pl. 38.

Fahlbusch, Klaus, 1964, *Die Stellung der Conodontida im biologischen System:* Palaeontographica, v. 123, Abt. A., p. 137-201, 9 text-fig., pl. 16-22.

Fåhraeus, L. E., 1966, *Lower Viruan (Middle Ordovician) conodonts from the Gullhögen Quarry, southern central Sweden:* Sver. Geol. Unders. Ser. C, no. 610, p. 1-40, text-fig. 1-3, pl. 1-4.

1975, *Conodontophorid ecology and evolution related to global tectonics:* Geol. Soc. Am. Abstr. Programs, v. 7, no. 6, p. 753-754.

———, & **Nowlan, G. S.**, 1978, *Franconian (Late Cambrian) to early Champlainian (Middle Ordovician) conodonts from the Cow Head Group, Western Newfoundland:* J. Paleontol., v. 52, p. 444-471, 6 text-fig., 3 pl.

Fay, R. O., 1952, *Catalogue of conodonts:* Univ. Kans. Paleontol. Contrib. (Vertebrata) Artic. 3, 206 p., 109 text-fig.

1957, *Comment on the Moore/Sylvester-Bradley "Parataxa plan":* Bull. Zool. Nomencl., v. 15, Doc. 1/26, p. 170-171.

1958, *A key to conodont genera and subgenera:*

Oklahoma Geol. Notes, v. 18, p. 103-120.
1959, *Generic and subgeneric homonyms of conodonts:* J. Paleontol., v. 33, p. 195-196.
Ferrigno, K. F., 1971, *Environmental influences on the distribution and abundances of conodonts from the Dundee Limestone (Devonian), St. Mary's, Ontario:* Can. J. Earth Sci., v. 8, p. 378-386, 5 text-fig.
Furnish, W. M., 1938, *Conodonts from the Prairie du Chien (Lower Ordovician) beds of the upper Mississippi Valley:* J. Paleontol., v. 12, p. 318-340, 2 text-fig., pl. 41, 42.
Glenister, B. F., & Klapper, Gilbert, 1966, *Upper Devonian conodonts from the Canning basin, Western Australia:* J. Paleontol., v. 40, p. 777-842, 3 text-fig., pl. 85-96.
Graves, R. W., & Ellison, S. P., 1941, *Ordovician conodonts of the Marathon Basin, Texas:* Univ. Mo. Sch. Mines Metall. Bull. Tech. Ser., v. 14, no. 2, 26 p., 2 text-fig., 3 pl.
Gross, Walter, 1954, *Zur Conodonten-Frage:* Senckenb. Lethaea, v. 35, p. 73-85, 2 text-fig., pl. 1-5.
1957, *Über die Basis der Conodonten:* Palaeontol. Z., v. 31, p. 78-91, pl. 7-9.
1960, *Über die Basis bei den Gattungen Palmatolepis und Polygnathus (Conodontida):* Palaeontol. Z., v. 34, p. 40-58, 9 text-fig.
Gunnell, F. H., 1931, *Conodonts from the Fort Scott limestone of Missouri:* J. Paleontol., v. 5, p. 244-252, pl. 29.
1933, *Conodonts and fish remains from the Cherokee, Kansas City, and Wabaunsee groups of Missouri and Kansas:* J. Paleontol., v. 7, p. 261-297, pl. 31-33.
Hadding, Assor, 1913, *Undre dicellograptusskiffern i Skåne jämte några därmed ekvivalenta bildningar:* Lunds Univ. Arsskr Avd. 2, v. 9, no. 15, p. 1-90, text-fig. 1-23, pl. 1-8.
Halstead, L. B., 1968, *The pattern of vertebrate evolution:* W. H. Freeman & Co. (San Francisco), 209 p.
Hamar, György, 1964, *The Middle Ordovician of the Oslo region, Norway. 17. Conodonts from the lower Middle Ordovician of Ringerrike:* Nor. Geol. Tidsskr., v. 44, p. 243-292, 6 text-fig., 6 pl.
1966, *The Middle Ordovician of the Oslo region, Norway. 22. Preliminary report on conodonts from the Oslo-Asker and Ringerrike districts:* Nor. Geol. Tidsskr., v. 46, p. 27-83, 6 text-fig., 7 pl.
Hardin, N. S., 1972, *Paleoecology of Ordovician conodonts of southwest Wisconsin:* Unpubl. MS thesis, Univ. Wisconsin (Madison).
Harley, J., 1861, *On the Ludlow bone-bed and its crustacean remains:* Q.J. Geol. Soc. London, v. 17, p. 542-552, pl. 17.
Harlton, B. H., 1933, *Micropaleontology of the Pennsylvanian Johns Valley shale of the Ouachita Mountains, Oklahoma, and its relationship to the Mississippian Caney shale:* J. Paleontol., v. 7, p. 3-29, pl. 1-7.
Harris, R. W., 1962, *New conodonts from Joins (Ordovician) Formation of Oklahoma:* Okla. Geol. Notes, v. 22, no. 8, p. 199-211, pl. 1.
1964, *Subgenera of the conodont genus Multioistodus in Simpson-Burgen (Ordovician Arbuckle) conodonts from Oklahoma:* Okla. Geol. Notes, v. 25, p. 34-47.
———, **& Harris, Beth,** 1965, *Some West Spring Creek (Ordovician Arbuckle) conodonts from Oklahoma:* Okla. Geol. Notes, v. 25, p. 34-47, pl. 1.
———, **& Hollingsworth, R. V.,** 1933, *New Pennsylvanian conodonts from Oklahoma:* Am. J. Sci., ser. 5, v. 25, no. 147, p. 193-204, 1 pl.
Hass, W. H., 1941, *Morphology of conodonts:* J. Paleontol., v. 15, p. 71-81, pl. 12-16.
1953, *Conodonts of the Barnett formation of Texas:* U.S. Geol. Surv. Prof. Pap. 243-F, p. 69-94, pl. 14-16.
1959, *Conodonts from the Chappel limestone of Texas:* U.S. Geol. Surv. Prof. Pap. 294-J, p. 365-400, pl. 46-50.
1962, *Conodonts:* in R. C. Moore (ed.), Treatise on Invertebrate Paleontology,- Part W, Miscellanea, p. *W*3-*W*69, text-fig. 1-42, Geological Society of America and University of Kansas (New York and Lawrence).
———, **& Lindberg, M. L.,** 1946, *Orientation of crystal units of conodonts:* J. Paleontol., v. 20, p. 501-504, 4 text-fig.
Hayashi, Shingo, 1968, *The Permian conodonts in chert of the Adoyama Formation, Ashio Mountains, Central Japan:* Earth Sci., v. 22, p. 63-77, pl. 1-4.
Heckel, P. H., & Baesemann, J. F., 1975, *Environmental interpretation of conodont distribution in Upper Pennsylvanian (Missourian) megacycles in eastern Kansas:* Bull. Am. Assoc. Pet. Geol., v. 59, no. 3, p. 486-509, 6 text-fig.
Helms, Jochen, 1959, *Conodonten aus dem Saalfelder Oberdevon (Thüringen):* Geologie, Jahrg. 8, no. 6, p. 634-677, text-fig. 1-3, pl. 1-6.
1963, *Zur "Phylogenese" und Taxionomie von Palmatolepis (Conodontida, Oberdevon):* Geologie, Jahrg. 12, no. 4, p. 449-485, text-fig. 1-3, pl. 1-4.
Higgins, A. C., 1975, *Conodont zonation of the late Visean–early Westphalian strata of the south and central Pennines of northern England:* Bull. Geol. Surv. G.B., no. 53, 90 p., 14 text-fig., 18 pl.
Hinde, G. J., 1879, *On conodonts from the Chazy and Cincinnati group of the Cambro-Silurian, and from the Hamilton and Genesee-shale divisions of the Devonian in Canada and the United States:* Q.J. Geol. Soc. London, v. 35, p. 351-369, pl. 15-17.
Hirschmann, Christa, 1959, *Über Conodonten aus dem Oberen Muschelkalk des Thüringer Beckens:* Freiberger Forschungsh. C, v. 76, p. 33-93, pl. 4-5.

Hofker, Jan, 1974, *Eine mögliche Tiergruppe, welche die Trägerin der sogenannten Conodonten war:* Palaeontol. Z., v. 48, p. 29-35, text-fig. 1-3.

Holmes, G. B., 1928, *A bibliography of the conodonts with descriptions of Early Mississippian species:* Proc. U.S. Natl. Mus., v. 72, art. 5, 38 p., 11 pl.

Huckriede, Reinhold, 1958, *Die Conodonten der Mediterranen Trias und ihr stratigraphischer Wert:* Palaeontol. Z., v. 32, p. 141-175, pl. 10-14.

Huddle, J. W., 1934, *Conodonts from the New Albany shale of Indiana:* Bull. Am. Paleontol., v. 21, no. 72, 136 p., 12 pl.

1968, *Redescription of Upper Devonian conodont genera and species proposed by Ulrich & Bassler in 1926:* U.S. Geol. Surv. Prof. Pap. 578, 55 p., 2 text-fig., 17 pl.

1970, *Revised descriptions of some late Devonian polygnathid conodonts:* J. Paleontol., v. 44, p. 1029-1040, 1 text-fig., pl. 137, 138.

1972, *Historical introduction to the problem of conodont taxonomy:* Geol. Palaeontol., Spec. Vol. 1, p. 3-16, 1 pl.

Hyman, L. H., 1940, *The invertebrates, vol. I (Protozoa through Ctenophora):* 726 p., McGraw-Hill (New York, London).

Igo, Hisayoshi, Koike, Toshio, & Yin, E. H., 1965, *Triassic conodonts from Kelantian, Malaya (Studies of Asiatic conodonts, Part III):* Mejiro Gakuen Women's Jr. Coll., Mem., v. 2, p. 5-20, pl. 1, 2.

James, U. P., 1884, *On conodonts and fossil annelid jaws:* Cincinnati Soc. Nat. Hist. J., v. 7, p. 143-149, pl. 7.

Jeppsson, Lennart, 1969, *Notes on some Upper Silurian multielement conodonts:* Geol. Foeren. Stockholm Foerh., v. 91, p. 12-24, 4 text-fig.

1971, *Elements arrangement in conodont apparatuses of Hindeodella type and in similar forms:* Lethaia, v. 4, p. 101-123, 5 text-fig.

1972, *Some Silurian conodont apparatuses and possible conodont dimorphism:* Geol. Palaeontol., v. 6, p. 51-69, text-fig. 1-5, pl. 1, 2.

1975, *Aspects of Late Silurian conodonts:* Fossils Strata, v. 6 (1974), 54 p., 6 text-fig., 12 pl.

Jones, D. J., 1941, *The conodont fauna of the Seminole formation of Oklahoma:* 55 p., 4 pl., University of Chicago Press (Chicago).

Khodalevich, A. N., & Tschernich, V. V., 1973a, *Konodonty iz zhivetskikh otlozheniy vostochnogo sklona yuzhnogo Urala:* Tr. Sverdl. Gorn. Inst., no. 93, p. 27-41, 1 text-fig., 4 pl. [*Conodonts from Givetian deposits of the eastern slope of the southern Urals.*]

1973b, *Novoe podsemeystvo Belodellinae (Konodonty):* Tr. Sverdl. Gorn. Inst., no. 93, p. 42-47, 1 text-fig., 1 pl. [*New subfamily Belodellinae (conodonts).*]

Kirk, S. R., 1929, *Conodonts associated with the Ordovician fish fauna of Colorado—A preliminary note:* Am. J. Sci., ser. 5, v. 18, no. 108, p. 493-496, 14 text-fig.

Klapper, Gilbert, 1966, *Upper Devonian and Lower Mississippian conodont zones in Montana, Wyoming, and South Dakota:* Univ. Kans. Paleontol. Contrib. Pap. 3, 43 p., 2 text-fig., 6 pl.

1971, *Patrognathus and Siphonodella (Conodonta) from the Kinderhookian (Lower Mississippian) of western Kansas and southwestern Nebraska:* Kans. State Geol. Surv. Bull. 202, pt. 3, 14 p., pl. 1, 2.

———, **& Furnish, W. M.,** 1963, *Conodont zonation of the early Upper Devonian in eastern Iowa:* Proc. Iowa Acad. Sci., v. 69, p. 400-410, 2 text-fig.

———, **& Murphy, M. A.,** 1975, *Silurian-Lower Devonian conodont sequence in the Roberts Mountains Formation of central Nevada:* Univ. Calif. Publ. Geol. Sci., v. 111 (1974), 62 p., 10 text-fig., 12 pl.

———, **& Philip, G. M.,** 1971, *Devonian conodont apparatuses and their vicarious skeletal elements:* Lethaia, v. 4, p. 429-452, 14 text-fig.

1972, *Familial classification of reconstructed Devonian conodont apparatuses:* Geol. Palaeontol., Spec. Vol. 1, p. 97-105, pl. 1-4.

———, **Sandberg, C. A., Collinson, Charles, Huddle, J. W., Orr, R. W., Rickard, L. V., Schumacher, Dietmar, Seddon, George, & Uyeno, T. T.,** 1971, *North American Devonian conodont biostratigraphy:* in W. C. Sweet & S. M. Bergström (eds.), Symposium on conodont biostratigraphy, Geol. Soc. Am. Mem. 127, p. 285-316, 6 text-fig.

Knüpfer, Jürgen, 1967, *Zur Fauna und Biostratigraphie des Ordoviciums (Gräfenthaler Schichten) in Thüringen:* Freiberger Forschungsh. C, no. 220, 199 p., 19 pl.

Kozitskaya, R. I., Kosenko, Z. A., Lipnyagov, O. M., & Nemirovskaya, T. I., 1978, *Konodonty Karbona Donetskogo Basseyna:* Naukova Dumka, Akad. Nauk Ukrainskoy SSR, 136 p., 1 text-fig., 32 pl. [*Conodonts of the Carboniferous of the Donets basin.*]

Kozur, Heinz, 1968, *Neue Conodonten aus dem Oberen Muschelkalk des germanischen binnenbeckens (Vorläufige Mitteilung):* Monatsber. Dtsch. Akad. Wiss. Berlin, v. 10, p. 130-142, pl. 1 (Berlin).

1972, *Die Conodontengattung Metapolygnathus Hayashi 1968 und ihr stratigraphischer Wert:* Geol. Paläontol. Mitteil. Innsbruck, v. 2, no. 11, p. 1-27, pl. 1-7.

1973, *Beiträge zur Stratigraphie und Paläontologie des Trias:* Geol. Paläontol. Mitt. Innsbruck, v. 3, no. 1, p. 1-30.

1975a, *Beiträge zur Conodontenfauna des Perm:* Geol. Paläontol. Mitt. Innsbruck, v. 5, no. 4, p. 1-44, pl. 1-4.

1975b, *Probleme der Triasgliederung und Parallelisiergung des germanischen und tethyalen Trias. Teil II. Anschluss der germanischen Trias an die internationale Trias Gliederung:* Frei-

berger Forschungsh. C, no. 304, p. 51-77.
1978, *Beiträge zur Stratigraphie des Perms. Teil II. Die Conodontenchronologie des Perms:* Freiberger Forschungsh. C, no. 334, p. 85-161, 8 pl.
———, & **Mock, Rudolf,** 1972, *Neue Conodonten aus der Trias der Slowakei und ihre Stratigraphische Bedeutung:* Geol. Palaeontol. Mitt. Innsbruck, v. 2, no. 4, p. 1-20.
1974, *Zwei Neue Conodonten-Arten aus der Trias des Slowakisschen Karstes:* Cas. Mineral. Geol., v. 19, p. 135-139.
———, & **Mostler, Helfried,** 1970, *Neue Conodonten aus der Trias:* Bericht Nat.-Med. Ver. Innsbruck, v. 58, p. 429-464.
1971, *Holothurien-Sklerite und Conodonten aus der Mittel- und Obertrias von Köveskal (Balatonhochland, Ungarn):* Geol. Palaeontol. Mitt. Innsbruck, v. 1, no. 10, p. 1-36.
1972, *Die Bedeutung der Conodonten fur stratigraphische und palaogeographische Untersuchungen in der Trias:* Mitt. Ges. Geol. Bergbaustud. Wien, v. 21, p. 777-810.
1976, *Neue Conodonten aus dem Jungpaläozoikum und der Trias:* Geol. Palaeontol. Mitt. Innsbruck, v. 6, no. 3, p. 1-33, 4 pls.
———, ———, & **Rahimi-Yazd, A.,** 1976, *Beiträge zur Mikrofauna permotriadischer Schichtfolgen. Teil II. Neue Conodonten aus dem Oberperm und der basalen Trias von Nord- und Zentraliran:* Geol. Palaeontol. Mitt. Innsbruck, v. 5, no. 3, p. 1-23, 7 pl.
Lamont, Archie, & Lindström, Maurits, 1957, *Arenigian and Llandeilian cherts identified in the Southern Uplands of Scotland by means of conodonts, etc.:* Trans. Edinburgh Geol. Soc., v. 17, pt. 1, p. 60-70, text-fig. 1, pl. 5.
Landing, Ed, 1977, *"Prooneotodus" tenuis (Müller, 1959) apparatuses from the Taconic allochthon, eastern New York: Construction, taphonomy and the protoconodont "supertooth" model:* J. Paleontol., v. 51, p. 1072-1084, 1 text-fig., 2 pl.
1979, *Conodonts and biostratigraphy of the Hoyt Limestone (Late Cambrian, Trempealeauan), eastern New York:* J. Paleontol., v. 53, p. 1023-1029, 3 text-fig.
Lane, H. R., 1967, *Uppermost Mississippian and Lower Pennsylvanian conodonts from the type Morrowan region, Arkansas:* J. Paleontol., v. 41, p. 920-942, 2 text-fig., pl. 119-123.
1968, *Symmetry in conodont element-pairs:* J. Paleontol., v. 42, p. 1258-1263, 3 text-fig.
———, **Merrill, G. K., Straka, J. J., II, & Webster, G. D.,** 1971, *North American Pennsylvanian conodont biostratigraphy:* in W. C. Sweet & S. M. Bergström (eds.), Symposium on conodont biostratigraphy, Geol. Soc. Am. Mem. 127, p. 395-414, 4 text-fig., pl. 1.
———, **Müller, K. J., & Ziegler, Willi,** 1979, *Devonian and Carboniferous conodonts from Perak, Malaysia:* Geol. Palaeontol., v. 13, p. 213-226, 2 text-fig., 2 pl.
———, & **Straka, J. J., II,** 1974, *Late Mississippian and early Pennsylvanian conodonts, Arkansas and Oklahoma:* Geol. Soc. Am. Spec. Pap. 152, 144 p., 44 text-fig. (text-fig. 32-44 are plates).
———, & **Ziegler, Willi,** 1979, *Gnathodus Pander, 1856 (Conodonta): Proposed designation of a type species under the plenary powers Z.N.(S) 2279,* Bull. Zool. Nomencl., v. 36, pt. 1, p. 57-62.
Lange, F. G., 1968, *Conodonten-Gruppenfunde aus Kalken des tieferen Oberdevon:* Geol. Palaeontol., v. 2, p. 37-57, 2 text-fig., 6 pl.
Lee, H. Y., 1970, *Conodonten aus der Choson-Gruppe (Unteres Ordovizium) von Korea:* Neues Jahrb. Geol., Palaeontol. Abh., v. 136, no. 3, p. 303-344, pl. 7, 8.
Legault, J. A., 1968, *Conodonts and fish remains from the Stonehouse Formation, Arisaig, Nova Scotia:* Contrib. Can. Paleontol., pt. 1, Geol. Surv. Can. Bull. 165, p. 1-23, 1 text-fig., pl. 1-3.
Lindström, Maurits, 1955, *Conodonts from the lowermost Ordovician strata of south-central Sweden:* Geol. Foeren. Stockholm Foerh., v. 76, h. 4 (1954), no. 21, p. 517-603, 6 text-fig., 10 pl.
1964, *Conodonts:* 196 p., 64 text-fig., Elsevier Publishing Co. (Amsterdam).
1970, *A suprageneric taxonomy of the conodonts:* Lethaia, v. 3, p. 427-445, 9 text-fig.
1971, *Lower Ordovician conodonts of Europe:* in W. C. Sweet & S. M. Bergström (eds.), Symposium on conodont biostratigraphy, Geol. Soc. Am. Mem. 127, p. 21-61, 20 text-fig., 1 pl.
1973, *On the affinities of conodonts:* in F. H. T. Rhodes (ed.), Conodont palaeozoology, Geol. Soc. Am. Spec. Paper 141, p. 85-102, 6 text-fig.
1974, *The conodont apparatus as a food-gathering mechanism:* Palaeontology, v. 17, p. 729-744, 4 text-fig.
———, **McTavish, R. A., & Ziegler, Willi,** 1972, *Feinstrukturelle Untersuchungen an Conodonten. 2. Einige Prioniodontidae aus dem Ordovicium Australiens:* Geol. Palaeontol., v. 6, p. 33-43, pl. 1-3.
———, & **Ziegler, Willi,** 1966, *Ein Conodontentaxon aus vier morphologisch verschiedenen Typen (Grenze Mittel-/Oberdevon, Rheinisches Schiefergebirge),* in Das Mitteldevon des Rheinischen Schiefergebirges, Ein Symposium: Fortschr. Geol. Rheinl. Westfalen, v. 9 (1965), p. 209-218, pl. 1, 2.
1971, *Feinstrukturelle Untersuchungen an Conodonten. 1. Die Überfamilie Panderodontacea:* Geol. Palaeontol., v. 5, p. 9-33, text-fig. 1-6, pl. 1-8.
Link, A. G., & Druce, E. C., 1972, *Ludlovian and Gedinnian conodont stratigraphy of the Yass basin, New South Wales:* Aust. Bur. Miner. Resour. Geol. Geophys. Bull. 134, 136 p., 67 text-fig., 12 pl.
Loomis, F. B., 1936, *Are conodonts gastropods?:* J. Paleontol., v. 10, p. 663-664, 1 text-fig.
Lys, Maurice, Serre, Bernadette, & Deroo, G., 1957,

Études micropaléontologiques dans le paléozoïque de la Montagne Noire (Foraminifères, Conodontes, etc.): Inst. Fr. Pét. Rev., v. 12, p. 783-833, 3 text-fig., 12 pl.

Macfarlane, J. M., 1923, *Fishes the source of petroleum:* 451 p., Macmillan Co. (New York).

McHargue, T. R., 1974, *Conodonts of the Joins Formation (Ordovician), Arbuckle Mountains, Oklahoma:* Unpubl. MA thesis, Univ. Missouri, 151 p., 8 pl. (Columbia).

McTavish, R. A., 1973, *Prioniodontacean conodonts from the Emanuel Formation (Lower Ordovician) of western Australia:* Geol. Palaeontol., v. 7, p. 27-58, 3 pl.

Mashkova, T. V., 1972, *Ozarkodina steinhornensis (Ziegler) apparatus, its conodonts and biozone:* Geol. Palaeontol., Spec. Vol. 1, p. 81-90, 2 text-fig., 2 pl.

1977, *Novye Konodonty zony amorphognathoides iz nizhnego Silura Podolii:* Paleontol. Zh., 1977, no. 4, p. 127-131, 2 text-fig. [*New conodonts of the amorphognathoides Zone from the Lower Silurian of Podolia.*]

1978, *Novye Knodonty zony pesavis-optima iz nizhnego Devona Urala i Zeravshana:* Paleontol. Zh., 1978, no. 2, p. 91-97, 2 text-fig. [*New conodonts of the pesavis-optima Zone from the Lower Devonian of the Urals and Zeravshan.*]

Matern, Hans, 1933, *Neue Conodonten aus Devon und Unterkarbon:* Senckenbergiana Lethaea, v. 15, p. 12-22, 18 text-fig.

Matthews, S. C., 1969a, *A Lower Carboniferous conodont fauna from east Cornwall:* Palaeontology, v. 12, p. 262-275, 1 text-fig., pl. 46-50.

1969b, *Two conodont faunas from the Lower Carboniferous of Chudleigh, south Devon:* Palaeontology, v. 12, p. 276-280, pl. 51.

———, **& Naylor, D.,** 1973, *Lower Carboniferous conodont faunas from south-west Ireland:* Palaeontology, v. 16, no. 2, p. 335-380, pl. 35-38.

———, **Sadler, P. M., & Selwood, E. B.,** 1972, *A Lower Carboniferous conodont fauna from Chillaton, south-west Devonshire:* Palaeontology, v. 15, p. 550-568, 1 text-fig., pl. 109-111.

Meischner, K. D., 1970, *Conodonten-Chronologie des deutschen Karbons:* Congr. Adv. Étud. Stratigr. Géol. Carbonifère, 6th, Sheffield, 1967, C.R., v. 3, p. 1169-1180, 3 text-fig.

Melton, W. G., & Scott, H. W., 1973, *Conodont-bearing animals from the Bear Gulch Limestone, Montana:* in F. H. T. Rhodes (ed.), Conodont paleozoology, Geol. Soc. Am. Spec. Pap. 141, p. 31-65, 20 text-fig.

Merrill, G. K., 1968, *Pennsylvanian conodont assemblages, natural assemblages and statistical assemblages* (abstr.): Geol. Soc. Am., 9th Central Section Mtg., Iowa City, Iowa, p. 198-199.

1972, *Taxonomy, phylogeny and biostratigraphy of Neognathodus in Appalachian Pennsylvanian rocks:* J. Paleontol., v. 46, p. 817-829, 1 text-fig., 2 pl.

1973, *Pennsylvanian conodont paleoecology:* in F. H. T. Rhodes (ed.), Conodont paleozoology, Geol. Soc. Am. Spec. Pap. 141, p. 239-274, 13 text-fig.

1975a, *Pennsylvanian conodont biostratigraphy and paleoecology of northwestern Illinois:* Geol. Soc. Am. Microform Pub., 3 p.

1975b, *Pennsylvanian conodonts of northwestern Illinois: Summary and new systematics:* Geology, v. 3, no. 12, p. 721-722, 1 text-fig.

———, **& King, C. W.,** 1971, *Platform conodonts from the lowest Pennsylvanian rocks of northwestern Illinois:* J. Paleontol., v. 45, p. 645-664, pl. 75, 76.

———, **& Merrill, S. M.,** 1974, *Pennsylvanian nonplatform conodonts, IIa: The dimorphic apparatus of Idioprioniodus:* Geol. Palaeontol., v. 8, p. 119-130, 6 text-fig.

Miller, A. K., & Youngquist, W. L., 1947, *Conodonts from the type section of the Sweetland Creek shale in Iowa:* J. Paleontol., v. 21, p. 501-517, pl. 72-75.

Miller, J. F., 1969, *Conodont fauna of the North Peak Limestone (Cambro-Ordovician), House Range, Utah:* J. Paleontol., v. 43, p. 413-439, pl. 63-66.

1975, *Conodont faunas from the Cambrian and lowest Ordovician of western North America:* Geol. Soc. Am. Abstr. Programs, v. 7, no. 7, p. 1200-1201.

1980, *Taxonomic revisions of some Upper Cambrian and Lower Ordovician conodonts, with comments on their evolution:* Univ. Kans. Paleontol. Contrib. Pap. 99, 44 p., 6 text-fig., 2 pl.

———, **& Rushton, A. W. A.,** 1973, *Natural conodont assemblages from the Upper Cambrian of Warwickshire, England:* Geol. Soc. Am. Abstr. Programs, v. 5, no. 4, p. 338-339.

Miller, S. A., 1889, *North American geology and palaeontology for the use of amateurs, students, and scientists:* 718 p., 1,265 text-fig., Western Methodist Book Concern (Cincinnati, Ohio).

Missarzhevsky, V. V., 1973, *Konodontoobraznye organizmy iz pogranichtsykh sloyebv Kembriya i Dokembiya Sibirskoy platformy i Kazakhstana:* in Problemy paleontologii i biostratigrafii nizhnego Kembriya i dalnego vostoka: Tr. Inst. Geol. Geofiz. Akad. Nauk SSSR, Sib. Otd., p. 53-57 (Moscow). [*Conodontlike organisms from the boundary beds of the Cambrian and Precambrian of the Siberian Platform and Kazakhstan:* in Problems of paleontology and biostratigraphy of the Lower Cambrian of Siberia and the Far East.]

Moore, R. C., 1962, *Conodont classification and nomenclature:* in R. C. Moore (ed.), Treatise on Invertebrate Paleontology, Part W, Miscellanea, p. *W*92-*W*98, text-fig. 49, Geological Society of America and University of Kansas Press (New York and Lawrence).

———, **& Sylvester-Bradley, P. C.,** 1957a, *Proposed*

insertion in the "Regles" of provisions recognizing "Parataxa" as a special category for the classification and nomenclature of discrete fragments or of life-stages of animals which are inadequate for identification of whole-animal taxa, with proposals of procedure for the nomenclature of "Parataxa": Bull. Zool. Nomencl., v. 15, sec. A, pt. 1/4, Doc. 1/1, p. 5-13.

1957b, *First supplemental application: Application for ruling of the International Commission directing that the classification and nomenclature of discrete conodonts be in terms of "parataxa":* Bull. Zool. Nomencl., v. 15, sec. A, pt. 1/4, Doc. 1/2, p. 14-34.

Mosher, L. C., 1967, *Are there post-Triassic conodonts?:* J. Paleontol., v. 41, p. 1554-1555.

1968, *Triassic conodonts from western North America and Europe and their correlations:* J. Paleontol., v. 42, p. 895-946, 14 text-fig., pl. 113-118.

1969, *Nomenclatural revisions for Triassic conodonts in Mosher, 1968:* J. Paleontol., v. 43, p. 1441.

1973, *Evolutionary, ecologic, and geographic observations on conodonts during their decline and extinction:* in F. H. T. Rhodes (ed.), Conodont paleozoology, Geol. Soc. Am. Spec. Pap. 141, p. 143-152, 1 text-fig.

———, **& Clark, D. L.,** 1965, *Middle Triassic conodonts from the Prida Formation of northwestern Nevada:* J. Paleontol., v. 39, p. 551-565, 2 text-fig., pl. 65, 66.

Moskalenko, T. A., 1967, *Novye Dannye po biostratigraphii nizhnego Paleozoya Sibirskoy platformy:* Tr. Inst. Geol. Geofiz. Akad. Nauk SSSR Sib. Otd., v. 58, p. 98-116, 16 text-fig., pl. 22-25. [*New data on the biostratigraphy of the Lower Paleozoic deposits of the Siberian Platform.*]

1970, *Konodonty krivolutskogo yarusa (sredniy ordovik) Sibirskoy platformy:* Tr. Inst. Geol. Geofiz. Akad. Nauk SSSR Sib. Otd., v. 61, 132 p., 14 pl. [*Conodonts from the middle Ordovician Krivolutskian stage of the Siberian platform.*]

1972, *Ordovician conodonts of the Siberian platform and their bearing on multielement taxonomy:* Geol. Palaeontol., Sp. Vol. 1, p. 47-55.

1973, *Konodonty srednego i verkhnego Ordovika Sibirskoy platformy:* Tr. Inst. Geol. Geofiz. Akad. Nauk SSSR Sib. Otd., v. 137, 143 p., 18 pl. [*Conodonts from the Middle and Upper Ordovician of the Siberian platform.*]

Mostler, Helfried, 1967, *Conodonten aus dem tieferen Silur der Kitzbühler Alpen (Tirol):* Naturhist. Mus. Wien, Ann., v. 71, p. 295-303, 5 text-fig., 1 pl.

Mound, M. C., 1965a, *A conodont fauna from the Joins Formation (Ordovician), Oklahoma:* Tulane Stud. Geol., v. 4, p. 1-46, pl. 1-4.

1965b, *Two new conodont genera from the Joins Formation (lower Middle Ordovician) of Oklahoma:* Biol. Soc. Wash. Proc., v. 78, p. 193-200, 13 text-fig.

1968, *Upper Devonian conodonts from southern Alberta:* J. Paleontol., v. 42, p. 444-524, text-fig. 1-14, pl. 65-71.

Müller, K. J., 1956a, *Zur Kenntnis der Conodonten-Fauna des europäischen Devons, 1. Die Gattung Palmatolepis:* Abh. Senckenb. Naturforsch. Ges., v. 494, 70 p., 1 text-fig., 11 pl.

1956b, *Triassic conodonts from Nevada:* J. Paleontol., v. 30, p. 818-830, pl. 95, 96.

1956c, *Taxonomy, nomenclature, orientation, and stratigraphic evaluation of conodonts:* J. Paleontol., v. 30, p. 1324-1340, pl. 145.

1959, *Kambrische Conodonten:* Z. Dtsch. Geol. Ges., v. 111, p. 434-485, pl. 11-15.

1962a, *Zur systematischen Einteilung der Conodontophorida:* Palaeontol. Z., v. 36, p. 109-117, 1 text-fig.

1962b, *Taxonomy, evolution, and ecology of conodonts:* in R. C. Moore (ed.), Treatise on Invertebrate Paleontology, Part W, Miscellanea, p. *W*83-*W*91, text-fig. 44-48. Geological Society of America and University of Kansas Press (New York and Lawrence).

1962c, *Supplement to systematics of conodonts:* in R. C. Moore (ed.), Treatise on Invertebrate Paleontology, Part W, Miscellanea, p. *W*246-*W*249, text-fig. 150-153, Geological Society of America and University of Kansas Press (New York and Lawrence).

1964, *Conodonten aus dem unteren Ordovizium von Südkorea:* Neues Jahrb. Geol. Palaeontol. Abh., v. 119, no. 1, p. 93-102, pl. 12, 13.

1969, *Bürstenbildung bei Conodonten:* Palaeontol. Z., v. 43, p. 64-71, pl. 5, 6.

1973, *Late Cambrian and Early Ordovician conodonts from northern Iran:* Geol. Surv. Iran Rep. 30, p. 1-77, text-fig. 1-11, pl. 1-11.

———, **& Andres, D.,** 1976, *Eine Conodontengruppe von Prooneotodus tenuis (Müller, 1959) in natürlichem Zusammenhang aus dem Oberen Kambrium vom Schweden:* Palaeontol. Z., v. 50, p. 193-200, 3 text-fig., pl. 22.

———, **& Clark, D. L.,** 1967, *Early late Devonian conodonts from the Squaw Bay Limestone in Michigan:* J. Paleontol., v. 41, p. 902-919, 7 text-fig., pl. 115-118.

———, **& Mosher, L. C.,** 1971, *Post-Triassic conodonts:* in W. C. Sweet & S. M. Bergström (eds.), Symposium on conodont biostratigraphy, Geol. Soc. Am. Mem. 127, p. 467-470.

———, **& Müller, E. M.,** 1957, *Early Upper Devonian (Independence) conodonts from Iowa, Part 1:* J. Paleontol., v. 31, p. 1069-1108, 8 text-fig., pl. 135-142.

———, **& Nogami, Yasuo,** 1971, *Über den Feinbau der Conodonten:* Mem. Fac. Sci. Kyoto Univ. Ser. Geol. Mineral., v. 38, p. 1-88, 31 text-fig., pl. 1-22.

1972a, *Growth and function of conodonts:* Int. Geol. Congr.. 24th Sess., Montreal, p. 20-27, text-fig. 1-10.

1972b, *Entöken und Bohrspuren bei den Conodontophorida:* Palaeontol. Z., v. 46, p. 68-86, 11 text-fig., pl. 14-16.

Munsell Color Company, 1971, *Munsell soil color charts:* Munsell Color Company, Inc. (Baltimore, Maryland).

Murchison, R. I., 1867, *Siluria. A history of the oldest rocks in the British Isles and other countries:* 566 p., 51 pl., J. Murray (London).

Murray, F. N., & Chronic, John, 1965, *Pennsylvanian conodonts and other fossils from insoluble residues of the Minturn Formation (Desmoinesian), Colorado:* J. Paleontol., v. 39, p. 594-610, pl. 71-73.

Nease, F. R., 1969, *Conodontophyta Chattanoogae, a non-petrified fossil plant from Chattanooga Shale:* Louisburg Coll. J. Arts Sci., v. 3, p. 10-23.

Newberry, J. S., 1875, *Descriptions of fossil fishes:* Ohio Geological Survey, rept. 2, pt. 2, p. 41-46, pl. 57.

Nicoll, R. S., 1975, *The effect of Late Carboniferous-Early Permian glaciation on the distribution of Permian conodonts in Australasia:* Geol. Soc. Am. Abstr. Programs, v. 7, p. 828-829.

———, **& Rexroad, C. B.,** 1969, *Stratigraphy and conodont paleontology of the Salamonie Dolomite and Lee Creek Member of the Brassfield Limestone (Silurian) in southeastern Indiana and adjacent Kentucky:* Indiana Geol. Surv. Bull. 40, 73 p., 4 text-fig., 7 pl.

Nieper, C. M., 1969, *Conodonts of Nora Fm., Toko Range, Glenormiston:* in Dorothy Hill, Geoffrey Playford, & J. T. Woods (eds.), Ordovician and Silurian fossils of Queensland, Queensland Palaeontolographical Society (Brisbane), p. 12-15, pl. 7.

Nogami, Yasuo, 1966, *Kambrische Conodonten von China. Teil 1. Conodonten aus den oberkambrischen Kushan-Schichten:* Kyoto Imp. Univ., Mem. Coll. Sci., Ser. B, v. 32, p. 351-367, pl. 9, 10.

1967, *Kambrische Conodonten von China. Teil 2. Conodonten aus den hoch oberkambrischen Yencho-Schichten:* Mem. Coll. Sci. Kyoto Univ. Ser. B, v. 33, p. 211-219, 3 text-fig., pl. 1.

Nohda, Susumu, & Steoguchi, Takeshi, 1967, *An occurrence of Jurassic conodonts from Japan:* Mem. Coll. Sci. Kyoto Univ. Ser. B, v. 33, no. 4, p. 227-237, pl. 2.

Norby, R. D., 1974, *Conodont apparatuses of the Heath Shale (Chesterian) of central Montana:* Geol. Soc. Am. Abstr. Prog., v. 6, p. 535.

Ochietti, Serge, & Cailleux, André, 1969, *Paléontologie: Comparaison des Conodontes et des mâchoires de Gnathostomulides:* C. R. Acad. Sci. Ser. D, v. 268, p. 2664-2666, 1 text-fig.

Orr, R. W., & Klapper, Gilbert, 1968, *Two new conodont species from Middle-Upper Devonian boundary beds of Indiana and New York:* J. Paleontol., v. 42, p. 1066-1075, 2 text-fig., pl. 139-140.

Owen, Richard, 1861, *Paleontology or a systematic summary of extinct animals and their geological relations:* 463 p., 174 text-fig., Adam & Charles Black (Edinburgh).

Packham, G. H., 1967, *The occurrence of shelly Ordovician strata near Forbes, New South Wales:* Aust. J. Sci., v. 30, no. 3, p. 106-107.

Pander, C. H., 1856, *Monographie der fossilen Fische des silurischen Systems der russisch-baltischen Gouvernements:* Akad. Wiss., St. Petersburg, 91 p., 1 text-fig., 7 pl.

Perlmutter, Barry, 1975, *Conodonts from the uppermost Wabaunsee Group (Pennsylvanian) and the Admire and Council Grove groups (Permian) in Kansas:* Geol. Palaeontol., v. 9, p. 95-115, pl. 1-3.

Philip, G. M., 1965, *Lower Devonian conodonts from the Tyers area, Gippsland, Victoria:* Proc. R. Soc. Victoria, v. 79, p. 95-117, 2 text-fig., pl. 8-10.

Pierce, R. W., & Langenheim, R. L., Jr., 1970, *Surface patterns of selected Mississippian conodonts:* Geol. Soc. Am. Bull., v. 81, p. 3225-3236.

1974, *Platform conodonts of the Monte Cristo Group, Mississippian Arrow Canyon Range, Clark County, Nevada:* J. Paleontol., v. 48, p. 149-169, text-fig. 1-3, pl. 1-4.

Pietzner, Horst, Vahl, Johanna, Werner, Hans, & Ziegler, Willi, 1968, *Zur chemischen Zusammensetzung und Mikromorphologie der Conodonten:* Palaeontographica, v. 128, p. 115-152, pl. 18-27.

Pilsbry, H. A., 1937, *Are conodonts molluscan teeth?:* Nautilus, v. 50, no. 3, p. 101.

Pollock, C. A., 1969, *Fused Silurian conodont clusters from Indiana:* J. Paleontol., v. 43, p. 929-935, 1 text-fig., pl. 110-112.

———, **Rexroad, C. B., & Nicoll, R. S.,** 1970, *Lower Silurian conodonts from northern Michigan and Ontario:* J. Paleontol., v. 44, p. 743-764, 3 text-fig., pl. 111-114.

Pomesano Cherchi, Antoinetta, 1967, *I Conodonti del Muschelkalk della Nurra (Sardegna nordoccidentale):* Riv. Ital. Paleontol. Stratigr., v. 73, no. 1, p. 205-272, 14 pl.

Poulsen, Valdemar, 1966, *Early Cambrian distacodontid conodonts from Bornholm:* Biol. Medd. K. Dan. Vidensk. Selsk., v. 23, p. 1-9, pl. 1.

Rabe, E. H., 1977, *Zur Stratigraphie des ostandinen Raumes von Kolumbien:* Giessener Geol. Schr., no. 11, 223 p.

Rexroad, C. B., 1957, *Conodonts from the Chester series in the type area of southwestern Illinois:* Illinois State Geol. Surv., Rep. Invest. 199, 43 p., 4 pl.

1958a, *The conodont homeomorphs Taphrognathus and Streptognathodus:* J. Paleontol., v. 32, p. 1158-1159.

1958b, *Conodonts from the Glen Dean formation (Chester) of the Illinois basin:* Ill. State Geol. Surv. Rep. Invest. 209, 27 p., 14 text-fig., 6 pl.

1967, *Stratigraphy and conodont paleontology of the Brassfield (Silurian) in the Cincinnati arch area:* Indiana Geol. Surv. Bull. 36, 64 p., 4 pl.

———, & **Burton, R. C.**, 1961, *Conodonts from the Kinkaid Formation (Chester) in Illinois:* J. Paleontol., v. 35, p. 1143-1158.

———, & **Collinson, Charles**, 1961, *Preliminary range chart of conodonts from the Chester Series (Mississippian) in the Illinois basin:* Ill. State Geol. Surv. Circ. 319, p. 3-11.

1963, *Conodonts from the St. Louis Formation (Valmeyeran series) of Illinois, Indiana, and Missouri:* Ill. State Geol. Surv. Circ. 355, 28 p., 5 text-fig., 2 pl.

———, & **Furnish, W. M.**, 1964, *Conodonts from the Pella formation (Mississippian), south-central Iowa:* J. Paleontol., v. 38, p. 667-676, 1 text-fig., pl. 111.

———, & **Nicoll, R. S.**, 1964, *A Silurian conodont with tetanus?:* J. Paleontol., v. 38, p. 771-773, 1 text-fig.

1971, *Summary of conodont biostratigraphy of the Silurian System of North America:* in W. C. Sweet & S. M. Bergström (eds.), Symposium on conodont biostratigraphy, Geol. Soc. Am. Mem. 127, p. 207-225, pl. 1, 2.

Rhodes, F. H. T., 1952, *A classification of Pennsylvanian conodont assemblages:* J. Paleontol., v. 26, p. 886-901, 4 text-fig., pl. 126-129.

1953a, *Some British Lower Palaeozoic conodont faunas:* Philos. Trans. R. Soc. London Ser. B, v. 237, p. 261-334, 20 text-fig., pl. 20-23.

1953b, *Nomenclature of conodont assemblages:* J. Paleontol., v. 27, p. 610-612.

1954, *The zoological affinities of the conodonts:* Biol. Rev. (Cambridge Philos. Soc.), v. 29, p. 419-452, 6 text-fig.

1955, *The conodont fauna of the Keisley limestone:* Q.J. Geol. Soc. London, v. 111, pt. 2, p. 117-142, pl. 7-10.

1957, *Comment on the Moore/Sylvester-Bradley "Parataxa Plan":* Bull. Zool. Nomencl., v. 15, Doc. 1/47, p. 305-312.

1962, *Recognition, interpretation, and taxonomic position of conodont assemblages:* in R. C. Moore (ed.), Treatise on Invertebrate Paleontology, Part W, Miscellanea, p. *W*70-*W*83, text-fig. 43, Geological Society of America and University of Kansas Press (New York and Lawrence).

1963, *Conodonts from the topmost Tensleep Sandstone of the eastern Big Horn Mountains, Wyoming:* J. Paleontol., v. 37, p. 401-408, 1 text-fig., pl. 47.

———, & **Austin, R. L.**, 1971, *Carboniferous conodont faunas of Europe:* in W. C. Sweet & S. M. Bergström (eds.), Symposium on conodont biostratigraphy, Geol. Soc. Am. Mem. 127, p. 217-352, pl. 1, 2.

———, ———, & **Druce, E. C.**, 1969, *British Avonian (Carboniferous) conodont faunas, and their value in local and intercontinental correlation:* Bull. Br. Mus. Nat. Hist. Geol., Suppl. 5, 313 p., 92 text-fig., 31 pl.

———, & **Müller, K. J.**, 1956, *The conodont genus Prioniodus and related forms:* J. Paleontol., v. 30, p. 695-699.

———, & **Wingard, Paul**, 1957, *Chemical composition, microstructure and affinities of the Neurodontiformes:* J. Paleontol., v. 31, p. 448-454, 2 text-fig.

Riedl, R. J., 1969, *Gnathostomulida from North America:* Science, v. 163, p. 445-452.

Rietschel, Siegfried, 1973, *Zur Deutung der Conodonten:* Nat. Mus., v. 103, p. 409-418, 11 text-fig.

Rodgers, John, 1969, *Gnathostomulida: Is there a fossil record?:* Science, v. 164, p. 855-856.

Rohon, J. V., & Zittel, K. A. von, 1887, *Ueber Conodonten:* Sitzungsber. Math. Naturwiss. Kl. Bayer. Akad. Wiss. Muenchen, v. 16, Jahrg. 1886, p. 108-136, 1 text-fig., 2 pl.

Rolle, Freidrich, 1882, in Adolf Kenngott (ed.), Handwörterbuch der Mineralogie, Geologie und Palaeontologie, v. 1, 550 p. (Breslau).

Roundy, P. V., 1926, *Introduction, the microfauna in Mississippian formations of San Saba County, Texas by Roundy and others:* U.S. Geol. Surv. Prof. Pap. 146, 63 p., 33 pl.

Sannemann, Dietrich, 1955, *Oberdevonische Conodonten (to II α):* Senckenbergiana Lethaea, v. 36, p. 123-156, 2 text-fig., pl. 1-6.

Schmidt, Hermann, 1934, *Conodonten-Funde in ursprunglichem Zusammenhang:* Palaeontol. Z., v. 16, p. 76-85, 8 text-fig., pl. 6.

1950, *Nachträge zur Deutung der Conodonten:* Decheniana, v. 104, p. 11-19, 7 text-fig.

———, & **Müller, K. J.**, 1964, *Weitere Funde von Conodonten-Gruppen aus dem oberen Karbon des Sauerlandes:* Palaeontol. Z., v. 38, p. 105-135, 11 text-fig.

Schönlaub, H. P., 1971, *Zur Problematik der Conodonten-Chronologie an der Wende Ordoviz/Silur mit besonderer Berücksichtigung der Verhältnisse im Llandovery:* Geol. Palaeontol., v. 5, p. 35-57, 2 text-fig., pl. 1-3.

Schopf, T. J. M., 1966, *Conodonts of the Trenton Group (Ordovician) in New York, southern Ontario and Quebec:* N.Y. State Mus. Sci. Serv. Bull. no. 405, 105 p., 7 text-fig., 6 pl.

Scott, A. J., 1961, *Three new conodonts from the Louisiana Limestone (Upper Devonian) of western Illinois:* J. Paleontol., v. 35, p. 1223-1227, 2 text-fig.

Scott, H. W., 1934, *The zoological relationships of the conodonts:* J. Paleontol., v. 8, p. 448-455, pl. 58, 59.

1942, *Conodont assemblages from the Heath formation, Montana:* J. Paleontol., v. 16, p. 293-300, 1 text-fig., pl. 37-40.

1969, *Discoveries bearing on the nature of the conodont animal:* Micropaleontology, v. 15, p. 420-426, 1 text-fig., 1 pl.

1973, *New Conodontochordata from the Bear Gulch Limestone (Namurian, Montana):* Mich. State Univ., Paleontol. Ser., v. 1, p. 81-100, 7 text-fig., pl. 1-3.

Seddon, George, 1970, *Frasnian conodonts from the Sadler Ridge–Bugle Gap area, Canning basin, Western Australia:* J. Geol. Soc. Aust., v. 16, p. 723-753, 9 text-fig., pl. 11-16.

———, **& Sweet, W. C.**, 1971, *An ecologic model for conodonts:* J. Paleontol., v. 45, p. 869-880, 3 text-fig.

Sergeeva, S. P., 1963, *Novyy ranneordovikskiy rod konodontov semeystva Prioniodinidae:* Paleontol. Zh., no. 4, p. 138-140. [*A new Early Ordovician conodont genus of the family Prioniodinidae.*]

1966, *Biostratigraficheskoe rasprostranenie konodontov v Tremadokskom yaruse (Ordovik) Leningradskoy Oblasti:* Dokl. Akad. Nauk SSSR, v. 67, p. 672-674. [*Biostratigraphic distribution of the conodonts of the Tremadocian Stage (Ordovician) from Leningrad region.*]

1974, *Nekotorbye norbye Konodonty iz ordovikskikh otlozhyeniy Leningradskoy oblasti:* Paleontol. Sb., v. 11, p. 79-85. [*Some new Ordovician conodonts from the Leningrad region.*]

Serpagli, Enrico, 1967, *I conodonti dell'Ordoviciano superiore (Ashgilliano) delle Alpi Carniche:* Boll. Soc. Paleontol. Ital., v. 6, p. 30-111, 8 text-fig., pl. 6-31.

1971, *Uppermost Wenlockian–Upper Ludlovian (Silurian) conodonts from western Sardinia:* Boll. Soc. Paleontol. Ital., v. 9 (1970), p. 76-96, 1 text-fig., pl. 21-24.

1974, *Lower Ordovician conodonts from Precordilleran Argentina (Province of San Juan):* Boll. Soc. Paleontol. Ital., v. 13, no. 1, 2, p. 17-98, pl. 7-31.

Shatzer, D. C., 1976, *Conodont biostratigraphy of the Fremont and Priest Canyon formations (Upper Ordovician) at Kerber Creek, Saguache County, Colorado:* Unpubl. MS thesis, The Ohio State Univ., 123 p., 4 pl. (Columbus).

Siewing, Rolf, 1976, *Probleme und neuere Erkenntnisse in der Gross-systematik der Wirbellosen:* Verh. Dtsch. Zool. Ges., 1976, p. 59-83, 13 text-fig.

Sinclair, G. W., 1953, *The naming of conodont assemblages:* J. Paleontol., v. 27, p. 489-490.

Smith, John, 1907, *On the occurrence of conodonts in the Arenig-Llandeilo formations of the Southern Uplands of Scotland:* Trans. Glasgow Nat. Hist. Soc., n.s., v. 7, pt. 3, p. 235-252, pl. 5-9.

Sparling, D. R., 1981, *Middle Devonian conodont apparatuses with seven types of elements:* J. Paleontol., v. 55, p. 295-316, 1 pl.

Staesche Ulrich, 1964, *Conodonten aus dem Skyth von Südtirol:* Neues Jahrb. Geol. Palaeontol. Abh., v. 119, p. 247-306, text-fig. 1-73, pl. 28-32.

Stauffer, C. R., 1930, *Conodonts from the Decorah shale:* J. Paleontol., v. 4, p. 121-128.

1935a, *Conodonts from the Glenwood beds:* Geol. Soc. Am. Bull., v. 46, p. 125-168, pl. 9-12.

1935b, *The conodont fauna of the Decorah shale (Ordovician):* J. Paleontol., v. 9, p. 596-620, pl. 71-75.

1936, *Pravognathus, a new name for Heterognathus Stauffer (not Girard):* J. Paleontol., v. 10, p. 79.

1938, *Conodonts of the Olentangy shale:* J. Paleontol., v. 12, p. 411-443, pl. 48-53.

1940, *Conodonts from the Devonian and associated clays of Minnesota:* J. Paleontol., v. 14, p. 417-435, pl. 58-60.

———, **& Plummer, H. J.**, 1932, *Texas Pennsylvanian conodonts and their stratigraphic relations:* Univ. Texas Bull. 3201, p. 13-50, pl. 1-4.

Stone, G. L., & Furnish, W. M., 1959, *Bighorn conodonts from Wyoming:* J. Paleontol., v. 33, p. 211-228, pl. 31, 32.

Sweet, W. C., 1955, *Conodonts from the Harding formation (Middle Ordovician) of Colorado:* J. Paleontol., v. 29, p. 226-262, pl. 27-29.

1970, *Uppermost Permian and Lower Triassic conodonts of the Salt Range and Trans-Indus Ranges, West Pakistan:* in Bernhard Kummel & Curt Teichert (eds.), Stratigraphic boundary problems, Permian and Triassic of West Pakistan, Univ. Kansas, Dep. Geol., Spec. Publ. 4, p. 207-275, pl. 1-5.

1979, *Late Ordovician conodonts and biostratigraphy of the western Midcontinent Province:* Brigham Young Univ. Geol. Stud., v. 26, no. 3, p. 45-86, 10 text-fig.

———, **& Bergström, S. M.**, 1962, *Conodonts from the Pratt Ferry Formation (Middle Ordovician) of Alabama:* J. Paleontol., v. 36, text-fig. 1-5, p. 1214-1252, pl. 168-171.

1969, *The generic concept in conodont taxonomy:* North Am. Paleontol. Conv., Proc., v. 1, p. 157-173, Allen Press (Lawrence, Kansas).

1971, eds., *Symposium on conodont biostratigraphy:* Geol. Soc. Am. Mem. 127, 499 p.

1972, *Multielement taxonomy and Ordovician conodonts:* Geol. Palaeontol., Spec. Vol. 1, p. 29-42, 4 text-fig.

1974, *Provincialism exhibited by Ordovician conodont faunas:* in C. A. Ross (ed.), Paleogeographic provinces and provinciality, p. 189-202, Soc. Econ. Paleontol. Mineral., Spec. Publ. 21.

———, **Ethington, R. L., & Barnes, C. R.**, 1971, *North American Middle and Upper Ordovician conodont faunas:* in W. C. Sweet & S. M. Bergström (eds.), Symposium on conodont biostratigraphy, Geol. Soc. Am. Mem. 127, p. 163-193, 2 pl.

———, **Mosher, L. C., Clark, D. L., Collinson, J. W., & Hasenmueller, W. A.**, 1971, *Conodont biostratigraphy of the Triassic:* in W. C. Sweet & S. M. Bergström (eds.), Symposium on conodont biostratigraphy, Geol. Soc. Am. Mem. 127, p. 441-465, 3 text-fig., 1 pl.

———, **& Schönlaub, H. P.**, 1975, *Conodonts of the genus Oulodus Branson & Mehl, 1933:* Geol. Palaeontol., v. 9, p. 41-59, pl. 1, 2.

———, **Thompson, T. L., & Satterfield, I. R.**, 1975, *Conodont stratigraphy of the Cape Lime-*

stone (Maysvillian) of eastern Missouri: Missouri Geol. Surv. Water Resour. Rep. Invest. 57, pt. 1, p. 1-60, pl. 1-3.

———, **Turco, C. A., Warner, Earl, & Wilkie, L. C.**, 1959, *The American Upper Ordovician Standard 1. Eden conodonts from the Cincinnati region of Ohio and Kentucky:* J. Paleontol., v. 33, p. 1029-1068, pl. 130-133.

Szaniawski, Hubert, 1971, *New species of Upper Cambrian conodonts from Poland:* Acta Palaeontol. Pol., v. 16, p. 401-413.

———, **& Matkowski, Krzysztof**, 1979, *Conodonts from the Kapp Starostin Formation (Permian) of Spitsbergen:* Acta Palaeontol. Pol., v. 24, no. 2, p. 231-264, pl. 4-10.

Szulczewski, Michal, 1971, *Upper Devonian conodonts, stratigraphy and facial development in the Holy Cross Mts.:* Acta Geol. Pol., v. 21, p. 1-129.

Tatge, Ursula, 1956, *Conodonten aus dem germanischen Muschelkalk:* Palaeontol. Z., v. 30, pt. 1, p. 108-127, 12 text-fig., pl. 5, 6.

Thomas, L. A., 1949, *Devonian-Mississippian formations of southeast Iowa:* Geol. Soc. Am. Bull., v. 60, p. 403-438, pl. 1-4.

Titus, Robert, 1974, *Paleoecology of the lower Trenton conodonts of New York State:* Geol. Soc. Am. Abstr. Programs, v. 5, p. 81.

Ulrich, E. O., 1878, *Observations on fossil annelids and description of some new forms:* Cincinnati Soc. Nat. Hist. J., v. 1, p. 87-91, pl. 4.

———, **& Bassler, R. S.**, 1926, *A classification of the toothlike fossils, conodonts, with descriptions of American Devonian and Mississippian species:* U.S. Nat. Mus. Proc., v. 68, art. 12, 63 p., 5 text-fig., 11 pl.

Van Wamel, W. A., 1974, *Conodont biostratigraphy of the Upper Cambrian and Lower Ordovician of north-western Öland, south-eastern Sweden:* Utrecht Micropaleontol. Bull., v. 10, p. 1-125, pl. 1-8.

Viira, Viive, 1966, *Distribution of conodonts in the Lower Ordovician sequence of Sukhramägi (Tallin):* Eesti NSV Tead. Akad. Toim., v. 15, no. 1, p. 150-155.

1975, *Konodonty Ordovika Pribaltiki:* Eesti NSV Tead. Akad., Geol. Inst. (1974), 142 p., 3 pl.

Votaw, R. B., 1971, *Conodont biostratigraphy of the Black River Group (Middle Ordovician) and equivalent rocks of the eastern Midcontinent, North America:* Unpubl. Ph.D. diss., 160 p., 3 pl., The Ohio State Univ. (Columbus).

Walliser, O. H., 1957, *Conodonten aus dem oberen Gotlandium Deutschlands und der Karnischen Alpen:* Notizbl. Hess. Landesamtes Bodenforsch. Wiesbaden, v. 85, p. 28-52, text-fig. 1-3, pl. 1-3.

1958, *Rhombocorniculum comleyense n. gen., n. sp. (incerte sedis, Unterkambrium, Shropshire):* Palaeontol. Z., v. 32, p. 176-180.

1964, *Conodonten des Silurs:* Abh. Hess. Landesamtes Bodenforsch., v. 41, p. 1-106, 10 text-fig., pl. 1-32.

1972, *Conodont apparatuses in the Silurian:* Geol. Palaeontol., Spec. Vol. 1, p. 75-80.

Webers, G. F., 1966, *The Middle and Upper Ordovician conodont faunas of Minnesota:* Minn. Geol. Surv. Spec. Publ. Ser., 4, 123 p., 2 text-fig., 15 pl.

Weddige, Karsten. 1977, *Die Conodonten der Eifel-Stufe im Typusgebiet und in benachbarten Faziesgebieten:* Senckenbergiana Lethaea, v. 58, p. 271-420, 9 text-fig., 6 pl.

Welles, S. P., 1947, *Vertebrates from the upper Moenkopi formation of northern Arizona:* Univ. Calif. Dept. Geol. Sci. Bull., v. 27, no. 7, p. 241-294, pl. 21, 22.

Wittekindt, H. P., 1966, *Zur Conodonten Chronologie des Mitteldevons:* Fortschr. Geol. Rhein. Westfalen, v. 9 (1965), p. 621-646, 1 text-fig., pl. 1-3.

Woodward, A. S., 1898, *Outlines of vertebrate palaeontology for students of zoology:* Cambridge Biological Series, 470 p., 228 text-fig.

Yochelson, E. L., 1975, *Discussion of early Cambrian "molluscs":* J. Geol. Soc. London, v. 131, p. 661-662.

Youngquist, Walter, 1945, *Upper Devonian conodonts from the Independence shale (?) of Iowa:* J. Paleontol., v. 19, p. 335-367, pl. 54-56.

———, **& Miller, A. K.**, 1949, *Conodonts from the Late Mississippian Pella beds of south-central Iowa:* J. Paleontol., v. 23, p. 617-622, pl. 101.

———, **Hawley, R. W., & Miller, A. K.**, 1951, *Phosphoria conodonts from southeastern Idaho:* J. Paleontol., v. 25, p. 356-364, pl. 54.

Ziegler, Willi, 1958, *Conodontenfeinstratigraphische Untersuchungen an der Grenze Mittel/Oberdevon und in der Adorfstufe:* Notizbl. Hess. Landesamtes Bodenforsch. Wiesbaden, v. 87, p. 7-77, text-fig. 1-7, pl. 1-12.

1959, *Ancyrolepis n. gen. (Conodonta) aus dem höchsten Teil der Manticoceras-Stufe:* Neues Jahrb. Geol. Paläontol. Abh., v. 108, p. 75-80, pl. 7.

1962, *Taxionomie und Phylogenie oberdevonischer Conodonten und ihre stratigraphische Bedeutung:* Abh. Hess. Landesamtes Bodenforsch., v. 38, p. 1-166, pl. 1-14.

1966, *Eine Verfeinerung der Conodontengliederung an der Grenze Mittel-/Oberdevon,* in Das Mitteldevon des Rheinischen Schiefergebirges, Ein Symposium: Fortschr. Geol. Rhein. Westfalen, v. 9 (1965), p. 647-676, 4 text-fig., 6 pl.

1969, *Eine neue Conodontenfauna aus dem höchsten Oberdevon:* Fortschr. Geol. Rheinl. Westfalen, v. 17, p. 343-360, 2 pl.

1970, *On microstructure and surface ornamentation of conodonts:* Geol. Soc. Am. Abstr. Programs, v. 2, p. 410.

1971, *Conodont stratigraphy of the European Devonian:* in W. C. Sweet & S. M. Bergström (eds.), Symposium on conodont biostratigraphy, Geol. Soc. Am. Mem. 127, p. 227-284, 3 text-fig.

1974, *Catalogue of conodonts:* v. I, 504 p., E. Schweizerbart'sche (Stuttgart).

1975, *Catalogue of conodonts:* v. II, 404 p., 26 text-fig., 25 pl., E. Schweizerbart'sche (Stuttgart).

———, & **Lindström, Maurits,** 1971, *Über Panderodus Ethington, 1959, und Neopanderodus n. g. (Conodonta) aus dem Devon:* Neues Jahrb. Geol. Palaeontol. Monatsh., Jahrg. 1971, no. 10, p. 628-640, 3 text-fig.

1972, *Microstructure and growth in some platform conodonts:* Geol. Soc. Am. Abstr. Programs, v. 4, p. 357.

1975, *Fortschrittsbericht Conodonten:* Palaeontol. Z., v. 49, no. 4, p. 565-595.

———, ———, & **McTavish, R. A.,** 1973, *Monochloracetic acid and conodonts—A warning:* Nature, v. 230, no. 5296, p. 584-585.

———, **Sandberg, C. A., & Austin, R. L.,** 1974, *Revision of Bispathodus group (Conodonta) in the Upper Devonian and Lower Carboniferous:* Geol. Palaeontol., v. 8, p. 97-112, 1 text-fig., pl. 1-3.

Zittel, K. A., & Rohon, J. V., 1886, *Über Conodonten:* Bayer. Akad. Wiss. Muenchen Math.-Phys. Kl. Sitzungsber., v. 16, p. 108-136, pl. 1, 2.

INDEX

Italicized names in the following index are considered to be invalid; those printed in roman type, including morphological terms, are accepted as valid. The names of all taxa above the rank of superfamily are distinguished by use of full capitals, and authors' names are set in large and small capitals. Page references having chief importance are in boldface type.